Khalid Bin Al-Waleed

A Biography of one of the Greatest Military Generals in History

Copyright

Akram, Ishaq, Kathir

Editor: Imam Ahmad

"What an excellent slave of Allah Khalid ibn Al-Waleed, one of the swords of Allah, unleashed against the unbelievers!"

The Prophet, peace be upon him

Chapter 1

(The Boy)

"The best of you in Jahiliyyah are the best of you in Islam, as long as they have understanding."
[Prophet Muhammad (SAWS)][1]

Khalid and the tall boy glared at each other. Slowly they began to move in a circle, the gaze of each fixed intently upon the other, each looking for an opening for his attack and each wary of the tricks that the other might use. There was no hostility in their eyes-just a keen rivalry and an unshakeable determination to win. And Khalid found it necessary to be cautious, for the tall boy was left-handed and thus enjoyed the advantage that all left-handers have over their opponents in a fight.

Wrestling was a popular pastime among the boys of Arabia, and they frequently fought each other. There was no malice in these fights. It was a sport, and boys were trained in wrestling as one of the requirements of Arab manhood. But these two boys were the strongest of all and the leaders of boys of their age. This match was, so to speak, a fight for the heavy-weight title. The boys were well matched. Of about the same age, they were in their early teens. Both were tall and lean, and newly formed muscles rippled on their shoulders and arms as their sweating bodies glistened in the sun. The tall boy was perhaps an inch taller than Khalid. And their faces were so alike that one was often mistaken for the other.

Khalid threw the tall boy; but this was no ordinary fall. As the tall boy fell there was a distinct crack, and a moment later the grotesquely twisted shape of his leg showed that the bone had broken. The stricken boy lay motionless on the ground, and Khalid stared in horror at the broken leg of his friend and nephew. (The tall boy's mother, Hantamah bint Hisham bin Al Mugheerah, was Khalid's first cousin.)

In course of time the injury healed and the leg of the tall boy became whole and strong again. He would wrestle again and be among the best of wrestlers. And the two boys would remain friends. But while they were both intelligent, strong and forceful by nature, neither had patience or tact. They were to continue to compete with each other in almost everything that they did.

The reader should make a mental note of this tall boy for he was to play an important role in the life of Khalid. He was the son of Al Khattab, and his name was Umar.

Soon after his birth Khalid was taken away from his mother, as was the custom among the better families of the Quraish, and sent to a Bedouin tribe in the desert. A foster mother was found for him, who would nurse him and bring him up. In the clear, dry and unpolluted air of the desert, the foundations were laid of the tremendous strength and robust health that Khalid was to enjoy throughout his life. The desert seemed to suit Khalid, and he came to love it and feel at home in it. From babyhood he grew into early childhood among the Arabs of the desert; and when he was five or six years old he returned to his parents' home in Makkah.

Some time in his childhood he had an attack of small pox, but it was a mild attack and caused no damage except to leave a few pock marks on his face. These marks did not, however, spoil his ruggedly handsome face, which was to cause a lot of trouble among the belles of Arabia - and some -to himself too.

The child became a boy; and as he reached the age of boyhood he came to realise with a thrill of pride that he was the son of a chief. His father, Al Waleed, was the Chief of the Bani Makhzum - one of the noblest clans of the Quraish - and was also known in Makkah by the title of AlWaheed- the Unique. Khalid's upbringing was now undertaken by the father who did his best (and with excellent success) to instill into Khalid all the virtues of Arab manhood-courage, fighting skill, toughness and generosity. Al Waleed took great pride in his family and his ancestors, and told Khalid that he was:

Khalid
son of Al Waleed
son of Al Mugheerah
son of Abdullah
son of Umar
son of Makhzum (after whom the clan was named)
son of Yaqza
son of Murra
son of Kab
son of Luwayy
son of Ghalib
son of Fihr
son of Malik
son of Al Nazr
son of Kinana
son of Khuzeima
son of Mudrika
son of Ilyas
son of Muzar
son of Nizar
son of Ma'add
son of Adnan
son of Udd
son of Muqawwam
son of Nahur
son of Teirah
son of Ya'rub
son of Yashjub
son of Nabit
son of Isma'il (regarded as the father of the Arabians)
son of Ibrahim (the prophet)
son of Azar
son of Nahur
son of Sarugh (or Asragh)
son of Arghu
son of Falakh
son of Eibar
son of Shalakh
son of Arfakhshaz

son of Saam
son of Noah (the prophet)
son of Lamk
son of Mattushalakh
son of Idris (the prophet)
son of Yard
son of Muhla'il
son of Qeinan
son of Anush
son of Sheis
son of Adam (the father of mankind)

1. Bukhari, from Abu Hurayrah. Sahih Al-Jami' Al-Saghir No. 3267

The great tribe of the Quraish that inhabited Makkah had evolved a clear-cut division of privilege and responsibility among its major clans. The three leading clans of the Quraish were the Bani Hashim, the Bani Abduddar (of which the Bani Umayyah was an offshoot) and the Bani Makhzum. The Bani Makhzum was responsible for matters of war. This clan bred and trained the horses on which the Quraish rode to war; it made arrangements for the preparation and provisioning of expeditions; and frequently it provided the officers to lead Quraish groups into battle. This role of the Bani Makhzum set the atmosphere in which Khalid was to grow up.

While still a child he was taught to ride. As a Makhzumi he had to be a perfect rider and soon acquired mastery over the art of horsemanship. But it was not enough to be able to handle trained horses; he had to be able to ride any horse. He would be given young, untrained colts and had to break them and train them into perfectly obedient and well-disciplined war horses. The Bani Makhzum were among the best horsemen of Arabia, and Khalid became one of the best horsemen of the Bani Makhzum. Moreover, no Arab could claim to be a good rider if he only knew horses; he had to be just as good on a camel, for both animals were vital for Arab warfare. The horse was used for fighting, and the camel for long marches, in which horses were tagged along unmounted.

Along with riding, Khalid learned the skills of combat. He learnt to use all weapons-the spear, the lance, the bow and the sword. He learnt to fight on horseback and on foot. While he became skillful in the use of all weapons, the ones for which he appears to have had a natural gift were the lance, used while charging on horseback, and the sword for mounted and dismounted dueling. The sword was regarded by the Arabs as the weapon of chivalry, for this brought one nearest to one's adversary; and in sword fighting one's survival depended on strength and skill and not on keeping at a safe distant from the opponent. The sword was the most trusted weapon.

As Khalid grew to manhood, he attained a great height-over six feet. His shoulders widened, his chest expanded and the muscles hardened on his lean and athletic body. His beard appeared full and thick on his face, with his fine physique, his forceful personality, and his skill at riding and the use of weapons, he soon became a popular and much-admired figure in Makkah. As a wrestler, he climbed high on the ladder of achievement, combining consummate skill with enormous strength.

The Arabs had large families, the father often having several wives to increase his offspring, Al Waleed was one of six brothers. (There may have been more, but the names of only six have been recorded.) And the children of Al Waleed that we know of were five sons and two daughters. The sons were Khalid, Waleed (named after the father), Hisham, Ammarah and Abdu Shams. The daughters were Faktah and Fatimah.

Al Waleed was a wealthy man. Thus Khalid did not have to work for a living and could concentrate on learning the skills of riding and fighting. Because of this wealthy background, Khalid grew up to disregard economy and became known for his lavish spending and his generosity to all who appealed to him for help. This generosity was one day to get him into serious trouble.

Al Waleed was a wealthy man. But the Quraish were a surprisingly democratic people and everybody was required to do some work or the other-either for remuneration or just to be a useful member of society. And Al Waleed, who hired and paid a large number of employees, would work himself. In his spare time he was a blacksmith [1] and butcher [2], slaughtering animals for the clan. He was also a trader, and along with other clans would organise and send trade caravans to neighbouring countries. On more than one occasion Khalid accompanied trade caravans to Syria and visited the great trading cities of that fair province of Rome. Here he would meet the Christian Arabs of the Ghassan, Persians from Ctesiphon, Copts from Egypt, and the Romans of the Byzantine Empire.

Khalid had many friends with whom, as with is brothers he would ride and hunt. When not engaged outdoors they would recite poetry, recount genealogical lines and have bouts of drinking. Some of these friends were to play an important part in Khalid's life and in this story; and the ones deserving special mention besides Umar, were Amr bin Al Aas and Abul Hakam. The latter's personal name was Amr bin Hisham bin Al Mugheerah, though he was to earn yet another name later: Abu Jahl. He was an elder cousin of Khalid. And there was Abul Hakam's son, Ikrimah, Khalid's favourite nephew and bosom friend.

Al Waleed was not only the father and mentor of his sons; he was also their military instructor, and from him Khalid got his first lesson in the art of warfare. He learnt how to move fast across the desert, how to approach a hostile settlement, how to attack it. He learned the importance of catching the enemy unawares, of attacking him at an unexpected moment and pursuing him when he broke and fled. This warfare was essentially tribal, but the Arabs well knew the value of speed, mobility and surprise, and tribal warfare was mainly based on offensive tactics.

On reaching maturity Khalid's main interest became war and this soon reached the proportions of an obsession. Khalid's thoughts were thoughts of battle; his ambitions were ambitions of victory. His urges were violent and his entire psychological make-up was military. He would dream of fighting great battles and winning great victories, himself always the champion-admired and cheered by all. He promised himself battle. He promised himself victory. And he promised himself lots and lots of blood. Unknown to him, destiny had much the same ideas about Khalid, son of Al Waleed.

1. Ibn Qutaibah: p. 575.
2. Ibn Rusta: p. 215.

Chapter 2

(The New Faith)

"It is He who has sent His Messenger with Guidance and the Religion of Truth, to make it prevail over all religion, and Allah is sufficient as a witness." [48:28]

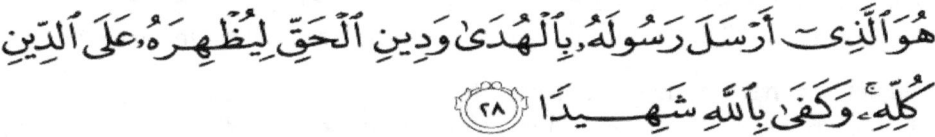

A certain Arab would walk the streets of Makkah at night, lost in thought. He was a member, no longer wealthy, of the noble clan of Bani Hashim. A strikingly handsome man of medium height with broad, powerful shoulders, his hair ended in curls just below his ears. His large, dark eyes, fringed with long lashes, seemed pensive and sad.

There was much in the way of life of the Arabs that caused him pain. Everywhere around him he saw signs of decay-in the injustice done to the poor and helpless, in the unnecessary bloodshed, in the treatment of women who were considered as no better than domestic animals. He would be deeply anguished whenever he heard reports of the live burial of unwanted female children.

Certain clans of the Arabs had made a horrible ritual of the killing of infant daughters. The father would let the child grow up normally until she was five or six years old. He would then tell her that he would take her for a walk and dress her up as if for a party. He would take her out of the town or settlement to the site of a grave already dug for her. He would make the child stand on the edge of this grave and the child, quite unaware of her fate and believing that her father had brought her out for a picnic, would look eagerly at him, wondering when the fun would start. The father would then push her into the grave, and as the child cried to her father to help her out, he would hurl large stones at her, crushing the life out of her tender body. When all movement had ceased in the bruised and broken body of his poor victim, he would fill the grave with earth and return home. Sometimes he would brag about what he had done.

This custom was not, of course, very widespread in Arabia. Among the famous families of Makkah-the Bani Hashim, the Bani Umayyah and the Bani Makhzum-there is not a single instance on record of a female child being killed. This happened only among some desert tribes, and only in some clans. But even the exceptional occurrence of this revolting practice was sufficient to horrify and sicken the more intelligent and virtuous Arabs of the time.

Then there were the idols of Makkah. The Kabah had been built by the Prophet Ibrahim as the House of God, but had been defiled with gods of wood and stone. The Arabs would propitiate these gods with sacrificial offerings, believing that they would harm a man when angered and be bountiful when pleased. In and around the Kabah there were 360 idols, the most worshipped of whom were Hubal, Uzza and Lat. Hubal, the pride of the Arab pantheon, was the largest of these gods and was carved of red agate. When the inhabitants of Makkah had imported this idol from Syria it was without a right hand; so they fashioned a new hand of gold and stuck it on to its arm.

In the religion of the Arabs there was a curious mixture of polytheism and belief in Allah-the true God. They believed that Allah was Lord and Creator, but they also believed in the idols, regarding them as sons and daughters of Allah. The position of the deity in the

Arab mind was like that of a divine council, God being the President of the council of which these other gods and goddesses were members, each having supernatural powers, though subservient to the President. The Arabs would swear by Hubal or by another god or goddess. They would also swear by Allah. They would name their sons Abdul Uzza, i.e. the Slave of Uzza. They would also name their sons Abdullah i.e. the Slave of Allah.

It would not be correct to suggest that everything was wrong with the Arab culture of the time. There was much in their way of life which was glorious and chivalrous. There were qualities in the Arab character which would be enviable today-courage, hospitality and a sense of personal and tribal honour. There was also an element of vindictiveness, in the blood feuds which were passed down from father to son, but this was understandable, and even necessary, in a tribal society where no central authority existed to enforce law and order. Violent tribal and personal retaliation was the only way to keep the peace and prevent lawlessness.

What was wrong with Arab culture lay in the fields of ethics and religion, and in these fields Arab life had hit an all-time low. This period became known in history as the Ignorance. During the Ignorance Arab actions were acts of ignorance; Arab beliefs were beliefs of ignorance. The Ignorance was thus not only an era but an entire way of life.

The Arab mentioned at the beginning of this chapter took to retiring to a cave in a hill not far from Makkah, for one month every year. In this cave he would spend his time in meditation and reflection, and he would wait-not knowing just what he was waiting for. Then one day, while he was meditating in the cave; he suddenly became conscious of a presence. He could see no one and there was no sound of movement, but he could feel that someone was there. Then a voice said, *"Read!"*

Alarmed by the phenomenon of the disembodied voice, the Arab exclaimed, *"What shall I read?"* The voice was louder as it repeated, *"Read!"* Again the Arab asked, *"What shall I read?"* The voice now seemed terrible as it called sternly, *"Read!"* Then the voice continued in a more gentle tone:

Read: in the name of your Lord who created,
Created man from a clot.
Read: and it is your Lord the Bountiful
Who taught by the pen;
Taught man that which he knew not. [Quran 95: 1-5]

This happened on a Monday in the month of August, 610 CE. The world would never be the same again, for Muhammad had received his first revelation. A new faith was born.

When Muhammad (SAWS) received this revelation, Khalid was 24 years old.

For three years the Prophet remained silent, receiving guidance through the Angel Jibril. Then he was ordered to start expounding the religion of Allah, and he started with his own family and clan. Most of them, however, scorned his teaching and made fun of the new faith.

One day the Prophet decided to collect his closer relatives and give them a good meal at his house. This would give him an opportunity to get them together and put them in a situation where they would have to listen to him. The meal was duly arranged and heartily eaten by the guests. The Prophet then addressed the assembled guests and said, *"O Bani Abdul Muttalib! By Allah, I do not know of any man among the Arabs who has come to you with anything better than I have brought you. I bring you the best of this world and the next. I have been ordered by Allah to call you to Him. Who will help me in this work and be my brother and deputy?"*

The response of the entire gathering was silence. No one replied, each watching the others to see if anyone would get up to support this man. And then a thin, under-sized boy with skinny legs, in his early teens, sprang up and piped in a voice which had not yet broken, *"I, O Prophet of Allah, will be your helper!"*

There was a roar of laughter from the guests at what appeared at the time to be a ridiculous sight-rude and contemptuous laughter-as they stood up and began to walk away. But the boy was impervious to such rudeness, for the next instant he had been clasped by the Prophet in a loving embrace. The Prophet declared, *"This is my brother and deputy."*1 The boy was the Prophet's cousin-Ali, son of Abu Talib. He was the first male to accept Islam at the hands of the Prophet .2

Gradually the truth began to spread; and a few individuals, mostly youths or weak, helpless people, accepted the new faith. Their number was small but their courage was high. And the Prophet's sphere of activity widened. In spite of the rebuffs and insults which were hurled at him by the Quraish, he continued to accost people at street corners and in the market place and to warn them of the Fire which awaited the evil-doer. He would decide their idols of wood and stone and call them to the worship of the true God. As his activities increased, the opposition of the Quraish became harder and more vicious. This opposition was directed mainly by four men: Abu Sufyan (whose personal name was Sakhr bin Harb, and who was the leader of the Bani Umayyah), Al Waleed (father of Khalid), Abu Lahab (uncle of the Prophet) and Abul Hakam. Of the first and the last we will hear a lot more in this story.

Abu Sufyan and Al Waleed were men of dignity and self-respect. While they directed the opposition against the Prophet, they did not demean themselves by resorting to violence or abuse. Al Waleed's initial reaction was one of ruffled dignity. *"Is the prophethood to be bestowed on Muhammad,"* he exploded, *"while I, the greatest of the Quraish and their elder, am to get nothing? And there is Abu Masud, the chief of the Saqeef. Surely he and I are the greatest of the two towns."*3 This grand old man lived in a world of his own where everything depended on nobility of birth and rank. He was, of course, being unfair to the Prophet, for the line of Muhammad joined his own six generations back, and the family of Muhammad was no less noble than his own. In fact, in recent history the Prophet's family had acquired greater prominence than any other family in Makkah. The Prophet's grandfather, Abdul Muttalib, had been the chief of all the Quraish in Makkah.

1. Tabari: Vol. 2, p. 63; Ibn Sad: Vol. 1, p. 171.
2. Ibn Hisham: Vol. 1, p. 245; Tabari: Vol. 2, p. 56. Masudi: Muruj; Vol. 2, p. 283.
3. Ibn Hisham: Vol. 1, p. 361.Quran 95:1-5

According to Ibn Hisham, it was in connection with this statement of Al Waleed that the Quranic verse was revealed: And they say: **If only this Quran had been revealed to some man from the, two great towns! [Quran 43:31]**. The two towns were Makkah and Taif. And another Quranic revelation believed to have referred to Al Waleed who, as we have stated in the preceding chapter, was known by the title of Al Waheed (the Unique), reads: **Leave Me (to deal) with him whom I created Waheed; and bestowed upon him ample means; and sons abiding in his presence; and made (life) smooth for him. Yet he desires that I should give more. Nay, for lo, he has been stubborn about Our revelations. On him I shall impose a fearful doom…Then he looked; then he frowned and showed displeasure; then he turned away in pride and said: This is nothing but magic from of old; this is nothing but the speech of a man. Him shall I fling into the fire: [Quran 74: 11-17 and 21-26]**

The most blood-thirsty and vindictive of these leaders was Abul Hakam-cousin and friend of Khalid. As a result of his violent opposition to Islam he was given by the Muslims the nickname of Abu Jahl, *the Ignorant One*, and it is by this name that posterity was to know him. A small, tough and wiry man with a squint, he has been described by a contemporary as: *"a man with a face of iron, a look of iron and a tongue of iron."*[1] And Abu Jahl could not forget that in their younger days, in a fierce wrestling match, Muhammad had thrown him badly, gashing his knee, the scar of which was to remain until his death.[2]

These prominent men of the Quraish, and some others, finding it impossible to stop the Prophet by either threat or inducement, decided to approach the aged and venerable Abu Talib, uncle of the Prophet and leader of the Bani Hashim. They would have killed the Prophet but for the strong sense of tribal and family unity which protected the Prophet. His killing would have led to a violent blood feud with the Bani Hashim, who would undoubtedly have taken revenge by killing the killer or a member of the killer's family.

The delegation of the Quraish now approached Abu Talib and said *"O Abu Talib! You are our leader and the best among us. You have seen what the son of your brother is doing to our religion. He abuses our gods. He vilifies our faith and the faith of our fathers. You are one of us in our faith. Either stop Muhammad from such activities or permit us to deal with him as we wish."* [3]

Abu Talib spoke gently to them, said that he would look into the matter, and dismissed them with courtesy. But beyond informing the Prophet of what the Quraish had said, he did nothing to stop him from spreading the new faith. Abu Talib was a poet. Whenever anything of this sort happened, he would compose a long poem and pour all his troubles into it.

In the house of Al Waleed, the actions of the Prophet became the most popular topic of conversation. In the evening Al Waleed would sit with his sons and other relatives and recount the actions of the day and all that the Quraish were doing to counter the movement of Muhammad. Khalid and his brothers heard their father describe the entire proceedings of the first delegation to Abu Talib. Some weeks later, they listened to him tell all about the second delegation to Abu Talib, which had no more effect than the first. The Prophet continued with his mission.

Then Al Waleed took a bold step. He decided to offer his own son, Ammarah, to Abu Talib in return for the person of Muhammad. Ammarah was a fine, strapping youth in whom men and women saw all the virtues and graces of young manhood. The Quraish delegation approached Abu Talib with Ammarah in tow. *"O Abu Talib"* said the delegates. *"Here is Ammarah, son of Al Waleed. He is the finest of youths among the Quraish, and the handsomest and noblest of all. Take him as your son. He will help you and be yours as any son could be. In return give us the son of your brother-the one who has turned against your faith and the faith of your fathers and has caused dissension in our tribe. We shall kill him. Is that not fair-a man for a man?"*

Abu Talib was shocked by the offer. *"I do not think that it is fair at all,"* he replied. *"You give me your son to feed and bring up while you want mine to kill. By Allah, this shall not be."*[4] The mission failed. We do not know how Ammarah reacted to the failure of the mission-with disappointment or relief!

1. Waqidi: Maghazi, p.20; Ibn Rusta p. 223.
2. Tabari: Vol. 1, p. 265; Ibn Sad: p. 186.
3. Ibn Hisham: Vol. 1, p. 265; Ibn Sad: p. 186.
4. Ibn Hisham: Vol. 1, p. 267; Ibn Sad: p. 186.

Now seeing no hope of persuading Abu Talib to stop the Prophet and despairing of persuading him themselves, the Quraish decided to make the life of Muhammad and his followers so wretched that they would be forced to submit to the wishes of the Quraish. They set the vagabonds of Makkah against him. These hooligans would shout and jeer at the Prophet wherever he passed, would throw dust into his face and spread thorns in his path. They would fling filth into his house, and in this activity they were joined by Abu Lahab and Abu Jahl. This ill treatment was soon to enter a more violent phase.

As the persecution of the Muslims gathered momentum, it also increased in variety of method. One man got the bright idea that he would hurt Muhammad's cause by challenging him to a wrestling match, and thus belittle and humiliate him in a public contest. This man was an unbelieving uncle of the Prophet by the name of Rukkana bin Abd Yazid, a champion wrestler who was proud of his strength and skill. No one in Makkah had ever thrown him. *"O son of my brother!"* he accosted the Prophet. *"I believe that you are a man. And I believe that you are not a liar. Come and wrestle with me. If you throw me I shall acknowledge you as a true prophet."* The man was delighted with himself at having thought up this unusual way of lowering the stock of Muhammad in the eyes of the Makkans. Muhammad would either decline, and thus look small, or accept and get the thrashing of a lifetime. But that is what he thought. His challenge was accepted, and in the wrestling match that ensued the Prophet threw him three times! But the scoundrel went back on his word.[1]

The Prophet himself was reasonably safe from physical harm, partly because of the protection of his clan and partly because he could give better than he took in a fight. But there were other Muslims who were in a vulnerable position-those who were not connected with powerful families or were physically weak. They included slaves and slave girls. There was one slave girl the news of whose conversion so infuriated Umar that he beat her. He continued to beat the poor girl until he was too tired to beat her any more. And Umar was a very strong man!

Many of the men and women were tortured by the Quraish, The most famous of these sufferers, of whom history speaks in glowing terms, was Bilal bin Hamamah-a tall, gaunt Abyssinian slave who was tortured by his own master, Umayyah bin Khalf. In the afternoon, during the intense heat of the Arabian summer, when the sun would dry up and bake everything exposed to it, Bilal would be stretched out on the burning sand with a large rock on his chest and left to the tender mercies of the sun. Every now and then his master would come to him, would look at his suffering, tormented face, his dry lips and his swollen tongue, and would say, *"Renounce Muhammad and return to the worship of Lat and Uzza."* But the faith of Bilal remained unshaken. Little did Umayyah bin Khalf know, while he was torturing Bilal, that he and his son would one day face his erstwhile slave in the Battle of Badr, and that Bilal would be his executioner and the executioner of his son.

Bilal and several other slaves, all victims of torture, were purchased by Abu Bakr, who was a wealthy man. Whenever Abu Bakr came to know of a Muslim slave being tortured, he would buy and free him.

In spite of all this persecution, the Prophet remained gentle and merciful towards his enemies; He would pray: *"O Lord! Strengthen me with Umar and Abul Hakam."* His prayer was answered in so far as it concerned Umar, who became the fortieth person to embrace Islam [2]; but Abu Jahl remained an unbeliever and died in his unbelief.

In 619, ten years after the first revelation, Abu Talib died [3]. The Prophet's position now became more delicate. The hostility of the Quraish increased, and so did the danger to the life of Muslims. The Prophet remained surrounded by a few faithful companions to whom he continued to preach, and among these companions were 10 who were especially close to him. These men became known as The Blessed Ten, and were held in especial esteem and affection by the Muslims as long as they lived.[4]

1. According to Ibn Hisham (Vol. 1, p. 390) the Prophet himself challenged Rukkana, but I have narrated Ibn-ul-Asir's version (Vol. 2, pp. 27-28), as the event is more likely to have happened this way.
2. This is Ibn Qutaibah's placing (p. 180). Tabari, however, places Umar as the 67th Muslim (Vol. 3, p. 270).
3. Ten years reckoning by the lunar year, which is, at an average, 11 days shorter that the solar year.
4. For the names of these 10 men, see the Companions page or Note 1 in Appendix B.

The Prophet remained in Makkah, bearing up against what became increasingly more unbearable. Then some men of Madinah (at the time known as Yathrib) met the Prophet and accepted Islam. Knowing the danger to which the Prophet was exposed, they invited him to migrate to their settlements and make his home with them. With this invitation came Allah's permission for the Muslims to migrate, and the Prophet sent most of them to Madinah.

In September 622, the Quraish finally made up their minds to assassinate Muhammad. On the eve of the planned assassination, during the night, the Prophet left his house and, accompanied by Abu Bakr, a slave and a guide, migrated to Yathrib. With his safe arrival at Yathrib, Madinah (as the place was now to be called) became the seat and center of the Muslim faith and the capital of the new Muslim State. The era of persecution was over.

Three months after the Prophet's departure from Makkah, Al Waleed called his sons to his death bed, He knew that he was dying. *"O my sons!"* he said. *"There are three tasks that I bequeath you. See that you do-not foil in carrying them out. The first is my blood feud with the Khuza'a. See that you take revenge. By Allah, I know that they are not guilty, but I fear that you will be blamed after this day. The second is my money, accruing from interest due to me, with the Saqeef, See that you get it back. Thirdly, I am due compensation or blood from Abu Uzeihar."*[1] This bad man married the daughter of Al Waleed and then put her away from him without returning her to her father's home.

Having made these bequests, Al Waleed died. He was buried with all the honour due to a great chief, a respected elder and a noble son of the Quraish.

The first of the problems was settled without too much difficulty; the Khuza'a paid blood money, and the matter was closed without violence. The second matter remained pending for many years, and was then shelved as unsettled. As for the third problem, i.e. the feud with the son-in-law of Al Waleed, Khalid's brother, Hisham, decided that he would be content with nothing less than the blood of Abu Uzeihar. He waited more than a year before he got his chance. Then he killed his man. The matter assumed an ugly aspect, and there was danger of further bloodshed between the two families; but Abu Sufyan intervened and made peace. No more blood was shed.

During the years following his father's death, Khalid lived peacefully in Makkah, enjoying the good life which his wealth made possible. He even travelled to Syria with a trade caravan, to a large town called Busra, which he was to approach many years later as a military objective.

We do not know how many wives or children he had at this time, but we know of two sons: the elder was called Sulaiman, the younger, Abdur-Rahman. The latter was born about six years before the death of Al Waleed, and was to achieve fame in later decades as a commander in Syria. But according to Arab custom, it was Sulaiman by whose name Khalid became known. Thus he was called variously: Khalid, his own name; Ibn Al Waleed, i.e. the son of Al Waleed; and Abu Sulaiman, i.e. the father of Sulaiman. Most people addressed him as Abu Sulaiman.

1. Ibn Hisham: Vol. 1, pp. 410-411.

Chapter 3

(The Battle of Uhud)

"Allah did indeed fulfil His Promise to you, when you were about to annihilate the enemy with His permission, until you flinched and fell to disputing about the command, and disobeyed after He showed you what you covet. Among you were some that hankered after this world and among you were some that desired the Hereafter. Then did He divert you from them in order to test you. But He has forgiven you, for Allah is full of grace to those who have faith."
[Quran 3:152]

Everybody in Makkah rejoiced at the arrival of the caravan from Palestine. The caravan had been in grave danger during the few days it moved along the coastal road near Madinah and very nearly fell into the hands of the Muslims. It was only the skill and leadership of Abu Sufyan, who led the caravan that saved it from capture. The caravan consisted of 1,000 camels and had taken goods worth 50,000 dinars, on which Abu Sufyan had made a cent per cent profit. Since every family of note in Makkah had invested in this caravan, its return with so much profit was a matter of jubilation for all Makkah. And it was spring in Arabia: the month of March, 624.

Even as the people of Makkah sang and danced, and the merchants rubbed their hands while awaiting their share of the profit, the battered and broken army of the Quraish picked its weary way towards Makkah. This army had rushed out in response to Abu Sufyan's call for help when he had first realised the danger from the Muslims. Before the Quraish army could come into action, however, Abu Sufyan had extricated the caravan and sent word to the Quraish to return to Makkah as the danger had passed. But Abu Jahl, who commanded the army, would have none of this. He had spent the past 15 years of his life in bitter opposition to the Prophet, and he was not going to let this opportunity slip away. Instead of returning, he had precipitated a battle with the Muslims.

Now this proud army was returning home in a state of shock and humiliation.

While the Quraish army was still on its way, a messenger from it sped to Makkah on a fast camel. As he entered the outskirts of the town, he tore his shirt and wailed aloud, announcing tragedy. The people of Makkah hastily gathered around him to seek news of the battle. They would ask about their dear ones and he would tell of their fate. Among those present were Abu Sufyan and his wife, Hind.

From this messenger Hind heard of the loss of her dear ones; of the death of her father, Utbah, at the hands of Ali and Hamza, uncle of the Prophet; of the death of her uncle, Sheiba, at the hands of Hamza; of the death of her brother, Waleed, at the hands of Ali; of the death of her son, Handhalah, at the hands of Ali. She cursed Hamza and Ali and swore vengeance.

The Battle of Badr was the first major clash between the Muslims and their enemies. A small force of 313 Muslims had stood like a rock against the onslaught of 1,000 infidels. After an hour or two of severe fighting the Muslims had shattered the Quraish army, and the Quraish had fled in disorder from the battlefield. The finest of the Quraish had fallen in battle or been taken prisoner.

A total of 70 infidels had been killed and another 70 captured by the Muslims, at a cost of only 14 Muslim dead. Among those killed were 17 members of the Bani Makhzum, most of them either cousins or nephews of Khalid. Abu Jahl had been killed. Khalid's brother, Waleed, had been taken prisoner.

As the messenger announced the names of those who had fallen and those who had killed them, the Quraish noted the frequency with which the names of Ali and Hamza were repeated. Ali had killed 18 men by himself and had shared in the killing of four others. Hamza had killed four men and shared with Ali in the killing of another four. The name of Ali thus dominated the proceedings of this sad assembly.

Two days later Abu Sufyan held a conference of all the leaders of the Quraish. There was not one amongst them who had not lost a dear one at Badr, Some had lost fathers, some sons, some brothers. The most vociferous at the conference were Safwan bin Ummayya and Ikrimah, son of Abu Jahl.

Ikrimah was the most difficult to restrain. His father had had the distinction of commanding the Quraish army at Badr and had fallen in battle. The son drew some comfort from the fact that his father had killed a Muslim at Badr and that he himself had killed another. Moreover, he had attacked and severed the arm of the Muslim who had mortally wounded his father; but that was not enough to quench his thirst for revenge. He insisted that as noble Quraish they were honour-bound to take revenge.

"And I have lost my son, Handhalah." said Abu Sufyan. *"My thirst for revenge is no less than yours. I shall be the first to prepare and launch a powerful expedition against Muhammad."*[1]

At this conference they all took the pledge of revenge; this time none would stay back. An expedition would be prepared such as had never assembled at Makkah before, and other local tribes would be invited to join the expedition and take part in the annihilation of the Muslims. The entire profit from the caravans, amounting to 50,000 dinars, would be spent on financing the expedition- Abu Sufyan was unanimously elected as the commander of the Quraish army.

Abu Sufyan now gave two decisions, the first of which was more or less universally accepted. This was to the effect that there should be no weeping and no mourning of any kind for those who had fallen at Badr. The idea behind this order was that tears would wash away the bitterness in their hearts, and that this bitterness should be kept alive until they had taken their revenge against the Muslims. However, those whose burden of sorrow was too heavy to carry wept secretly.

The second decision related to the prisoners who were in Muslim hands. Abu Sufyan forbade all efforts to get them released for fear that if these efforts were made immediately, the Muslims might put up the price. This decision, however, was not followed by everyone. Within two days a man left Makkah secretly at night to ransom his father; and when others came to know about this, they took the matter into their own hands and got their dear ones released. Abu Sufyan had no choice but to revoke his decision.

The rate of ransom varied. The top rate was 4,000 dirhams and there was a graduated scale down to 1,000 dirhams for those who could not afford to pay more. A few prisoners who were too poor to pay but were literate, earned their freedom by teaching a certain number of Muslim children to read and write. Some destitute ones were released by the Prophet without ransom on condition that they would never again take up arms against Muslims.

Among those who went to negotiate the release of the prisoners were Ikrimah, Khalid (who had missed the battle of Badr on account of his absence from the Hijaz) and Khalid's brother, Hisham. Khalid and Hisham arranged the release of their brother, Waleed. When Hisham heard that the ransom would be 4,000 dirhams, he began to haggle for a lower sum but was rebuked by Khalid. The sum of 4,000 dirhams was duly paid for the release of Waleed, where after the three brothers left Madinah and camped for the night at a place called Zhul Halifa, a few miles away. Here, during the night, Waleed slipped away from the camp, returned to Madinah, reported to the Prophet and became a Muslim. He thereafter proved a devout Muslim and became very dear to the Prophet; and in spite of his new faith, his relations with Khalid remained as warm and loving as ever.

While at the Quraish conference the main theme of the discussion had been revenge, another factor which drove the Quraish to war with the Muslims was economic survival. The main route of the Quraish caravan to Syria and Palestine lay along the coastal road which now, after the Battle of Badr, was no longer open to them. In November, Safwan bin Ummayya felt the need for more trade, and dispatched a caravan towards Syria on another route which he thought might be safe. This caravan left Makkah on the road to Iraq, and after travelling some distance turned north-west towards Syria, bypassing Madinah at what Safwan considered a safe distance. But the Holy Prophet came to know of this caravan and sent Zaid bin Harithah with 100 men to capture it, which Zaid did.

Safwan then went to Abu Sufyan, and both leaders agreed that since the economic well-being and prosperity of the Quraish depended on their profitable trade with Syria, the sooner the Muslims were crushed the better. Ikrimah also was impatient and pressed for speed. Abu Sufyan, however, as a wise old chief, knew that it would take time to prepare the expedition and purchase the camels, the horses and the weapons. He promised to do his best.

The preparations for the expedition now began in right earnest. While they were in progress, an unbeliever of doubtful character approached Abu Sufyan with a proposal. This man was Abu Amir of Madinah. He had taken exception to the arrival of the Holy Prophet at Madinah and to the speed with which members of his own clan, the Aws, had begun to embrace Islam. Consequently he had left Madinah and sworn never to return as long as Muhammad remained in power. At Makkah he took to inciting the Quraish against the Muslims. In the old days Abu Amir had been known as the Monk, but the Holy Prophet had given him the nickname of the Knave! Thus the Muslims knew this man as Abu Amir the Knave.[2]

1. Waqidi: *Maghazi*, pp. 156-7
2. Ibn Hisham: Vol. 2, p. 67.

"I have 50 members of my clan with me", he said to Abu Sufyan. *"I have much influence with my clan, the Aws. I propose that before the battle begins I be permitted to address the Aws among the Muslims, and I have no doubt that they will all desert Muhammad and come over to my side."* [1] Abu Sufyan gladly accepted the arrangement. The Aws were one of the two major tribes of Madinah and would comprise more than a third of the Muslim army.

Parleys were begun with neighbouring tribes, and strong contingents were received from the Kinana and the Thaqeef. Early in March 625, the assembly of the expedition began at Makkah. At this time Abbas, uncle of the Prophet, wrote to him from Makkah to inform him of the preparations being made against him.

In the second week of March, the Quraish set out from Makkah with an army of 3,000 men, of whom 700 were armoured. They had 3,000 camels and 200 horses. With the army went 15 Quraish women in litters, whose task it was to remind the Quraish of the comrades who had fallen at Badr and to strengthen their spirits. Among these women was Hind, who acted as their leader, and the role came naturally to her. Others were the wife of Ikrimah, the wife of Amr bin Al Aas and the sister of Khalid. One of the women, whom we shall hear of again later, was Amrah bint Alqama, and there were also some songstresses who carried tambourines and drums.

As the expedition moved towards Madinah, one of the leaders of the Quraish, Jubair bin Mut'im, spoke to his slave, who was known as the *Savage*-Wahshi bin Harb. *"If you kill Hamza, the uncle of Muhammad, in revenge for the killing of my uncle at Badr, I shall free you."* 2 The Savage liked the prospect. He was a huge, black Abyssinian slave who always fought with a javelin from his native Africa. He was an expert with this weapon and had never been known to miss.

After travelling a little further, the Savage saw one of the litter-carrying camels move up beside him. From the litter Hind looked out and spoke to the Savage. *"O Father of Blackness!"* she addressed him. *"Heal, and seek your reward."* 3 She promised him that if he would kill Hamza in revenge for his killing her father, she would give him all the ornaments that she was wearing.

The Savage looked greedily at her ornaments-her necklace, her bracelets, the rings that she wore on her fingers. They all looked very expensive and his eyes glittered at the prospect of acquiring them.

The Holy Prophet had been warned by Abbas of the Quraish preparations before they left Makkah. While they were on their way, he continued to receive information of their progress from friendly tribes. On March 20, the Quraish arrived near Madinah and camped a few miles away, in a wooded area west of Mount Uhud. On this very day the Prophet sent two scouts to observe the Quraish, and these scouts returned to give their exact strength.

On March 21, the Prophet left Madinah with 1,000 men, of whom 100 were armoured. The Muslims had two horses, of which one was the Prophet's. They camped for the night near a small black hillock called Shaikhan, a little over a mile north of Madinah.

The following morning, before the march was resumed, the Hypocrites, numbering 300 under the leadership of Abdullah bin Ubayy, left the Prophet on the plea that fighting the Quraish outside Madinah had no prospect of success, and that they would not take part in an operation which in their view was doomed to failure. The Hypocrites returned to Madinah. The Prophet was now left with 700 men; and with this strength he marched from the camp. The Prophet had not actually intended to fight outside Madinah. It had been his wish that the Muslims should await the arrival of the Quraish on their home ground and fight the battle in Madinah; but most of the Muslims had insisted that they go out to meet the Quraish, and so the Prophet, submitting to their demand, had marched out to give battle to the Quraish outside Madinah. But although he was going out to meet his enemy in the open, he would nevertheless fight the battle on ground of his own choice. He moved to the foot of Mount Uhud and deployed for battle.

1. Waqidi: *Maghazi*, p. 161
2. Ibn Hisham: Vol. 2, pp. 61-2.
3. *Ibid.*

Uhud is a massive feature lying four miles north of Madinah (the reference point in Madinah being the Prophet's Mosque) and rising to a height of about 1,000 feet above the level of the plain. The entire feature is 5 miles long. In the western part of Uhud, a large spur descends steeply to the ground, and to the right of this spur, as seen from the direction of Madinah, a valley rises gently and goes up and away as it narrows at a defile about 1,000 yards from the foot of the spur. Beyond this defile it shrinks into nothingness as it meets the main wall of the ridge. At the mouth of this valley, at the foot of this spur, the Prophet placed his army. The valley rose behind him.

He organised the Muslims as a compact formation with a front of 1,000 yards. He placed his right wing at the foot of the spur and his left wing at the foot of a low hill, about 40 feet high and 500 feet long, called Ainain. The Muslim right was safe, but their left could be turned from beyond Ainain; so, to meet this danger, the Prophet placed 50 archers on Ainain, from which they could command the approaches along which the Quraish could manoeuvre into the Muslim rear. These archers, under the command of Abdullah bin Jubair, were given instructions by the Prophet as follows; *"Use your arrows against the enemy cavalry. Keep the cavalry off our backs. As long as you hold your position, our rear is safe. On no account must you leave this position. If you see us winning, do not join us; if you see us losing, do not come to help us."* [1] The orders to this group of archers were very definite. Since Ainain was an important tactical feature and commanded the area immediately around it, it was imperative to ensure that it did not fall into the hands of the Quraish.

Behind the Muslims stood 14 women whose task it was to give water to the thirsty, to carry the wounded out of battle and to dress their wounds. Among these women was Fatimah, daughter of the Prophet and wife of Ali. The Prophet himself took up his position with the left wing of his army.

The Muslim dispositions were intended to lead to a frontal positional battle and were superbly conceived. They gave the Muslims the benefit of fully exploiting their own sources of strength-courage and fighting skill. They also saved them from the dangers posed by the Quraish strength in numbers and in cavalry-the mobile manoeuvre arm which the Muslims lacked. It would have suited Abu Sufyan to fight an open battle in which he could manoeuvre against the Muslim flanks and rear with his cavalry and bring his maximum strength to bear against them. But the Prophet neutralised Abu Sufyan's advantages, and forced him to fight on a restricted front where his superior strength and his cavalry would be of limited value. It is also worth noting that the Muslims were actually facing Madinah and had their backs to Mount Uhud; the road to Madinah was open to the Quraish.

Now the Quraish moved up. They established a battle camp a mile south of the spur, and from here Abu Sufyan led his army forward and formed it in battle array facing the Muslims. He organised it into a main body of infantry in the centre with two mobile wings. On the right was Khalid and on the left Ikrimah, each with a cavalry squadron 100 strong. Amr bin Al Aas was appointed in over-all charge of the cavalry, but his task was mainly that of co-ordination. Abu Sufyan placed 100 archers ahead of his front rank for the initial engagement. The Quraish banner was carried by Talha bin Abi Talha, one of the survivors of Badr. Thus the Quraish deployed with their backs to Madinah, facing the Muslims and facing Mount Uhud. In fact they stood between the Muslim army and its base at Madinah.

Just behind the Quraish main body stood their women. Before battle was joined, these women, led by Hind, marched back and forth in front of the Quraish, reminding them of those who had fallen at Badr. Thereafter, just before the women withdrew to their position in the rear of the army, the clear, strong voice of Hind rose as she sang:

O you sons of Abduddar!
Defenders of our homes!
We are the daughters of the night;
We move among the cushions.
If you advance we will embrace you.
If you retreat we will forsake you
With loveless separation. 2

1. Ibn Hisham: Vol. 2, pp. 65-66; Waqidi: *Maghazi*, p. 175.
2. Ibn Hisham: Vol. 2, p. 68. Waqidi: *Maghazi*, p. 176.

It was the morning of Saturday, March 22, 625 (the 7th of Shawwal, 3 Hijri)-exactly a year and a week after Badr. 1 The armies faced each other in orderly ranks, 700 Muslims against 3,000 unbelievers. This was the first time that Abu Sufyan had commanded in the field against the Prophet, but he had able lieutenants and felt certain of victory. The Muslims repeated to themselves the Quranic words: **"Sufficient for us is Allah, and what a good protector He is." [Quran 3: 173]** And they awaited the decision of Allah.

The first event, after the forming up of the two armies, was the attempt by the Knave to subvert the Aws. This man stepped forward ahead of the front rank of the Quraish, along with his 50 followers and a large number of the slaves of the Quraish. He faced the Aws and called, *"O people of the Aws! I am Abu Amir. You know me!"* The reply from the Aws was unanimous: *"No welcome to you, O Knave!"* This was followed by a shower of stones hurled with great delight by the Aws at the Knave and his group, under which the group hastily withdrew through the ranks of the Quraish. Observing the look of derision on the faces of the Quraish, the Knave assumed a prophetic posture and observed, *"After me my people will suffer."* 2 But the Quraish were not impressed!

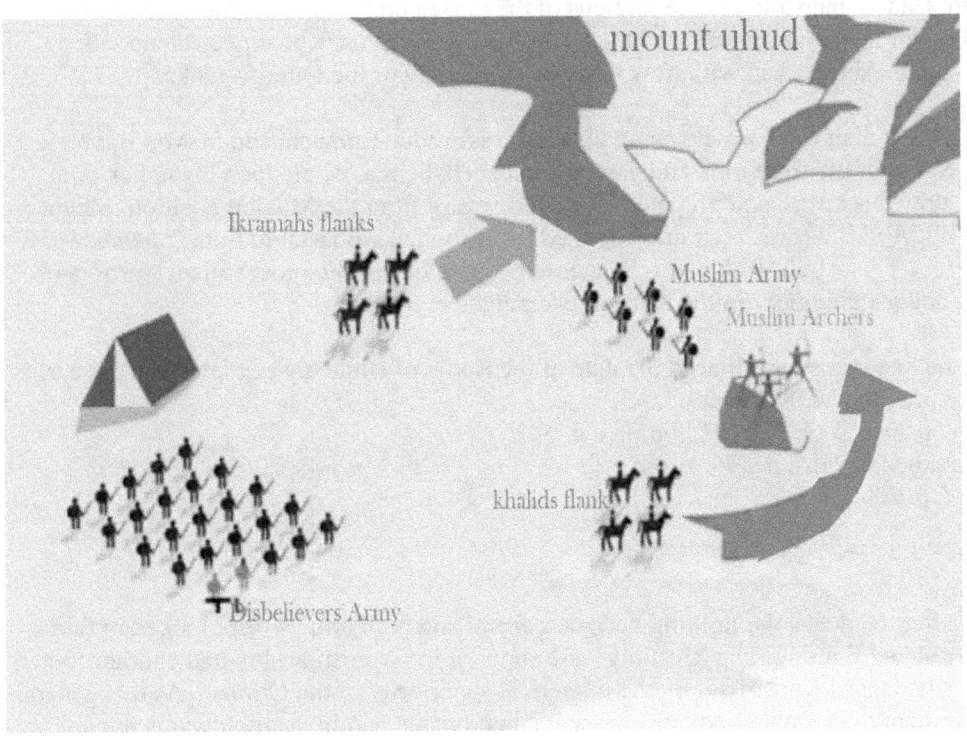

After the encounter of the Knave, the archers opened up from both sides. This was a kind of artillery duel between the 100 archers of the Quraish and the Muslim archers, who were dispersed along the front rank of the Muslims. Many salvoes were fired. Under cover of the Quraish archers Khalid advanced with his squadron to attack the left wing of the Muslims, but was forced back by accurate fire from the Muslim archers. As the archers' engagement ended, the song of the Quraish women was again heard on the battlefield: *"We are the daughters of the night..."*

The next phase was the phase of duels by the champions of the two armies. Talha, the standard bearer of the Quraish, stepped out of the front rank and called *"I am Talha, son of Abu Talha. Will anyone duel?"* [3] On his challenge, Ali strode out and before Talha could deliver a single blow, Ali struck him with his sword and felled him. Talha was only wounded, and as Ali raised his sword to strike again, Talha begged for mercy. Ali promptly turned away. Later, however, while the general engagement was in progress, the wounded Talha was despatched by the Muslims. On the fall of Talha, another infidel came forward and picked up the Quraish standard. This man was killed by Hamza, As Hamza killed him, he was noticed by the Savage who stood behind the Quraish ranks. Stealthily the Savage began to move towards the right in order to approach Hamza from a flank. Hamza was easily recognisable by a large ostrich feather which he wore in his turban.

Now the duels became more general. One after the other the relatives of Talha picked up the standard, and one after the other they were killed by the Muslims, the largest number falling before Ali's sword, Abu Sufyan also rode up to duel and was faced by Handhalah bin Abu Amir, who was dismounted. Before Abu Sufyan could use his lance or draw his sword, Handhalah struck at the forelegs of the horse and brought it down. Abu Sufyan shouted for help and was assisted by one of his companions, who engaged and killed Handhalah. Abu Sufyan withdrew hastily to the safety of the Quraish ranks.

Another Quraish warrior who came forward was Abdur-Rahman, son of Abu Bakr. He stepped out of the front rank and gave the usual challenge, whereupon his father, Abu Bakr, drew his sword and prepared to move forward from the Muslim position to fight him. But Abu Bakr was restrained by the Holy Prophet, who said to him, *"Sheathe your sword,"* 4 This Abdur-Rahman was later to become one of the most valiant warriors of Islam and acquire glory in the Muslim campaigns in Syria.

1. Some historians have placed the date of the Battle of Uhud a week later, but the earlier date is probably more correct.
2. Ibn Hisham: Vol. 2, p. 67; Ibn Sad: p. 543.
3. Waqidi: *Maghazi*, p. 176.
4. *Ibid*: p. 200.

Soon after the duels, the fighting became general and both armies were locked in fierce hand-to-hand fighting. The Muslims were superior in swordsmanship and courage, but these advantages were offset by the numerical superiority of the Quraish. As this general engagement of the main body progressed, Khalid made another sally towards the left wing of the Muslims, where the Prophet stood, but was again driven back by the Muslim archers on Ainain.

The Prophet himself participated in this action by firing arrows into the general mass of the Quraish. Beside him stood Sad bin Abi Waqqas, who was an arrow-maker by profession and was among the best archers of his time. The Prophet would indicate targets to Sad and Sad would invariably score a hit.

Hamza was fighting near the left edge of the Muslim force. By now he had killed two men and found a third one approaching him-a man named Saba bin Abdul Uzza, whom Hamza knew well. *"Come to me!"* shouted Hamza, *"O son of the skin-cutter!"* 1 (The mother of Saba used to perform circumcision operations in Makkah!) The colour rose in Saba's face as he drew his sword and rushed at Hamza.

As the two men began to duel with sword and shield, the Savage, crawling behind rocks and bushes, approached Hamza, At last he got within javelin range and with an experienced eye measured the distance between himself and his victim. Then he stood up and raised his javelin for the throw. Hamza struck a mortal blow on the head of Saba, and Saba fell in a heap at Hamza's feet. At this very moment the Savage hurled his javelin. The cruel weapon, thrown with unerring aim, struck Hamza in the abdomen and went right through his body. Hamza turned in the direction of the Savage and, roaring with anger, took a few steps towards him. The Savage trembled as he waited behind a large rock, but Hamza could only take a few steps before he fell.

The Savage waited until all movement had ceased in Hamza's body, and then walked up to the corpse and wrenched out his javelin. He then casually walked away from the scene of fighting. He had done his job. The Savage would fight more battles in his life, but there would be no more battles for the noble Hamza- "Lion of Allah and of His Prophet!" [2]

Soon after this, the Quraish army began to waver and the Muslims pressed harder in their assault. When several Quraish standard-bearers had been either killed or wounded, their standard was picked up by a slave who continued to fight with it until he too was killed and the standard fell again. As it fell, the Quraish broke and fled in disorder.

There was now complete panic in the ranks of the Quraish. The Muslims pursued them, but the Quraish outran their pursuers. The Quraish women wailed when they saw what had befallen their men. They also took to their heels; and raising their dresses in order to be able to run faster, gave a fine view of their flashing legs to the delighted Muslims. All the women ran except Amra, who remained where she had stood, close behind the original Quraish battle line.

The Muslims got to the Quraish camp and began to plunder it. There was complete confusion in the camp with women and slaves milling around, hoping not to be killed, while the Muslims rifled everything they could find and shouted with glee. There was now no order, no discipline, no control, for the Muslims felt that the battle was won. The first phase of the battle was indeed over. The casualties had been light, but the Quraish had been clearly defeated. This should have marked the end of the Battle of Uhud, but it did not.

As the Quraish fled and the Muslims, following in their footsteps, entered the Quraish camp, the two mobile wings of the Quraish stood firm. Both Khalid and Ikrimah moved back a bit from their previous positions but kept their men under complete control, not permitting a single rider to retreat. And Khalid now watched this confused situation, looking now at the fleeing Quraish, now at the plundering Muslims, now at the archers on Ainain. He did not quite know what to do; but he was capable of a high degree of patience and waited for an opportunity which would give him a line of action. Soon his patience was rewarded.

1. Ibn Hisham: Vol. 2, p. 70.
2. Waqidi: *Maghazi*, p. 225.

When the archers on Ainain saw the defeat of the Quraish and the arrival of the Muslims at the Quraish camp, they became impatient to take part in the plunder of the camp. The Quraish camp looked very tempting. They turned to their commander, Abdullah bin Jubair, and asked for permission to join their comrades, but Abdullah was firm in his refusal. *"You know very well the orders of the Messenger of Allah"*, he said. *"We are to remain on this hill until we receive his orders to leave it."* *"Yes, but that is not what the Messenger of Allah intended,"* the archers replied. *"We were to hold this hill during battle. Now the battle is over, and there is no point in our remaining here."* And in spite of the protests of their commander, most of the archers left the hill and ran towards the Quraish camp shouting, *"The booty! The booty!"* [1] Abdullah was left with nine archers on the hill. This movement was observed by the keen eyes of Khalid, who waited until the archers had reached the Quraish camp.

Then Khalid struck. He launched a mounted attack against the few archers who remained on the hill, with the intention of capturing this position and creating for himself room for manoeuvre, Ikrimah saw the movement of Khalid and galloped across the plain to join Khalid's squadron. As Khalid's squadron reached the top of the hill, Ikrimah's squadron was just behind while Ikrimah himself came ahead and began to take part in the assault on the Muslim archers.

The faithful archers who had remained on the hill resisted gallantly. Some were killed while the remainder, all wounded, were driven off the hill by the assault of Khalid. Abdullah bin Jubair, defending to the last the position which the Prophet had entrusted to him, suffered many wounds and was then slain by Ikrimah. Now Khalid's squadron, followed by Ikrimah's, swept forward and came in behind the line that had been held by the Muslims an hour ago. Here the two squadrons wheeled left and charged at the Muslims from the rear. Ikrimah with a part of his squadron assaulted the group which stood with the Holy Prophet, while Khalid's squadron and the remainder of Ikrimah's squadron attacked the Muslims in the Quraish camp.

Khalid drove into the rear of the unsuspecting Muslims, confident that having taken them unawares he would soon tear them to pieces. But the Muslims refused to be torn to pieces. As the Quraish cavalry reached the camp, there was an uproar in the ranks of the Muslims, and a few of them lost their heads and fled. Most of them, however, stayed and fought. As long as the Prophet lived, these men were not going to acknowledge defeat. But as the Muslims turned to fight the Quraish cavalry, Amra rushed towards the Quraish standard which lay on the ground. She picked up the standard and waved it above her head in the hope that the main body of the Quraish would see it.

By now Abu Sufyan had regained control over most of the infantry, He saw the movement of the cavalry. He saw the Quraish standard waving in the hands of Amra and he got his men back into action. Knowing that the Muslims had been taken in the rear by the cavalry, the Quraish rushed into battle once again, shouting their war cry: *"O for Uzza! O for Hubal!"* 2

The Muslims were now caught between two fires, the Quraish cavalry attacking from the rear and the bulk of the Quraish infantry attacking from the front. Abu Sufyan himself charged into battle and killed a Muslim. The situation soon became desperate for the Muslims, who broke up into small groups, each fighting on its own to repel the attacks of the cavalry and infantry. The confusion increased, and in the dust a few of the Muslims even began to fight each other. There was some alarm, but still no panic. Losses began to mount among the Muslims, but they held out-determined to fight to the last. At about this time, Khalid killed his first man-Abu Aseera-with his lance and knocked down another Muslim. Believing him dead, Khalid rode on; but the second man was only wounded and got up to fight again.

1. Waqidi: *Maghazi*, pp. 178-179.; Ibn Sad: pp. 545, 551.
2. Waqidi: *Maghazi*, p. 188; Ibn Sad: p. 545.

The battle was now divided into two separate actions. There was the main body of the Muslims holding out against the main part of the Quraish army, and there was the group with the Holy Prophet holding out against part of Ikrimah's squadron and some of the; Quraish infantry which had returned to attack him.

Now began the ordeal of the Prophet.

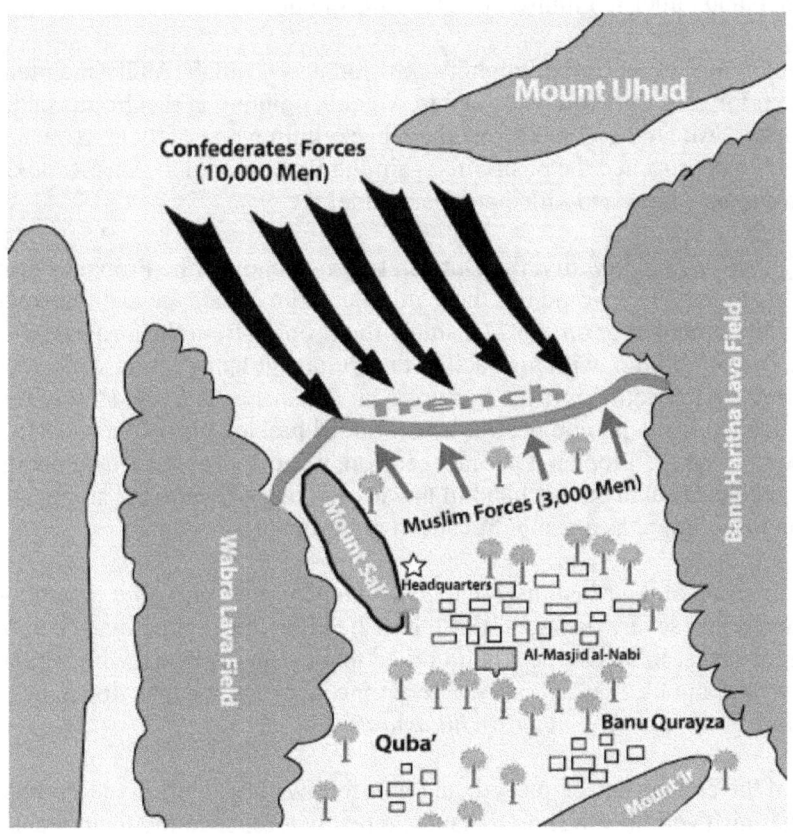

When the Muslims left their positions in pursuit of the Quraish, the Holy Prophet remained at his battle location. Here he had with him 30 of his Companions who stuck to him and refused to be tempted by the prospect of plunder. Among these 30 were some of the closest of his followers, including, Ali, Abu Bakr, Sad bin Abi Waqqas, Talha bin Ubaidullah, Abu Ubaidah, Abdur-Rahman bin Auf, Abu Dujanah and Mus'ab bin Umair. With the group were also present two women who had busied themselves with carrying water to the Muslims and had now joined the Prophet.

As Khalid captured the archers' position and the Quraish cavalry began to wheel round to attack the Muslims in the rear, the Prophet realised the seriousness of the predicament in which the Muslims were placed. He could do nothing to control and direct the actions of the main body, for it was too far away; and he knew that his own group would soon be under attack. His present position was utterly untenable, so he decided to move to the foot of the spur immediately behind him (not the spur at the foot of which the Muslim right wing had been placed), and with this intention he started to move backwards. But he had not gone more than about a quarter of a mile with his 30 Companions when Ikrimah with his horsemen moved up and barred his way. The Prophet determined to stand and fight where he stood; and it was not long before a Quraish infantry group also arrived to attack the Prophet.

The Prophet's group found itself assailed from front and rear. The Muslims formed a cordon around the Prophet to defend him and the fighting gradually increased in intensity. The Prophet himself used his bow to effect and continued to use it until it broke. Thereafter he used his own arrows to augment those of Sad, whose superb archery gave a great deal of trouble to the Quraish. Every Muslim took on an opposing group of three or four men and either fell himself or drove his opponents back.

The first of the Quraish to reach the Prophet's position was Ikrimah. As Ikrimah led a group of his men forward the Prophet turned to Ali and, pointing at the group, said, *"Attack those men."* Ali attacked and drove them back, killing one of them. Now another group of horsemen approached the position. Again the Prophet said to Ali, *"Attack those men."* 1 Ali drove them back and killed another infidel.

As the fighting increased in severity, the Quraish began to shower the Prophet's group with arrows and stones, They would use these missiles from a distance and then charge with swords, either mounted or on foot. To shield the Prophet from the arrows, Abu Dujanah stood in front of him, with his back to the Quraish infantry, from which came most of the arrows. After some time the back of Abu Dujanah was so studded with arrows that he looked like a porcupine, but he continued passing his own arrows to Sad. Talha also stood beside the Prophet. On one occasion, when an arrow seemed about to hit the Prophet in the face, Talha put his hand in the arrow's line of flight and stopped it with his hand. Talha lost a finger as a result, but saved the Prophet.

Against the main body of the Muslims, Khalid was launching assault after assault with his squadron and doing severe damage. About now he killed his second man-Sabt bin Dahdaha-with his lance. In this battle Khalid relied mainly on his lance, with which he would run down and impale his adversary. Every time he brought a man down, he would shout, *"Take that! And I am the Father of Sulaiman!"* 2

The first rush of the counter-attack passed, and was followed by a lull in the Prophet's sector, as the Quraish withdrew a short distance to rest before resuming their attacks. During this lull, one of the Muslims, noticed that the Prophet was looking cautiously over his shoulder. The man asked the reason for this, and the Prophet replied casually, **"I am expecting Ubayy bin Khalf. He may approach me from behind. If you see him coming, let him get near me".** He had hardly said this when a man detached himself from Ikrimah's squadron and slowly advanced towards the Prophet, mounted on a large, powerful horse. The man shouted, *"O Muhammad! I have come! It is either you or me!"* At this some of the Companions asked the Prophet for permission to deal with the man, but the Prophet said, **"Let him be!"** 3 The Companions moved aside, and left the way open for the rider to approach.

1. Tabari: Vol. 2, p. 197.
2. Waqidi: *Maghazi*, p.198.
3. *Ibid*: pp. 195-6; Ibn Hisham: Vol. 2, p. 84.

At the Battle of Badr, a young man by the name of Abdullah bin Ubayy (not to be confused with the Abdullah bin Ubayy who was the leader of the Hypocrites) was taken prisoner by the Muslims. His father, Ubayy bin Khalf, came to release his son and paid 4,000 dirhams as ransom. Once the ransom had been paid and the young man released, while still in Madinah, Ubayy had been insolent to the Prophet. He had said, *"O Muhammad! I have a horse which I am strengthening with a lot of fodder, because in the next battle I shall come riding that horse and I shall kill you"* The Prophet had then replied, **"No, you shall not kill me. But I shall kill you while you are on that horse, if Allah wills it."** 1 The man had laughed scornfully as he rode away with his son.

And now Ubayy bin Khalf was approaching the Prophet on his horse. He saw the Companions move out of the way. He saw the Prophet waiting for him, and grudgingly he admired the man he had set out to kill. The Prophet was wearing two coats of mail. He wore a chain helmet, the side-flaps of which covered his cheeks; His sword rested in its sheath, tucked into a leather belt, and in his right hand he held his spear. Ubayy noticed the powerful, broad shoulders of Muhammad; notice the large, hard hands-hands strong enough to break a spear in two. The Prophet looked a magnificent sight.

It is known to few people today that Prophet Muhammad was one of the strongest Muslims of his time. Add to his great personal strength the fact of divine selection, and one can imagine what a formidable opponent he would prove to anybody. But Ubayy was undaunted, He had just killed a Muslim, and his spirits were high.

The Prophet could easily have told his Companions to slay Ubayy. They would have fallen upon him and torn him to pieces. Or he could have given Ali the simple order, *"Kill that man"*, and that man would be as good as dead, for when Ali set out to kill a man nothing could save him. But the Prophet had ordered his Companions to stand aside. This time he wanted no help from anyone. This was a matter of personal honour-a matter of chivalry. Muhammad would fight alone as a chivalrous Arab. He would keep his rendezvous with a challenger.

As Ubayy reached the Prophet, he pulled up his horse. He was in no hurry. Not for a moment doubting that Muhammad would await his attack, he took his own time over drawing his sword. And then suddenly it was too late, for the Prophet raised his spear and struck at the upper part of Ubayy's chest. Ubayy tried to duck, but was not quick enough. The spear struck him on the right shoulder, near the base of the neck. It was a minor wound, but Ubayy fell off his horse, and in the fall broke a rib. Before the Prophet could strike again, Ubayy had risen and turned tail, running screaming towards his comrades. They stopped him and asked how he had fared, to which Ubayy replied in a trembling voice, *"By Allah, Muhammad has killed me."*

The Quraish examined his wound, and then told him not to be silly because it was a superficial wound which would soon heal. Ubayy's voice rose higher as he said, *"I shall die!"* When the Quraish tried to console him further, Ubayy lost all control over himself and in a frantic voice screamed, *"I tell you I shall die! Muhammad had said that he would kill me. If Muhammad were to just spit on me, I would die!"* [2] Ubayy remained inconsolable.

When the Quraish returned to Makkah, he went with them. While they were camped at a place called Saraf, not far from Makkah, the wretched man died. The cause of his death was certainly not the physical effect of the wound. And Allah knows best!

1. Ibn Sad: p. 549; Ibn Hisham: Vol. 2, p. 84.
2. Ibn Hisham: Vol. 2, p. 84.

The situation gradually became more desperate as the Muslims held on grimly and showed no sign of breaking up. Abu Sufyan and Khalid both wanted a quick decision, for the battle had gone on long enough. The Quraish therefore decided to press harder and if possible get at the Prophet, as his death would have the probable effect of ending resistance.

A strong group of Quraish infantry consequently advanced against the Prophet. The Muslim defenders continued to fight, and many of them were cut down. Three of the Quraish managed to break through the cordon and got within stone throwing distance of the Prophet. These three men were: Utbah bin Abi Waqqas, Abdullah bin Shahab and Ibn Qamiah. They all began to hurl stones at the Prophet.

The first (a brother of Sad) landed four stones on the Prophet's face, broke two of the Prophet's lower teeth and cut his lower lip. Abdullah managed to land one stone which gashed the Prophet's forehead, while Ibn Qamiah with one stone cut the Prophet's cheek and drove two links of the Prophet's chain helmet into his cheek bone.

The Prophet fell to the ground as a result of these blows and was helped up by Talha. At this moment the few Muslims left with the Prophet counter-attacked fiercely and drove the Quraish back. Sad dropped his bow, drew his sword and rushed at his brother; but the latter outran him and took shelter in the Quraish ranks. Sad was later to say that he had never wanted to kill a man so badly as he wanted to kill his brother, Utbah, for wounding the Prophet.

There was again a little respite in which the Prophet wiped the blood from his face. As he did so he said, **"How can a people prosper who colour the face of their Prophet with blood, while he calls them to their Lord!"** [1] Abu Ubaidah, who was a bit of a surgeon, tried to pull out the two links which had dug into the Prophet's cheek bone. Finally he had to use his teeth to pull them out, and in the process lost two of his teeth. He later became known among the Arabs as *Al Asram*, i.e. the one without the incisors.

During this respite the Prophet regained his strength and recovered from the physical shock of his wounds. A black lady by the name of Umm Eiman, who had once nursed the Prophet in his childhood, stood near him. From the Quraish ranks a man by the name of Haban bin Al Arqa slowly walked up to within bow-range, and fitting an arrow to his bow, shot it in the direction of the lady who was standing with her back to him. The arrow struck Umm Eiman in her backside. Haban found this terribly funny and roared with laughter as he turned and began to walk back towards the Quraish. The Prophet saw what had happened and was deeply angered. He took an arrow from his quiver and gave it to Sad. **"Shoot that man"**, [2] he ordered. Sad fitted the Prophet's arrow to his bow and, taking careful aim, fired it at the infidel, hitting him in the neck. This time the Prophet laughed!

The Quraish now started their last onslaught with violent assaults against the Prophet from all directions. The cordon formed by the Companions was able to hold the attack at practically all points; but at one place it was breached, and Ibn Qamiah broke through again and rushed towards the Prophet. This man was one of those who had struck the Prophet with stones in the previous phase of the attack. Near the Prophet and a bit to his right, stood Mus'ab bin Umair and a lady by the name of Umm Ammarah. This lady had given up her task of carrying water to the wounded, and picking up a sword and a bow from one of the dead, had actually taken part in the recent fighting. She had brought down one horse and wounded one unbeliever.

Ibn Qamiah mistook Mus'ab for the Prophet and rushed at him. Mus'ab was waiting for him with drawn sword and they began to duel. After a few passes, Ibn Qamiah struck Mus'ab bin Umair and killed him with a deadly blow.

As he fell, Umm Ammarah rushed at Ibn Qamiah and struck him on the shoulder with her sword. Ibn Qamiah wore a coat of mail, and since the blow lacked the power of muscle behind it, it did no damage. In return Ibn Qamiah struck the lady on her shoulder with his sword, but as it was a hasty blow it did not kill her. It just made a deep gash in her shoulder as a result of which the lady fell and was unable to move for some time.

1. Ibn Hisham: Vol. 2, p. 80: Waqidi: *Maghazi* p. 191.
2. Waqidi: *Maghazi*, p. 189.

As soon as Umm Ammarah fell, the infidel saw the Prophet standing by himself and rushed at him. He raised his sword and struck a savage blow at the Prophet's head. The sword cut a few links in the Prophet's chain helmet but was unable to penetrate it. Deflected by the helmet, the sword continued in its thrust and landed on the Prophet's right shoulder. The violence of the blow was such, and the power of muscle behind it so great, that the Holy Prophet fell into a shallow ditch just behind him. From here he was later lifted up by Ali and Talha.

Seeing the Prophet fall, Ibn Qamiah turned and rushed back to the Quraish, shouting at the top of his voice: *"I have killed Muhammad! I have killed Muhammad!"* [1] This shout carried across the battlefield and was heard by Quraish and Muslim alike. It broke the spirit of the Muslims, and most of them turned and fled towards Mount Uhud. A few Muslims, however, decided that if the Messenger of Allah was dead there was no point in their living on. They rushed at the Quraish cavalry-determined to sell their lives as dearly as possible, but were cut down in no time by Khalid and Ikrimah. Here Khalid killed his third man-Rafa'a bi Waqsh.

As the main body of the Muslims fled to the hills, most of the Quraish turned to loot the dead, and the Muslims defending the Holy Prophet now found that none of the Quraish remained near them. The temptation of loot proved as strong for the Quraish as it had proved a little while before for the Muslims. Finding his way clear, the Prophet, surrounded by the survivors of his group, withdrew towards the defile in the valley. In this withdrawal a few of the Quraish followed the Prophet but were beaten off and one or two of them were killed by the Companions. Khalid saw the movement of the Prophet's group towards the mountain pass, but made no attempt to intercept it, for he was busy pursuing the main body of the Muslim infantry. Thus the Prophet had no difficulty in reaching the defile, and the group climbed the steep slope of the spur, where it formed a rocky bluff about 400 feet high, on the east edge of the defile. Here the Prophet stopped, in a cleft in the rock, to survey the tragic panorama which stretched before him.

Of the group of 30 who had fought with the Prophet in the preceding few actions, only 14 remained and most of these were wounded. Sixteen of them had fallen-in defence of the Prophet and in the way of Allah.

Thus the Muslims abandoned the field of battle. Some fled in panic far away; some returned to Madinah; some did not rejoin the Prophet till two days later. But those who intended to seek refuge in the hills moved in small groups, fought their way through the Quraish cavalry and reached the foot of Mount Uhud. Here they dispersed, some taking shelter in the foothills, some climbing up to the ridge, others hiding in the re-entrants.

None of them knew what he would do next. The Quraish were in complete command of the battlefield.

On arrival at the defile the Prophet had some time to see to his wounds. Here his daughter, Fatimah, joined him. Ali brought water in his shield from a nearby pool, and Fatimah cried softly as she washed the blood from her father's face and dressed his wounds. In the shelter of this difficult pass, where the Quraish could not attack in strength, the Prophet rested his weary body.

Of the Muslims who had taken shelter on Mount, Uhud, some were moving about aimlessly, not knowing where to go or what to do. One of them, a man named Kab bin Malik, wandering towards the defile, saw the Prophet and recognised him. This man had a powerful voice. He climbed onto a large rock, and facing the direction where he knew most of the Muslims had taken shelter, he shouted, *"Rejoice, O Muslims! The Messenger of Allah is here!"*2 As he shouted, he pointed with his hand towards the Prophet. As a result of this call, which was not heard by the Quraish, many groups of Muslims moved over the hills and joined the Prophet. These included Umar, whose delight at seeing the Prophet again was boundless.

Meanwhile Abu Sufyan was looking for the body of the Prophet. He wandered over the battlefield and looked at each dead face, hoping that he would see the face of his enemy. Every now and then he would ask his men, *"Where is Muhammad?"* While he was so wandering, he came across Khalid and asked him the question. Khalid told him that he had seen Muhammad, surrounded by his Companions, moving towards the defile. Khalid pointed out the rocky bluff to Abu Sufyan, and the latter asked him to take his horsemen to attack the position.

1. Ibn Hisham: Vol. 2, p. 78.
2. Tabari: Vol. 2, p. 200; Waqidi: *Maghazi*, p. 185.

Khalid looked at the boulder-strewn valley which led to the spur, and then at the steep slope of the spur itself. He had misgivings about the manoeuvre, for he knew that in this sort of terrain his cavalry would be at a serious disadvantage. But he hoped that some opportunity might present itself, as it had done soon after the initial defeat of the Quraish. Khalid was an irrepressible optimist. He began to move his squadron towards the spur.

The Prophet saw this movement and prayed: ***"O Lord, let not those men get here."*** 1 Thereupon Umar took a group of Muslims and moved some distance down the slope to face the Quraish cavalry. As Khalid came up with his squadron, he saw Umar and other Muslims waiting for him on higher ground. Khalid realised that the situation was hopeless-that not only was his enemy better placed, but his own cavalry would be unable to manoeuvre in this difficult terrain. He withdrew. And this was the last tactical manoeuvre in the Battle of Uhud.

Abu Sufyan and Khalid, among many others, now saw a sight which they would never forget and of which they did not approve. The battlefield where the Muslim martyrs lay was invaded by Hind and the Quraish women. Hind found the body of Hamza and, knife in hand, fell upon it.

Hind was a large, heavily built woman and had no difficulty in mutilating the corpse. She cut open the belly and pulled out Hamza's liver. Slicing off a piece of it she put it in her mouth; and she swallowed it! She then cut off Hamza's nose and ears, and made the other women do the same to many of the other corpses.

The Savage now approached Hind. She turned to him, took off all her ornaments and gave them to him. *"And when we get to Makkah,"* she said, *"I shall give you 10 dinars."* 2 Having disposed of her own jewelry, she made a necklace and anklets of the ears and noses of the martyrs who had been mutilated, and she put on these grisly ornaments! Having done so, this extraordinary woman sang:

We have repaid you for the day of Badr-
One bloody day after another.
I could not bear the loss of Utbah,
Or of my uncle, my brother, my son.
Now my heart is cooled, my vow fulfilled;
And the savage has driven the pain from my heart.
The savage shall I thank as long as I live,
Until my bones turn to nothing in my grave. 3

Soon after this gruesome drama had been enacted, Abu Sufyan walked up the valley. He was still hoping that Muhammad might be dead; that Khalid had made a mistake. He climbed on to a large rock some distance from the Prophet's position and shouted, *"Is Muhammad among you?"* The Prophet motioned to his Companions to remain silent. Abu Sufyan repeated the question twice, but there was no reply.

Then thrice Abu Sufyan asked, *"Is Abu Bakr among you?"* And thrice he asked, *"Is Umar among you?"* There was nothing but silence from the spur.

Abu Sufyan now turned towards the Quraish, who stood not far from him, and shouted, *"These three are dead. They will trouble you no more."* At this Umar could no longer restrain himself and roared at Abu Sufyan, *"You lie, O Enemy of Allah! Those whom you have counted are alive, and there are enough of us left to punish you severely."*

Abu Sufyan's response was loud and contemptuous laughter. He knew that the Muslims were in no condition at the moment to punish anybody. But he called to Umar, *"May Allah protect you, O Son of Al Khattab! Is Muhammad really alive?"*

1. Ibn Hisham: Vol. 2, p. 86.
2. Waqidi: *Maghazi*, p. 222.
3. Ibn Hisham: Vol. 2, p. 91.

"By my Lord, yes. And even now he hears what you say."

"You are more truthful than Ibn Qamiah", replied Abu Sufyan.

Then took place a last dialogue between Abu Sufyan and the Prophet. The Prophet did not speak personally to his enemy, but would tell Umar what to say and Umar would shout the reply back at Abu Sufyan.

Abu Sufyan: *Glory to Hubal! Glory to Uzza!* 1
The Prophet: *Glory to Allah, Most High and Mighty!*
Abu Sufyan: *We have Uzza and Hubal. You have no Uzza and no Hubal.*
The Prophet: *We have Allah as Lord. You have no Lord.*
Abu Sufyan: *The deed is done. This was our day for your day of Badr. The destiny of war is not constant. We shall meet at Badr again next year.*
The Prophet: *At Badr we shall meet. You have our pledge.*
Abu Sufyan: *You will find among your dead some who have been mutilated. I neither ordered this nor approved of it. Do not blame me for this.* 2

Having made this last statement, Abu Sufyan turned away and walked back to his army.

The Quraish left the battlefield and gathered in their old camp of the day before. As they left, the Holy Prophet sent Ali as a scout to see how the Quraish were mounting- mounting camels or horses. Ali carried out his reconnaissance and returned to the Prophet to report that the Quraish were mounting camels and were leading their horses. The Prophet observed, **"That means that they intend to return to Makkah and will not attack Madinah. Had they wished to attack Madinah, they would have mounted their horses for battle. In that case, by my Lord, I would have gone this very instant to fight them again,"** 3

The Quraish spent the night in Hamrat-ul-Asad, 10 miles, from Madinah. 4 The Muslims returned to Madinah, except for some stragglers who were to turn up the following day and the day after.

The next morning the Holy Prophet got up and put on his armour. His face showed clear signs of the damage which it had suffered in the battle. His cheek, forehead and lip that had been badly cut were still swollen. The loss of his two teeth caused him pain, and his right shoulder hurt badly where the sword of Ibn Qamiah had landed. This shoulder was to trouble him for a whole month.

The Prophet sent for Bilal, his *Muazzin,* 5 and ordered him to call the Faithful to battle. Only those would be permitted to join this morning's expedition who had taken part in the battle of the day before. The thundering voice of Bilal rang across the streets of Madinah and carried the message into every Believer's home.

The Muslims rose from their mats as they heard the Prophet's orders to assemble for battle. Most of them were wounded, some more severely than others. They had spent a sleepless night in pain and suffering. All night long the women had been busy nursing the soldiers, washing and dressing their wounds. Not many of the Muslims were in fit shape for battle; but they got up from their mats. There were no groans or cries of pain.

Some limped, others used hastily improvised crutches, yet others put their arms around their comrades to get support as they walked. They came, limping and staggering, towards the Prophet. They saw the Prophet and they cried *Labbeik*- Present, Sir! And these tired, wounded Muslims, led by a tired, wounded Prophet, set out to fight the infidel. They numbered about 500.

As the Muslims were assembling for battle, a wild argument was taking place in the Quraish camp. Ikrimah, no less aggressive than he had been the day before, was insisting on a return to battle for the reason that the Muslims were in a bad way as a result of the battle and now was the time to seek them again and completely crush them before they recovered from the setback.

1. god and goddess in the Arab pantheon.
2. Ibn Hisham: Vol. 2, pp. 93-4; Waqidi: *Maghazi*, pp. 229-30; Ibn Sad: p. 551.
3. Ibn Hisham: Vol. 2, p. 94.
4. This place was near the present Bir Ali, on the main road to Makkah.
5. The one who call the Adhan-the Muslim call to prayer.

"Enough is enough", replied Safwan bin Umayyah. *"We have won the battle, and this victory should be sufficient for us. If the Muslims are in a bad way, we too are not in perfect condition. Most of our horses and many of our men are wounded. In the next battle, if we fight it with our present strength, we might not be as lucky as we were yesterday."* 1

By now the Quraish leaders had also heard of the defection of the 300 Hypocrites. The fear that troubled them was the possibility of the return of these 300 in a repentant mood to the Prophet, for this would considerably augment the strength of the Muslims with fresh troops. While this argument was in progress, the Quraish soldiers discovered and caught two Muslim scouts who had been sent by the Prophet to seek information of the Quraish. These scouts were promptly killed, but their presence confirmed the fears of Safwan and Abu Sufyan that the Muslims were in an aggressive mood and sought battle. Abu Sufyan promptly gave orders for the move to Makkah; and the Quraish army rode away.

In the afternoon the Muslims arrived at Hamrat-ul-Asad and found it deserted. They set up camp. After four nights at Hamrat-ul-Asad, they returned to Madinah.

The campaign of Uhud was over. A total of 70 Muslims had fallen in battle. Abu Sufyan had killed one. Safwan bin Ummayya, Khalid and Ikrimah had each killed three Muslims. On the Quraish side, 22 unbelievers had been killed including six by Ali and three by Hamza, It was a defeat for the Muslims, but not a decisive one.

This was the second major battle in the history of Islam. It was the first battle in which Abu Sufyan commanded an army against the Muslims, and the first battle in the life of Khalid. The Holy Prophet lost this battle, and the blame for this rests squarely on the shoulders of the fickle archers who disobeyed the orders of the Prophet and of their own immediate commander, In fact, in leaving their position these archers momentarily ceased to be Muslims and became tribal Arabs, bent on plunder.

Several writers have expressed the opinion that the Arabs of this period were ignorant about regular warfare; that militarily they were nothing better than raiders, and that they knew nothing about regular battles. It has been suggested by many of these writers that the Arabs learnt the art of war from the Romans and the Persians with whom they came into military contact after the Prophet's death. This is just not true. We have already considered the dispositions adopted by the Prophet and the sound military reasons underlying his deployment. It should also be noted that in selecting the battlefield the Prophet left Madinah open to assault by the Quraish, Madinah was the base of the Muslims, but the route to that base, which ran south of the Muslim position, was open to Abu Sufyan, The Muslims were not in the way of Abu Sufyan had he decided to move to

Madinah. In this decision, the Prophet guessed rightly that Abu Sufyan would not dare to move to Madinah, because in doing so he would expose his flank and rear to attack by the Muslims. And this is just what happened. Abu Sufyan did not move to Madinah for fear of the Muslims who stood on the flank of the route. This was a classic example, repeated time and again in military history, of a force defending its base not by sitting on it for a frontal action, but by threatening from a flank any enemy movement towards that base.

While Abu Sufyan was forced to fight the battle under conditions not favourable to him, the disposition of his forces was sound, following the normal pattern, as practised by the Romans and the Persians, of having a main body of infantry in the centre and mobile wings for manoeuvre against the enemy's flanks and rear. So far as the selection of the battlefield and the dispositions are concerned, it is doubtful if any Roman or Persian general commanding these forces could have acted differently and deployed the forces in another manner than done by the Prophet and Abu Sufyan. Certainly no critic has offered us a better solution!

Another important fact which this battle brings out is the military judgement and skill of Khalid. When the main body of the Quraish fled, its smaller parts-the cavalry squadrons-remained firm on the battlefield. Generally when the bulk of an army flees its parts do not remain. In this we see the unusual courage of Khalid (and Ikrimah) in keeping their squadrons under control on the battlefield, although reason could suggest no possible advantage in doing so. We see the patience of Khalid and his refusal to accept defeat. It was only the keen eye of Khalid which observed the opening left by the archers when they abandoned their position. He saw the opening and took an immediate decision to exploit the opportunity with a rapid riposte which would get him into the vulnerable rear of the Muslims. It was this brilliant manoeuvre by Khalid which turned the near-complete victory of the Muslims into their near-complete defeat.

1. Ibn Hisham: Vol. 2, p. 104; Waqidi: *Maghazi*, pp. 231-2, 263.

We also see the determination and doggedness of Khalid in the relentless pressure which he maintained against the stubborn Muslims until they broke. His killing three men showed the personal courage and fighting skill of the man. Possessing the boldness and dash of youth, and the patience and judgement of age, Khalid showed promise of great military achievements.

This was the first battle of Islam in which a fine manoeuvre was carried out. Henceforth manoeuvres and stratagem would achieve more prominence in Muslim battles. Some of the names that have been mentioned in this account would achieve undying fame within the next two decades as victor and conquerors... Khalid, Amr bin Al Aas, Abu Ubaidah, Sad bin Abi Waqqas.

Chapter 4

(The Battle of the Ditch)

"You have indeed in the Messenger of Allah a beautiful example for anyone whose hope is in Allah and the Last Day, and remembers Allah much.

When the Believers saw the Confederate forces, they said, 'This is what Allah and His Messenger promised us, and Allah and His Messenger spoke the truth.' It added only to their faith and obedience.

Among the Believers are men who have been true to their covenant with Allah. Of them, some have completed their vow and some wait, but they have never wavered in the least.

That Allah may reward the men of Truth for their Truth, and punish the hypocrites if He wills, or turn to them in mercy, for Allah is Oft-Forgiving, Most Merciful.

And Allah turned back the Unbelievers for all their fury – no advantage did they gain, and enough is Allah for the Believers in their fight. And Allah is full of Strength, full of Might.

He took those of the People of the Book who aided them, down from their strongholds, and cast terror into their hearts: some you slew, and some you captured. He made you heirs of their lands, their houses, and their goods, and of a land which you had not frequented before. And Allah has Power over all things."

[Quran 33:21-27]

For several days after his return to Makkah, the Battle of Uhud occupied the mind of Khalid. He thought time and again of how the opportunity had arisen when the archers abandoned their position, and how quickly and accurately he had grasped the possibilities of manoeuvre. Khalid was to repeat such counterstrokes in later battles of his career. But the one fact that weighed heavily on his mind, and which he found difficult to explain, was the courage and tenacity of the Muslims. It did not seem natural that a small force, so vastly outnumbered and attacked from all directions, should hold out with such rocklike determination and be prepared to fight to the end in defence of its leader and its faith. After all, the Muslims were the same stock as the Quraish and other Arabs. Perhaps there was something that the new faith did to its votaries which other faiths could not do. Perhaps there was something about the personality of Muhammad which other men lacked. Such thoughts would occupy the mind of Khalid, but so far he was not in any way inclined towards the new faith. In fact he looked forward to facing the Muslims again, but without bitterness or rancor. He thought of the next battle as a sportsman might think of his next match.

And Khalid continued to enjoy the good life with the vigor and enthusiasm which were characteristic of the man.

For the next two years there was no direct military clash between the Muslims and the Quraish. There was, however, an incident known as the Incident of Rajee-a brutal and horrible affair which further embittered relations between Makkah and Madinah.

This incident took place in July 625. Some Arabs came to the Prophet as a delegation from their tribe, expressed their desire to embrace Islam and asked him to send some men, well versed in the Quran and the ways of Islam, to explain the faith and its obligations to their tribe. The Prophet nominated six of his Companions for this task, and these men, proud of being selected to spread the true faith, set off with the delegation, entirely unaware of the trap that awaited them. When these men, with their guides, reached a place called Rajee, not far from Usfan, they were ambushed by 100 warriors from the tribe which had invited them. The Muslims drew their swords, but they never had a chance. Three of them were killed and three captured. The prisoners were led to Makkah, en route to which one of them was able to free himself from his bonds and attacked his captors, but he too was killed. The two captives who eventually got to Makkah were Khubaib bin Adi and Zaid bin Al Dasinna. Both of them had killed infidels in battle; and their captors now took them to Makkah and sold them at a high price to the relatives of the dead infidels, who bought them eagerly with the intention of killing them in revenge for those whom they had lost.

For some days no action was taken against the prisoners, as this was the holy month of Safar. As soon as the month ended, the two captives were taken to Tan'eem, a place by the north-western edge of Makkah, where the entire population of the town had gathered, including, slaves, women and children. Two wooden stakes had been dug in the ground, and to these the captives were led. They asked to be allowed to say a final prayer and the request was granted. When the prayer was over, the captives were tied to the stakes.

Each of them was now given the option of returning to the idol worship of the Quraish or death. Both the Muslims chose the option of death. Next Abu Sufyan went up to each captive and said, *"Do you not wish that you were safe in your home and Muhammad were here in your place?"* Each of them vehemently rejected the suggestion and said that no amount of suffering could put such an idea into his mind. Vexed and angered, Abu Sufyan turned away and remarked to his friends, *"I have never seen men love their leader as the men of Muhammad love Muhammad."* [1]

Zaid was the first to die, and his death was quick and easy. A slave walked up to him and drove a spear through his chest. Next came the turn of Khubaib, and this was to be a show. This is what the people of Makkah had come to watch with joyful anticipation.

At a signal, 40 boys carrying spears rushed to the stake where Khubaib was tied and began to prick him with their spears. Sometimes they would move away and then come rushing at him again with raised spears as if to kill him, but would withhold the blow at the last moment and just prick lightly?sufficient to cut and pierce the skin but not to kill. Some of the boys were clumsy and cut deeper than others, and soon the body of Khubaib was covered with blood that flowed from hundreds of shallow wounds. As each spear pricked him he would wince, but not a sound escaped his lips. And the spectators were thrilled by the spectacle of Khubaib's suffering.

1. Ibn Hisham: Vol. 2, p. 172

When this had gone on for some time, a man with a spear walked up to Khubaib and dispersed the boys. Perhaps by now the boys had tired of the fun. Perhaps the audience had tired of the game. This man now raised his spear and drove it through the heart of Khubaib, putting an end to his agony. The two bodies were left to rot at the stake.

The man who organised this show and prepared the boys for the part which they had to play was none other than Ikrimah, son of Abu Jahl. Little did Ikrimah know, when he arranged this horrible and gory entertainment that he could be forgiven his savage opposition to Islam and the Muslim, blood that he had shed at Badr and Uhud, but *this* he could not be forgiven. On this day Ikrimah became a *war criminal*.

It will be remembered that before leaving the battlefield of Uhud, Abu Sufyan had thrown a challenge to meet the Muslims, again at Badr in a year's time, and the Prophet had accepted the challenge. This would mean a rendezvous during March 626, but as the time of the rendezvous approached, Abu Sufyan felt disinclined to meet the Muslims. The winter rains had been even more scant than usual, and as the winter passed there was a sudden increase in temperature. The weather was hot and dry and the year promised to be an unusually bad one. Abu Sufyan decided to postpone the operation and sent an agent to Madinah to spread the rumor that the Quraish were assembling in vast numbers, and would this time come in much greater strength than at Uhud. His intention was to frighten the Muslims into remaining at Madinah, but when these reports reached the Prophet, he declared, *"I shall keep the rendezvous with the infidel even if I have to go alone"* 1

In late March, the Muslims marched from Madinah. They numbered 1,500 men, of whom 50 had horses. The army arrived at Badr on April 4, 626 (the 1st of Dhul Qad, 4 Hijri), but there was no sign of the Quraish.

When Abu Sufyan received news of the movement of the Muslims from Madinah, he got the Quraish together and rode out of Makkah. The army consisted of 2,000 men and a hundred horses, and stalwarts like Khalid, Ikrimah and Safwan again rode with the army. When the Quraish got to Usfan, however, Abu Sufyan decided that he was not under any circumstances going to fight this campaign. He turned to his subordinates and said, *"This is a terrible year in which to engage in warfare. There is drought in the land and we have seldom known such heat. These conditions are not suitable for battle. We shall fight again in a year of abundance."* 2 Having given these reasons for not continuing the movement, he ordered a return to Makkah. Safwan and Ikrimah protested vehemently against this decision but their protests were of no avail. The Makkans returned to Makkah.

The Muslims remained at Badr for eight days. Then, on hearing of Abu Sufyan's return to Makkah, they struck camp and went home to Madinah.

After the return of the Quraish to Makkah, peace may have prevailed between the Muslims and the Quraish had it not been for the machinations of certain Jews. To understand the reasons for this activity, we must go back to the days when the Prophet arrived at Madinah after his flight from Makkah.

When the Prophet got to Madinah, in what was later to be numbered as the first year of the Hijra, the Muslims formed into two groups, *viz.* the Emigrants *(Muhajireen)* those who had migrated from Makkah, and the Helpers (Ansar)?the newly converted Muslims of Madinah who had invited the Prophet to come and live with them. A third small group among the Muslims became known as the Hypocrites *(Munafiqeen)*, and these were inhabitants of Madinah who had accepted the Prophet and his faith in order to conform to the general trend of events but were not Muslims at heart. Their leader was Abdullah bin Ubayy, a man who commanded a position of prestige in Madinah and felt that the arrival of the Prophet had somehow reduced him in status and influence. These Hypocrites were the people who had abandoned the Muslim army on the eve of Uhud. They were to continue to create obstacles in the path of the Prophet, and without openly opposing him or his faith, would make every effort to weaken the resolution of the Muslims whenever they had to go to battle.

1. Ibn Sad: p. 563
2. *Ibid*.

An important element in the population of Madinah consisted of Jews, comprising three tribes known as Bani Qainqa, Bani Nazir and Bani Quraizah. When the Prophet arrived at

Madinah, these Jews accepted him without reservation and could see no possible threat to their position from the new faith. Each of the tribes entered into a pact with the Prophet which could be described as a friendship pact or a non?aggression pact. The pact included a clause under which one party would not in any way assist the enemies of the other party, should the other party be engaged in hostilities.

While the Prophet had been in Makkah, the revelations of the Quran had dealt mainly with spiritual and religious matters. Thus the character of Islam then was essentially spiritual and religious, dealing with man's relationship with Allah. When the Prophet migrated to Madinah, Islam took on a more dynamic and vital role in the affairs of men, entering the fields of society, politics and economics. It began to deal with man as a member of society and society as an instrument for the achievement of a more virtuous, more progressive and more prosperous way of life for mankind. This new dynamism which entered the character of Islam was bound to bring it into conflict with the older faiths. A clash was inevitable sooner or later; and the nearest of the older religions with which Islam came in conflict was Judaism. The Jews first became conscious of the threat to their position when the Muslims won a resounding victory at the Battle of Badr. Then the Bani Qainqa broke their pact and came out in open opposition to the Muslims. The Prophet besieged this tribe in its strongholds and forced it into submission. As punishment for violating their pledge, the Bani Qainqa were banished from Madinah, and they migrated to Syria.

The next Jewish tribe to break its pledge was the Bani Nazir, which happened soon after the Battle of Uhud. This tribe received the same punishment from the Muslims. Some of its members migrated to Syria, while others settled down in the area of Khaibar, north of Madinah. In the operations against both these tribes, Abdullah bin Ubayy, the chief of the Hypocrites, first sided with the Jews, secretly inciting them to fight the Prophet and promising active help from his followers. Later, when he saw the fortunes of war turning in favour of the Muslims, he abandoned the Jews to their fate.

The third Jewish tribe, the Bani Quraizah, continued to live peacefully in Madinah. Its relations with the Muslims were perfectly normal and entirely peaceful, each side respecting and observing the terms of the pact. But the Jews of the Bani Nazir who had settled at Khaibar did not forgive the Muslims the banishment which they had suffered. After Uhud they came to know of the agreement between the Muslims and the Quraish to fight another battle, and they waited patiently, hoping that in that battle the Muslims would be crushed. But when they found a year later that there was not going to be another battle, they decided to take direct action to bring on an attack against the Muslims.

As the summer of 626 came to an end, a delegation of the Jews of Khaibar set out for Makkah. Their leader was Huyaiy bin Akhtab, who had been the chief of the Bani Nazir in Madinah. On arrival at Makkah this delegation conferred with Abu Sufyan, and set about to organise an expedition against the Prophet. It was necessary for Huyaiy to work on the fears and emotions of the Quraish, and he started off by outlining the danger the Quraish faced from the spread of Islam in Arabia. If the Muslims reached Yamamah, the Quraish trade routes to Iraq and Bahrain would be blocked.

"Tell me, O Son of Akhtab", asked Abu Sufyan. *"You are one of the People of the Book. Is it your opinion that the new religion of Muhammad is better than our religion?"* Without batting an eye Huyaiy replied, *"As one who knows the Book, I can assure you that your religion is better than Muhammad's. You are in the right."* [1] This pleased the

Quraish no end, and they agreed to fight Muhammad if other Arab tribes would join them.

The Jewish delegation then went to the Ghatfan and the Bain Asad with whom it had similar talks and achieved similar results. These and various other tribes all agreed to take part in a massive expedition to fight and destroy the Muslims.

1. Ibn Hisham: Vol. 2, p. 214.

After Uhud the Quraish had accepted the loss of trade with Syria as inevitable. Since the Muslims remained in power at Madinah, the coastal route to Syria could not be used by the Makkans. So the Makkans increased their trade with Iraq, Bahrain and the Yemen, and thus more or less made up for the loss which they had suffered in the stoppage of their trade with Syria. As a result of the conference with the Jewish delegation, however, Abu Sufyan became more conscious of the danger to the Meccan trade by the further spread of Islam. If the Muslims reached Yamamah, the Quraish trade would have to be confined to the Yemen, for the routes to Iraq and Bahrain would then be in Muslim hands. And this further curtailment of their trade would be an economic blow which the Quraish could never survive. Abu Sufyan had also been needled a great deal by Safwan bin Umayyah for his lack of spirit in the last expedition. Both these factors combined to make Abu Sufyan determined and zealous to take out another expedition to Madinah.

Preparations for the expedition were begun. Tribal contingents began to concentrate in early February 627. The Quraish provided the largest force, consisting of 4,000 men, 300 horses and 1,500 camels. Next came the Ghatfan with 2,000 men under Uyaina bin Hisn, while the Bani Sulaim sent 700 warriors. The Bani Asad contributed a contingent, whose strength is not known, under Tulaiha bin Khuwailid. While the Quraish and some lesser tribes assembled at Makkah, the Ghatfan, Bani Asad and Bani Sulaim concentrated in their tribal settlements north, north?east and east of Madinah respectively, whence they would march direct to Madinah. The total strength of the force, including smaller tribes which have not been mentioned, was 10,000, and Abu Sufyan assumed command of the expedition. This became known as *the collection of tribes*. For want of a better name, we shall call them the Allies.

On Monday, February 24, 627 (the 1st of Shawal, 5 Hijri), the Allies, converging from their separate tribal regions, arrived near Madinah and established their camps. The Quraish camped in the area of the stream junction south of the wood, west of Mount Uhud, where they had camped for the Battle of Uhud. The Ghatfan and other tribes camped at Zanab Naqnia, about 2 miles east of Mount Uhud. Having established their camps, the Allies advanced on Madinah.

Hardly had the concentration of the Allies begun when agents brought word of it to Madinah. As more and more tribal contingents gathered, the reports became increasingly alarming. Finally the Prophet received the information that 10,000 warriors bent on destroying the Muslims were marching on Madinah. There was alarm and despondency among the Muslims as this unpleasant intelligence was received. The Muslims had, of course, always been numerically inferior to their enemies. The ratio of relative strengths at Badr and Uhud had been one to three and one to four respectively, and although the number of Muslims at Madinah had now increased to 3,000 able-bodied men, many hundreds among them were Hypocrites on whom no reliance could be placed. And

10,000 seemed a terribly large figure. Never before in the history of the Hijaz had such a vast army assembled for battle.

Then came light in the form of a suggestion by Salman the Persian. He explained that when the Persian army had to fight a defensive battle against superior odds, it would dig a ditch, too wide and too deep to cross, in the way of the enemy. To the Arabs this was an unfamiliar method of warfare, but they saw its virtue and the proposal was accepted.

The Prophet ordered the digging of the ditch. Many of the Arabs who could not comprehend such tactics seemed unwilling to get down to the arduous labour of digging, and the Hypocrites as usual went about dissuading the people from taking all this trouble. But the Prophet got down to digging with his own hands, and after this no self-respecting Muslim could keep away from the task. The ditch was sited and its entire length divided among the Muslims at the scale of 40 cubits per group of 10 men. As the Muslims sweated at this backbreaking task, Hassaan bin Thabit walked about reciting his poetry and infusing fresh spirit into the Muslims. Hassaan was a poet, and perhaps the greatest poet of his day. He could extemporise verses on any subject and on any occasion, and do it so beautifully that his listeners could hardly believe that the composition was extemporaneous. He could move people to a frenzy of emotion. But if Hassaan was one of the greatest poets of his age, that is where his talents ended. To such manly pursuits as fighting, Hassaan was in no way inclined, as we shall see later.

The ditch ran from Shaikhan to the hill of Zubab, and thence to the Jabal Bani Ubaid. All these hills were included in the area protected by the ditch, and on the west the ditch turned south to cover the left flank of the western of the two hills known as Jabal Bani Ubaid. East of Shaikhan and south-west of Jabal Bani Ubaid stretched vast lava fields- areas of broken, uneven ground covered by, and at times formed of, large black boulders, impassable for major military movement. A little south of the centre of the ditch stood the prominent hill of Sil'a, about 400 feet high, a mile long and a little less across, running generally north-south but with spurs extending in all directions. In fact the little hill of Zubab lay just off the north-eastern spur of Sil'a.

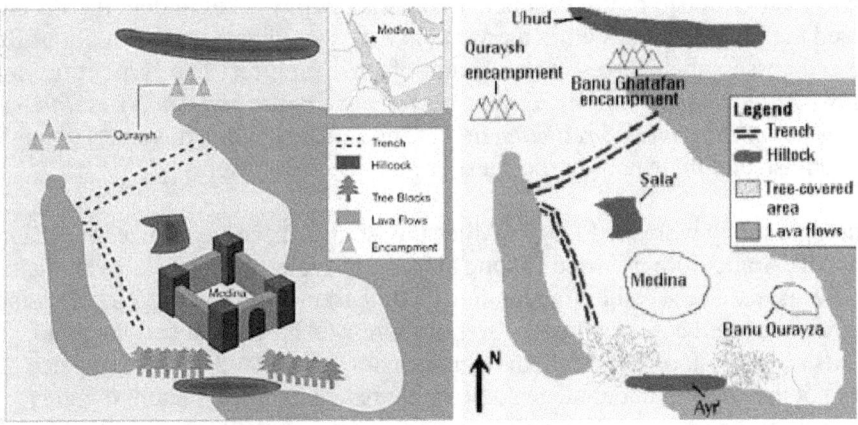

Once the digging of the ditch was complete, the Muslims established their camp just ahead of the hill of Sil'a. Their total strength was 3,000 which included Hypocrites whose fighting value and reliability were uncertain. The Prophet's plan was to keep the bulk of his army uncommitted to strike at any spot where the enemy managed to get a foothold across the ditch. To guard against surprise, the ditch was lightly covered along its entire length by 200 men, most of whom were placed as picquets on the hills commanding the ditch. A mobile force of 500 men was employed to patrol the various settlements of Madinah and deal with any infiltrators who might enter unseen, and also give some protection to areas not covered by the ditch. (Madinah was not then a city as it is now but consisted of a group of settlements and forts. The centre of Madinah, physical and spiritual, was the Prophet's Mosque.) The women and children were placed in forts and houses away from the main front, which faced north and north-west.

The winter that was now passing had been a severe one. It was also to prove a long winter.

When the Quraish saw the ditch they were first dismayed and then moved to indignation. They had come in such strength that victory had seemed certain. Abu Sufyan had joyfully expected to fight a victorious battle, and now here was this blessed ditch in the way!

"By Allah!" Abu Sufyan exploded. *"Such stratagems are not the way of the Arab!"* [2] In the simple mind of the average Arab there was no room for such tactics. To the chivalrous Arab this was definitely 'not cricket'?

However, the Allies moved up their camp, deployed along the ditch on the north and north-west, and settled down to a siege that was to last 23 days. By day the Allies would come up to the ditch which the Muslims covered lightly from the home side. There would be an exchange of archery which would go on for most of the day, and for the night the Allies would return to their camp. Mostly by day and sometimes by night, Allied patrols would move up and down the ditch trying to find a place at which a crossing could be attempted. They were eventually to find one such place, but more of that later.

For 10 days the siege continued with no decision and no let-up on either side. The morale of both sides came under considerable strain, but tended to harden rather than weaken. The Muslims began to feel the pangs of hunger. There were no large stocks of food in Madinah, and the Muslims were now on half rations. The Hypocrites became louder and more open in their criticism of the Prophet. While the ditch was being dug, the Prophet had promised the Muslims that within a few years they would destroy the might of Rome and Persia and possess themselves of the wealth of those empires. The Hypocrites now began to say, *"Muhammad promises us the treasures of Caesar and Chosroes, but he cannot get us out of this simple predicament!"* [3] The true Believers, however, remained firm and steadfast, and their faith in their leader remained unshaken.

The situation gradually worsened for the Allies too, so that discontent raised its head in their ranks. The Arabs were not used to long sieges and preferred a quick, lively battle to this form of warfare. The weather had remained unpleasant and began to cause a good deal of distress among the Allies. Food also ran short, as Abu Sufyan had made no arrangements for provisions to tide them over such a long period of time. But since the Allies were not themselves under siege some measures were hastily taken to gather provisions from outlying areas. The men began to grumble and Abu Sufyan had to think hard to find some way out of this impasse. Finally, he consulted Huyaiy the Jew, and between them they hit upon a new plan which showed every promise of success.

1. The western end of the ditch is also reported to have ended at Mazad. This too is correct, for the three western hills, the two southern ones of Jabal Bani Ubaid and the little one to their north-are also called Mazad.
2. Ibn Hisham: Vol. 2, p. 224.
3. *Ibid*: Vol. 2, p. 212.

On the night of Friday, March 7, Huyaiy stole into the settlement of the Bani Quraizah. He knocked at the door of their leader, Kab bin Asad; but the latter, guessing that Huyaiy had come as a Jew and probably intended to incite his fellow Jews against the Prophet, refused to see him. After some wrangling, however, Huyaiy was allowed in, and he gently and cleverly began to work on Kab, pressing him to join the Allies in the war against the Muslims. At first Kab refused. *"Muhammad has kept his pact with us, and we have no reason to complain"*, he said. *"In any case you have no certainty of victory. If we join you and the campaign fails, your idol-worshippers will go back in peace to their homes and we will have to bear the brunt of the wrath of Muhammad."* [1] But the visitor continued to press, now threatening, now tempting, now begging, and eventually got Kab to agree to a pact with the Allies. According to the terms of this pact there would be a simultaneous attack by the Allies and the Bani Quraizah. These Jews had their settlement and their forts two miles south-east of Madinah, and they would attack from this direction and draw some of the Muslims away from the ditch while the Allies attacked frontally. In case the attack failed, the Allies would leave a strong garrison in the Jewish forts to defend the Jews against the Muslims who were bound to turn against them in revenge. The Bani Quraizah asked for 10 days to prepare themselves before the attack was begun, during which period the Allies could continue minor operations from the north.

Thus the last of the Jews of Madinah, following in the footsteps of their co-religionists, broke their pact with the Muslims. Little did they know how heavily they would pay for their perfidy!

It was not long before the Prophet came to know about this pact. He got the intelligence through one of his agents who entered the camp of the Allies one night and unknown to them, overheard certain conversations. Then rumors of the pact also spread, and the report was ultimately confirmed by the incident of 'Safiyyah and the Jew'.

Safiyyah was an aunt of the Prophet, and along with other women and children had moved to a small fort in the south-eastern part of Madinah. Present in the fort was Hassaan the Poet, and he was the only man there! One day Safiyyah, looking down from the fort, saw a fully armed Jew moving stealthily beneath the wall as if seeking a way around the fort. Safiyyah at once concluded that he was a scout of the Bani Quraizah who had been sent to reconnoiter a route which the Jews might take in their attack. This Jew would act as a guide, leading his tribe into the unprotected rear of the Muslims.

Safiyyah went to the poet and said, *"O Hassaan! There is a Jew who is seeking a way by which he can lead the Bani Quraizah to attack our settlements from the rear. You know that the Messenger of Allah and all the men are busy at the front and cannot detach forces to protect us. This man must be killed. Go and kill him at once!" "May Allah bless you, O Daughter of Abdul Muttalib,"* replied Hassaan, *"you know that such work is not for me."* Throwing a glance of contempt at the poet, Safiyyah picked up a club, tied a waist-band around her waist and went down to meet the Jew. The brave lady killed the Jew. Leaving him lying with a crushed skull in a pool of blood, she returned to the fort and said to Hassaan, *"I have killed him, O Hassaan! Now go and take the booty from his body, for it is not right for a woman to undress a man." "May Allah bless you, O Daughter of Abdul Muttalib,"* replied Hassaan, *"I have no need for such booty!"* [2]

When the news of this incident reached the Muslims, there was no doubt left in their minds about the treachery of the Bani Quraizah. The situation now became more tense, and the Hypocrites became more outspoken. From half rations the Muslims came down to quarter rations. (Later it was to become no rations!) Their resolution was still unshaken; but if the siege continued very much longer, sheer starvation would force the Muslims to submit. And the Muslims could find no direct military solution to the problem.

1. Ibn Hisham: Vol. 2, p. 221; Waqidi: *Maghazi* p. 292.
2. Ibn Hisham: Vol. 2, p. 228.

The Prophet now decided to use diplomacy to achieve results which were not attainable by force of arms. He started secret negotiations with Uyaina, the commander of the Ghatfan contingent. (Uyaina was a brave and simple soul. A one-eyed man possessing more brawn than brain, he was to earn from the Prophet the nickname of 'the willing fool' [1]). The aim of the negotiations was to create a rift between the two major Allies, the Ghatfan and the Quraish-by drawing the Ghatfan away from the siege. If this were achieved, other tribes might also pull away from the Quraish; but even if they did not, the absence of the powerful Ghatfan contingent of 2,000 warriors would reduce the Allied strength to manageable proportions, where after military action could be taken to drive the Allies away from Madinah.

"If the Ghatfan secede from the alliance and return to their homes, they shall be given one-third of the date produce of Madinah", were the terms offered by the Prophet. This offer was accepted by Uyaina who had by now lost all hope of military victory. The pact

was drawn up, but before it could be signed and sealed (without which it would not be binding), the Prophet decided to mention the matter to some of the Muslim leaders. These Muslims protested vehemently. *"Dates!"* they exclaimed. *"Let the infidels get nothing from us but the sword!"* [2] This disagreement with the Prophet was so general and so strong that he decided to submit to the wishes of the Muslims, and the negotiations were dropped.

These stout hearted Believers could not understand the seriousness of the military situation or the intricacies of diplomacy as well as the Prophet did. He knew that the only solution to the problem lay in breaking the siege by diplomatic manoeuvre, and he now began to look about for another opening. Soon an opening presented itself. Among the Ghatfan was a man by the name of Nuaim bin Masud who had become a Muslim but had kept his conversion a secret. A prominent figure in the region, he was well known to all the three major partners in the alliance-the Quraish, the Ghatfan and the Jews of Bani Quraizah. He was also a very capable man.

Nuaim left the Ghatfan camp one night and slipped into Madinah. He came to the Prophet, explained his position and expressed his desire to be of service to the Muslims. *"Send me where you will"*, he said [3]. This was just the opportunity for which the Prophet had prayed. In a conference with Nuaim the Prophet went over the entire situation and laid down the course of action which Nuaim was to take.

The same night Nuaim stole into the settlement of the Bani Quraizah and visited Kab. He outlined the dangers of the situation as they applied to the Jews. *"Your situation is not like the situation of the Quraish and the Ghatfan"*, he explained. *"You have your families and your homes here, while their homes and families are at a safe distance from Madinah. They have no great stake in this battle. If they do not succeed in defeating Muhammad, they will return to their homes and leave you to face the wrath of the Muslims. You must take no action in collaboration with them unless they give you hostages from their best families. Thus you will have an assurance of their good faith."*

Nuaim next went to the Quraish and spoke to Abu Sufyan, who knew him well and had respect for his judgement. *"You have made a pact"*, he said, *"with a people who are treacherous and unreliable. I have come to know through friends in Madinah that the Bani Quraizah have repented and entered into a fresh pact with Muhammad. To prove their loyalty to Muhammad, they are going to ask you for hostages from your best families, whom they will promptly hand over to Muhammad, who will put them to death. The Jews will then openly come out as allies of the Muslims and both will make a joint attack against us. On no account must you give hostages to the Jews!"*

He then went to the Ghatfan where he painted the same picture. By the time Nuaim had finished, the seeds of doubt and discord had been firmly planted in the minds of the Allies.

The uncertainty began to tell on Abu Sufyan, who had relied unquestioningly on the alliance with the Jews. He decided to hasten the course of battle and put the intentions of the Jews to test. During the night of Friday, March 14, following the visit of Nuaim, he sent a delegation headed by Ikrimah to the Bani Quraizah. *"This is a terrible situation"*, explained Ikrimah. *"This cannot be allowed to continue any longer. We attack tomorrow. You have a pact with us against Muhammad. You must join in the attack from the direction of your settlement."*

1. Ibn Qutaiba: p. 303.
2. Ibn Hisham: Vol. 2, p. 223.
3. *Ibid*: Vol. 2, p. 229.

The Jews hummed and hawed for a while and then came out with their terms. *"Our position is more delicate than yours. If you have no success you may abandon us, and then we will be left alone to face the wrath of Muhammad. To make sure that this does not happen, you must give us hostages from your best families who will stay with us until the battle has been fought to a satisfactory conclusion. Anyway, tomorrow is Saturday and Jews are forbidden to fight on the Sabbath. Those who break the Sabbath are turned by Allah into pigs and monkeys."* Ikrimah returned empty-handed. Abu Sufyan then decided to make one more attempt at persuading the Jews to join battle on the morrow, and sent another delegation to Kab; but the stand of the two sides remained the same:

Quraish: *No hostages; fight tomorrow!*

Jews: *No fighting on the Sabbath; anyway, hostages first!*

All three groups now said, *"Nuaim was right. How wise he was in his advice to us!"* 1 Nuaim had done his work well. The Bani Quraizah had been neatly detached from the alliance.

The next morning, Saturday, March 15, Khalid and Ikrimah, tiring of the delay and seeing no hope of concerted action by the Allies, decided to take matters into their own hands and try to force a decision one way or another. They moved forward with their cavalry squadrons to a place just west of Zubab, where the ditch was not as wide as in other places and where it could be cleared on horseback or by men scrambling across on foot. This place was right in front of the Muslim camp, which nestled at the foot of Sil'a.

Ikrimah's squadron moved up first and a small group jumped the ditch, the horses landing neatly on the Muslim side. There were seven men in the group, including Ikrimah and an enormous man who urged his enormous horse ahead of the group and began to survey the Muslims, who were surprised by the sudden appearance of the Quraish. The stage was now set for one of the most remarkable duels of history, which, because of its unusual course, is here described in full detail.

This huge man was of a tremendous height and bulk, and while on his feet would tower above his fellow men. Sitting on his great horse, he looked positively unreal. Big, strong and fearless, he had a fierce countenance-an aspect which thrilled his comrades and dismayed his enemies.

This was Amr bin Abdu Wud. (We shall call him the Giant!) Horse and rider stood motionless as he let his gaze wander scornfully over the ranks of the Muslims.

Suddenly the Giant raised his head and roared, *"I am Amr bin Abdu Wud. I am the greatest warrior in Arabia. I am invincible. I... I. . ."* He certainly had a high opinion of himself. *"Is there anyone among you who has the courage to meet me in personal combat?"*

The challenge was received by the Muslims in silence. They looked at one another. They looked at the Holy Prophet. But no one moved, for the Giant was famous for his strength and skill, and though wounded several times, had never yet lost a duel, nor spared an opponent. It was said that he was equal to 500 horsemen; that he could lift a horse bodily and hurl it to the ground; that he could pick up a calf with his left hand and use it as a shield in combat; that he could... The stories were endless. The vivid Arab imagination had created around this formidable warrior a legend of invincibility.

So the Muslims remained silent, and the Giant laughed with contempt-a laugh in which the Quraish also joined, for they stood quite close to the ditch and could see and hear all that went on.

1. Ibn Hisham: Vol. 2, pp. 230 - 231; Ibn Sad: p. 574.

"So is there none among you who has the courage of a man? And what of your Islam? And your Prophet?" At this blasphemous taunt, Ali left his position in the front rank of the Muslims, approached the Holy Prophet and sought permission to engage the challenger and silence his insolent tongue once and for all. The Prophet replied, *"Sit down. This is Amr!"* Ali returned to his position.

There was another burst of scornful laughter, more taunts, and another challenge. Again Ali went up to the Prophet. Again the Prophet declined permission. More laughter, more taunts. Again the challenge from Amr, and this time more insulting than before. *"Where is your paradise?"* He shouted, *"Of which you say that those who lose in battle will enter it? Can you not send a man to fight me?"*

When for the third time Ali moved towards the Prophet, the latter saw in Ali's eyes a look which he knew well; and he knew that Ali could no longer be restrained. He looked at Ali fondly, for Ali was dearer to him than any other man. He took off his turban and wound it around Ali's head. He next took off his sword and girded it at Ali's waist. And he prayed: *"O Lord! Help him!"* [1]

This sword which the Prophet now gave to Ali had once belonged to an infidel by the name of Munabba bin Hajaj. This man had been killed at the Battle of Badr, and the sword had come to the Muslims as part of the spoils of war. The Prophet had taken the sword for himself. Now in Ali's hand this was to become the most famous sword in Islam, killing more men in fair combat than any sword in history. This was the *Zulfiqar*.

Ali hastily collected a small group of Muslims and strode out towards the unbelievers. The group stopped at some distance from the Giant, and Ali stepped forward and got to within dueling distance of the challenger. The Giant knew Ali well. He had been a friend of Ali's father, Abu Talib. He now smiled indulgently at Ali as a man might smile at a boy.

"O Amr!" called Ali. *"It is believed that if any man of the Quraish offers you two proposals, you always accept at least one of them."*

"True."

"Then I have two proposals to offer you. The first is: accept Allah and His Messenger and Islam."

"I have no need of them."

"Then dismount from your horse and fight me."

"Why, O son of my brother? I have no desire to kill you."

"But I", replied Ali, *"Have a great desire to kill you!"* [2]

The Giant's face flushed with anger. With a cry of rage he sprang off his horse, displaying a degree of agility surprising in so huge a monster. He hamstrung his horse, drew his sword and rushed at Ali. The fight was on.

Amr struck at Ali many times, but Ali remained unharmed. He would parry the blow with his sword or shield or nimbly step aside to let the Giant's sword whistle past him harmlessly. At last the Giant stood back, panting and baffled. He wondered how this could be. Never before had any man survived so long in personal combat against him. And now this boy was looking at him as if he was playing a game!

Then things happened so fast that no one could quite follow the sequence-neither the Muslims nor the Quraish nor the Giant himself. Ali dropped his sword and shield to the ground; his body shot through the air like a missile and his hands grasped the Giant's throat; with a wrestler's kick he knocked the Giant off balance, and the Giant came crashing to the ground-all in a matter of seconds. Now the Giant lay on his back with Ali sitting astride his chest. The two armies gasped and murmured, then held their breath.

1. Ibn Sad: p. 572.
2. Ibn Hisham: Vol. 2, p. 225.

The bewilderment on the Giant's face changed to fury. At last he had been thrown, and by this young upstart who was less than half his size! But although he was down, he was not finished. He would still win the battle and re-establish his position as the greatest warrior in Arabia. He would toss this youngster into the air as a leaf is tossed by the wind.

The Giant's face went purple, the veins stood out on his neck and his huge biceps and forearms trembled as he strained to break Ali's grip. But he could not move it an inch. There was the quality of steel in the muscles of Ali.
"Know, O Amr", said Ali gently, *"that victory and defeat depend upon the will of Allah. Accept Islam! Thus not only will your life be spared, but you will also enjoy the blessings of Allah in this life and the next."* Ali drew a sharp dagger from his waistband and held it close to Amr's throat.

But this was more than the Giant could take. Was he whom Arabia considered her greatest champion to live the rest of his life under the shadow of defeat and disgrace? Was it to be said of him that he saved his life in personal combat by submitting to the conditions of his opponent? No! He, Amr bin Abdu Wud, had lived by the sword. He would perish by the sword. A life spent in violence must end with violence. He gathered the spittle in his mouth and spat into the face of Ali!

He knew what would happen. He knew that there would be a sharp intake of breath, that Ali's right arm would shoot into the air and then plunge the dagger into his throat. Amr was a brave man and could face death without flinching. He arched his back and raised his chin to offer his throat to Ali, for he knew what was to come. At least he thought he knew!

But what happened next left him even more bewildered. Ali rose calmly from Amr's chest, wiped his face, and stood a few paces away, gazing solemnly at his adversary. *"Know, O Amr, I only kill in the way of Allah and not for any private motive. Since you spat in my face, my killing you now may be from a desire for personal vengeance. So I spare your life. Rise and return to your people!"*

The Giant rose. But there was no question of his returning to his people a loser. He would live a victor, or not at all. Intending to make one last attempt at victory, he picked up his sword and rushed at Ali. Perhaps he would catch Ali unawares.

Ali had just enough time to pick up his sword and shield and prepare for the fresh assault. The blow which the Giant now delivered in furious desperation was the most savage blow of the encounter. His sword shattered Ali's shield, but in doing so lost its force and impetus, and could then do no more than inflict a shallow cut on Ali's temple. The wound was too slight to worry Ali. Before the Giant could raise his sword again, the *Zulfiqar* flashed in the sunlight, and its tip slashed open the Giant's throat. The blood of the Giant gushed forth like a fountain.

For a moment the Giant stood motionless. Then his body began to sway as if he was drunk. And then he fell on his face with a crash and lay still.

The earth did not shake with the impact of that colossal body. The earth is too big. But the hill of Sil'a shook with the cry of *Allah-o-Akbar* that thundered from 2,000 Muslim throats. The triumphant cry echoed through the length and breadth of the valley before it faded away into the stillness of the desert.

The Muslim group now rushed at the six remaining Quraish. In the sword fighting that ensued, one more Quraish was killed and one Muslim fell. A few minutes later the Quraish group turned and hastily withdrew across the ditch. Ikrimah dropped his spear as he jumped the ditch, on which Hassaan the Poet wrote many a rude verse. A man known as Nofal bin Abdullah, a cousin of Khalid's, was not successful in clearing the ditch and fell into it. Before he could rise, the Muslims were on the bank and hurling stones at him. Nofal wailed, *"O Arabs! Surely death is better than this!"* [1] Thereupon Ali obliged the man by descending into the ditch and cutting off his head.

1. Tabari: Vol. 2, p. 240.

The Muslim group now returned to camp, and a strong guard was placed at the crossing.

On the afternoon of the next day, Khalid moved up with a squadron, intending to succeed where Ikrimah had failed. He tried to cross the ditch, but this time the Muslim guard at the crossing saw him advance and deployed in sufficient time to prevent his crossing. There was a heavy exchange of archery in which one Muslim and one Quraish were killed, but Khalid was unable to cross.

Since the opposition at the moment appeared too strong to overcome, Khalid decided to resort to stratagem. He moved his squadron back, as if he had given up his intention of crossing the ditch, and placed it at some distance from the ditch. The Muslims took the bait, and believing that Khalid had abandoned his attempt to cross the ditch, withdrew and began to relax, waiting for the peace and quiet of the night. Suddenly Khalid galloped back with his squadron; and before the Muslim guard had time to re-deploy, a few of the Quraish, led by Khalid, managed to cross the ditch. But they had not advanced far from the ditch when the Muslims formed up again and held Khalid within the small bridgehead which he had occupied. Khalid tried hard to break through, but the Muslim resistance was too strong, and he had no success. There was some hand-to-hand fighting between the Quraish group and the Muslim guard in which Khalid killed one Muslim. The Savage also was there; and with the same javelin that he had used against Hamza at Uhud, he killed a Muslim in this sally across the ditch. Before long, however, seeing the situation as hopeless, Khalid broke contact and withdrew across the ditch. This was the last major military action in the Battle of the Ditch.

For the next two days there was no activity except for a certain amount of sporadic archery which did no damage to either side. The Muslims now ran out of food; but their courage was hardened by desperation and they were determined to starve rather than surrender to the hated infidel. In the Allied camp tempers rose and spirits fell. Everyone knew that the expedition, which had been expected to lead to a glorious victory, had ended in fiasco. There was widespread grumbling, and what made the situation intolerable was the fact that no one could find a way out of the impasse.

Then on Tuesday night, March 18, the area of Madinah was struck by a storm. Cold winds lashed at the Allied camp and howled across the valley. The temperature dropped sharply. The Allied camp was more exposed than the Muslim camp and the storm appeared to strike the Allies with a vengeance. It put out fires, knocked down cooking pots, carried away tents. The Allies sat huddled under their blankets and cloaks as the storm raged around them, waiting for an end that would not come.

Abu Sufyan could take no more. He leapt to his feet, and raising his voice against the storm, shouted to his men: *"This is no proper abode for us. Men and animals have suffered grievously from exposure. The Bani Quraizah have turned out to be pigs and monkeys and have betrayed us in our hour of need. The storm has ravaged our camp, put out our fires, knocked down our tents. Let us return to Makkah. Lo, I am one who goes!"*
[1]

Having made this last speech, Abu Sufyan jumped on to his camel and rode out with his men, hoping to get away from the pitiless storm. But the demons of the storm were to pursue him the whole night. The Ghatfan now came to know of the movement of the Quraish and so did the other tribes. Without further delay they mounted their camels and departed for their settlements and pastures. In the rear of the Quraish army rode Khalid and Amr bin Al Aas with their cavalry squadrons acting as a rear guard in case the Muslims should come out of Madinah and attempted to interfere with the Quraish movement. It was a bitter and disillusioned Abu Sufyan who led his army back to Makkah. The burden of failure lay heavy on his heart.

The next morning the Muslims found the Allies gone, and returned to their homes. This was the last attempt by the Quraish to crush the Muslims; henceforth they would remain on the defensive.

1. Ibn Hisham: Vol. 2, p. 232.

The Battle of the Ditch was over. Each side had lost four men. It was a victory for the Muslims in that they achieved their aim of defending themselves and their homes against the Allies, while the Allies failed in their attempt to crush the Muslims. In fact the Allies failed to do any damage at all. The siege had lasted 23 days and had imposed a terrible strain on both sides. It had been ended by the storm, but the storm was not the cause of the raising of the siege. It was the last straw. Strictly speaking, this operation was a siege and a confrontation rather than a battle, for the two armies never actually came to grips.

This was the first instance in Muslim history of the use of politics and diplomacy in war, and it shows the interplay of politics and arms in the achievement of the national aim. The use of armed force is one aspect of war-a violent and destructive aspect-to be used only when political measures fail to achieve the aims of the State. When a shooting war becomes inevitable, politics, with diplomacy as its principle instrument, prepares the ground for the use of armed force. It sets the stage, weakens the enemy, and reduces his strength to a state where armed force can be employed against him with the maximum prospect of success.

And this is just what the Prophet did. He used the instrument of diplomacy to split and weaken the enemy, not only in numbers but also in spirit. Most of the Muslims could not understand this, but they were learning from their leader. The Prophet's words, **"War is stratagem"** 1 , were to be remembered and frequently quoted in later Muslim campaigns.

1. Ibn Hisham: Vol. 2, p. 229; Waqidi: *Maghazi*, p. 295.

Chapter 5

"Verily We have granted you a manifest Victory. That Allah may forgive you your faults of the past and those to follow, fulfil His Favour to you, and guide you on the Straight Path.
And that Allah may help you with powerful help. It is He who sent down Tranquility into the hearts of the Believers, that they may add Faith to their Faith; for to Allah belong the Forces of the heavens and the earth, and Allah is full of Knowledge, full of Wisdom."
[Quran 48:1-4]

The Truce of Hudaibiya was signed in early April 628 (late Dhul Qad, 6 Hijri). The signing of such a truce was not the intention of the Prophet as he set out for Makkah in the middle of March. His intention was to perform the pilgrimage-the off-season pilgrimage known as Umra-and he took with him 1,400 fully armed Muslims and a large number of sacrificial animals.

The Quraish, however, feared that the Muslims were coming to fight a battle and subdue the Quraish in their home town, for the initiative had now passed to the Muslims. Consequently, the Quraish moved out of Makkah and concentrated in a camp nearby, from where Khalid was sent forward with 300 horsemen on the road to Madinah to intercept the Muslim army. Khalid did not see how he could stop such a large force with only 300 men, but he decided to do whatever was possible to delay the advance of the Muslims. He arrived at Kuraul Ghameem, 15 miles from Usfan, and took up a blocking position in a pass through which the road crossed this hilly region.

When the Muslims arrived at Usfan, their advance was preceded by a detachment of 20 horsemen who had been sent forward as a reconnaissance element. This detachment made contact with Khalid at Kuraul Ghameem, and informed the Prophet at Usfan of the position and strength of Khalid's force.

The Prophet decided that he would not waste time in fighting an action at this place. He was in any case anxious to avoid bloodshed, as his intention was the pilgrimage and not battle. He ordered his forward detachment to remain in contact with Khalid, and draw Khalid's attention to itself; and with Khalid so engaged, the Prophet moved his army from the right, travelling over little used tracks through difficult hilly country, which he crossed, not far from the coast through a pass known as Saniyat-ul-Marar 2 . The march proved a strenuous one, but it was successfully accomplished and Khalid's position bypassed. It was not till the outflanking movement was well under way that Khalid saw in the distance the dust of the Muslim column, and realising what had happened, hastily withdrew to Makkah. The Muslims continued the march until they had got to Hudaibiya, 13 miles west of Makkah, and pitched camp.

At Hudaibiya battle seemed imminent for some time in spite of the Prophet's wish to avoid bloodshed. Some skirmishes took place, but there were no casualties. After a few days, however, the Quraish realised that the Muslims had indeed come for pilgrimage and not for war. Thereafter envoys travelled back and forth between the two armies, and finally a truce was agreed upon, which became known as the Truce of Hudaibiya. It was signed on behalf of the Muslims by the Prophet and on behalf of the Quraish by Suhail bin Amr. Its terms were as follows:

a. For 10 years there would be no war, no raids, no military action of any sort between the Muslims and the Quraish.

b. The following year the Muslims would be permitted to perform the pilgrimage. They would be allowed three days in Makkah.

c. Any member of the Quraish who deserted to the Muslims would be returned; any Muslim who deserted to the Quraish would **not** be returned.

1. This Kura-ul-Ghameem is not the Kura marked on today's maps. The latter lies by an inlet of the Red Sea, while the former was in a hilly region with the hills extending westwards from it to the sea. It was southeast of Usfan.
2. This pass was also called Zat-ul-Hanzal (Abu Yusuf: p. 209).

Other tribes could join the truce on either side and would be bound by the same terms.

Some Muslims were incensed at the third clause, dealing with deserters, especially the hot-headed Umar who protested vehemently against it; but all protests were overruled by the Prophet. The truce actually gave certain distinct long-term and solid advantages to the Muslims, although these were not at the time apparent to everyone. It would be to the Muslims' advantage to be generous in their terms, as this would have a favourable psychological impact on the Arab tribes and would show the confidence that the Muslims enjoyed in their dealings with the infidels. Moreover, if some Muslims were not permitted to leave Makkah, they would act as the eyes and ears of the Muslims in the midst of the enemy, and could in certain ways influence the people in Makkah. Their presence within the Quraish camp would in fact be a source of strength to the Muslims. *"Anyway"*, said the Prophet, *"when anyone wishes to join us, Allah will devise means for him to do so."* [1]

As a result of the last clause of the truce, two tribes living in and around Makkah joined the main participants: the Khuza'a as allies of the Muslims and the Bani Bakr as allies of the Quraish. These two tribes were mutually hostile and had been feuding since the Ignorance.

After a stay of over two weeks at Hudaibiya, the Muslims returned to Madinah. The following year, in March 629 (Dhul Qad, 7 Hijri), the Muslims, led by the Prophet, performed the pilgrimage. The Quraish evacuated Makkah and lived in the surrounding countryside for three days, and did not return to their homes until after the Muslims had departed for Madinah.

For some time a change had been taking place in the mind of Khalid. At first he thought mainly of military matters and military objectives. Conscious of his own ability and military prowess, he felt that he was truly deserving of victory, but somehow victory always eluded him. At the Battle of Uhud, despite his masterly manoeuvre, the Muslims had been able to avoid a major defeat. He admired the Prophet's dispositions and the way the Prophet had forced battle on the Quraish with the odds in his favour. At the Battle of the Ditch again victory had eluded the Quraish. They had gone to battle after such careful preparations and in such strength that victory had seemed certain; yet the simple expedient of the ditch had snatched victory from their grasp. The Quraish army had gone forth like a lion and come back like a mouse. In the expedition of Hudaibiya, when he had tried to intercept the Muslims, the Prophet had neatly outmaneuvered him while his attention was riveted to the small Muslim detachment in front of him. Khalid was looking for *the Man*, and he could not help admiring Muhammad-his generalship, his character, and his personality-qualities which he could find in no one else.

Above all Khalid wanted the clash of battle and the glory of victory. His martial spirit sought military adventure, and with the Quraish there was only misadventure. He could see no hope of fighting successful battles on the side of the Quraish. Perhaps he should join the Prophet, with whom there were unlimited prospects of victory and glory.
There was plenty of military activity at Madinah. Every now and then expeditions would be sent out against the unbelieving tribes, either to break up hostile concentrations before they became too large or to capture camels and other live-stock. Between the Battle of Uhud and the pilgrimage, 28 expeditions were taken out by the Muslims, some led by the Prophet in person and others by officers appointed by him. With very few exceptions these expeditions had ended in complete success for the Muslims. The greatest of these had been the Campaign of Khaibar, in which the last resistance of the Jews was crushed. These expeditions had not only enlarged the political boundaries of Islam, but had also resulted in a great increase in wealth. Whenever reports of Muslim military successes arrived at Makkah, Khalid would think wistfully of the 'fun' that the Muslims were having. Now and then he would wish that he were in Madinah, for that is 'where the action was'!

After the Prophet's pilgrimage serious doubt entered Khalid's mind regarding his religious beliefs. He had never been deeply religious and was not unduly drawn towards the gods of the Kabah. He had always kept an open mind. Now he began to ponder deeply on religious matters, but did not share his thoughts with anyone. And then suddenly it flashed across his mind that Islam was the true faith. This happened about two months after the Prophet's pilgrimage.

1. Waqidi: *Maghazi*, p. 310.

Having made up his mind about Islam, Khalid met Ikrimah and some others and said, *"It is evident to the intelligent mind that Muhammad is neither a poet nor a sorcerer, as the Quraish allege. His message is truly divine. It is incumbent on all sensible men to follow him."*

Ikrimah was stunned by the words of Khalid. *"Are you abandoning our faith?"* he asked incredulously.

"I have come to believe in the true Allah."

"It is strange that of all the Quraish you should say so."

"Why?"

"Because the Muslims have killed so many of your dear ones in battle. I for one shall certainly not accept Muhammad, nor shall I ever speak to you again unless you give up this absurd idea. Do you not see that the Quraish seek the blood of Muhammad?"

"That is a matter of Ignorance", replied Khalid.

When Abu Sufyan heard from Ikrimah of Khalid's change of heart, he sent for both the stalwarts. *"Is it true what I hear?"* he asked Khalid.

"And what do you hear?"

"That you wish to join Muhammad."

"Yes. And why not? After all Muhammad is one of us. He is a kinsman."

Abu Sufyan flew into a rage and threatened Khalid with dire consequences, but was restrained by Ikrimah. *"Steady, O Abu Sufyan!"* said Ikrimah. *"Your anger may well lead me also to join Muhammad. Khalid is free to follow whatever religion he chooses."* [1] Ikrimah, the nephew and bosom friend, had stood up for Khalid in spite of their religious differences.

That night Khalid took his armour, his weapons and his horse, and set out for Madinah. On the way he met two others travelling in the same direction: Amr bin Al Aas and Uthman bin Talha (son of the Quraish standard bearer at Uhud) and there was mutual astonishment when they found that each was travelling to Madinah with the same purpose, for each had regarded the other two as bitter enemies of Islam! The three seekers arrived at Madinah on May 31, 629 (the 1st of Safar, 8 Hijri), and went to the house of the Prophet. Khalid entered first and made his submission. He was followed by Amr and then Uthman. All three were warmly welcomed by the Prophet; their past hostility was forgiven, so that they could now start with a clean sheet. Khalid and Amr bin Al Aas were the finest military minds of the time and their entry into Islam would spell victory for Muslim arms in the following decades.

Khalid, now 43 and in the prime of life, was glad to be in Madinah. He met old friends and found that he was welcomed by all. The old feuds were forgotten. There was a new spirit in Madinah-the spirit of the pioneer. There was activity, anticipation, enthusiasm, optimism, and this atmosphere entered the heart of Khalid. He breathed the clear air of the new faith and was happy.

He also met Umar and they were friends again. There remained a little of the old rivalry between the two, but this existed as a subconscious undercurrent rather than a deliberate feeling or intention. Khalid now realised that in his rivalry with Umar he was at a disadvantage, for he was a new convert while Umar was an emigrant who had left his home in Makkah. Umar had been the fortieth person to become a Muslim. While the Muslims were at Makkah he could take no great pride in having this position, for then the Muslims were few in number; but now thousands had entered Islam and with this large number, being the fortieth amounted to having a very important position. Now Khalid was competing not only against a man of equal strength, will and ability, but also against Muslim No. 40!

1. Waqidi: *Maghazi*, p. 321.

Khalid took to visiting the Prophet frequently. He would listen for hours to the talks of the Prophet. He would drink at the fountain of wisdom and virtue that was Muhammad, Messenger of Allah. One day Khalid and Fadhl bin Abbas (cousin of the Prophet) visited him in the house of his wife, Maimuna, who was an aunt of Khalid. Just then a Bedouin friend had sent a cooked dish as a gift to the Prophet, and as was his custom, the Prophet asked the guests to stay and share his meal. A cloth was spread on the ground and around it they all sat-the Prophet, his wife, and the two guests.

As the Prophet extended his hand towards the dish, Maimuna asked, *"O Messenger of Allah, do you know what this is?"*

"No."

"This is roast lizard!"

The Prophet withdrew his hand. ***"This meat I shall not eat"***, he said.

"O Messenger of Allah", asked Khalid, *"is it forbidden?"*

"No."

"Can we eat it?"

"Yes, you may do so."

Maimuna also abstained from the food, but Khalid and Fadhl ate their fill of the dish. Roast lizard was a favourite among the desert Arabs. Apparently it was a favourite with Khalid too, for he ate heartily! 1

1. This little-known incident is taken from Ibn Sad: p. 381.

Chapter 6

"What an excellent slave of Allah: Khalid ibn al-Walid, one of the swords of Allah, unleashed against the unbelievers!"
[Prophet Muhammad (SAWS)]1

Three months after his arrival at Madinah, Khalid got his chance to show what he could do as a soldier and a commander for the faith which he had just embraced. The Prophet had sent an envoy to the Ghassan 2 Chieftain of Busra, with a letter inviting him to join Islam. While passing through Mutah this envoy was intercepted and killed by a local Ghassan chieftain by the name of Shurahbil bin Amr. This was a heinous crime among the Arabs, for diplomatic envoys held traditional immunity from attack no matter how hostile a power they represented. The news of this outrage inflamed Madinah.

An expedition was immediately prepared to take punitive action against the Ghassan, and the Prophet appointed Zaid bin Harithah as the commander of the force. If he were killed, the command was to be taken over by Jafar bin Abi Talib. If he were killed, the command would devolve upon Abdullah bin Rawahah. Having appointed these officers in the chain of command, the Prophet said, *"If all three of these are killed, let the men select a commander from among themselves."* 3

The expeditionary force consisted of 3,000 men, one of whom was Khalid, serving as a soldier in the ranks. The mission the Prophet gave to Zaid was to seek out and kill the person responsible for the murder of the Muslim envoy, and to offer Islam to the people of Mutah. If they accepted Islam, they were not to be harmed. At the time this force was sent out the Muslims had no knowledge of the enemy strength that they would have to deal with.

Spirits were high as the expeditionary force began its march from Madinah. When the force arrived at Ma'an, reports were received for the first time that Heraclius, the Eastern Roman Emperor, was in Jordan with "100,000 Romans" and had been joined by "100,000 Christian Arabs"-mainly from the Ghassan. The Muslims remained in Ma'an for two days debating their next move. There was a certain amount of hesitation and nervousness. Some suggested that the Prophet be informed of the large strength of the enemy so that he could give them fresh orders on what course of action they should adopt; but Abdullah bin Rawahah (the third-in-command) did not agree with this suggestion, as it would entail unnecessary delay and would give the impression that the Muslims were afraid. He recited a few verses and made a stirring speech to raise the spirits of the men. He concluded by saying, *"Men fight not with numbers or weapons but with faith. By going into battle we have a choice of two glorious alternatives: victory and martyrdom."* 4 This speech dispelled all doubt from the minds of the Muslims, and they promptly resumed their march towards Syria.

The Muslims reached a place near the frontier of Balqa-a district in the east of what is now Jordan-where they made contact with a large force of Christian Arabs. Not finding this place suitable for battle, the Muslim commander withdrew his force to Mutah. The Christian Arabs followed the Muslims, and the two forces again met at Mutah. Both sides now decided to fight. It was the second week of September 29 (the third week of Jamadi-ul-Awwal, 8 Hijri).

Zaid deployed his force in the normal pattern of a centre and two wings. The right wing was commanded by Qutba bin Qatadah and the left wing by Ubaya bin Malik. Zaid himself commanded the centre, and in the centre, too, was Khalid. The battlefield lay to the east of, and stretched up to about a mile from, the present village of Mutah. The ground here was even, but had a slight undulation, and the gentle slope of a low ridge rose behind the Muslims as they faced the Christian Arabs to the north. 5

1. Tirmidhi and Ahmad from Abu Hurayrah, Sahih Al-Jami' Al-Saghir No. 6776.
2. A large and powerful tribe inhabiting Syria and Jordan.
3. Ibn Sad: p. 636.
4. Ibn Hisham: Vol. 2, p. 375.
5. A new mosque is being built by the Jordanian Government to mark the site of this battle.

The Christian Arabs, who were commanded by Malik bin Zafila, formed themselves into a deep mass confronting the Muslims. Some historians have given their strength as 100,000, while others have doubled that figure. These estimates are clearly mistaken. The enemy probably consisted of between 10 and 15 thousand men. In this battle the Muslims failed to gain a victory. If the enemy had been only twice their strength, they would undoubtedly have thrashed him; and an enemy had to be many times their strength to, inflict a defeat on them. It is largely on this basis that the above estimate of the enemy's strength is made.

The battle began, and both armies very quickly got to grips with each other. This was essentially a battle of guts and stamina rather than military skill. The commander himself fought at the head of his men with his standard, and after a short while Zaid was killed. As the standard fell from his hands, the second-in-command, Jafar, picked it up and continued fighting at the head of the army. After his body had been covered with scores of wounds, Jafar also fell; and the standard went down for the second time. This distressed the Muslims, for Jafar was held in great esteem and affection as a cousin of the Prophet. A certain amount of confusion became noticeable among the Muslims, but soon the third-in-command, Abdullah bin Rawahah, picked up the standard and restored order. He continued to fight until he also was killed.

Now there was real disorder in the ranks of the Muslims. A few of them fled from the scene of battle, but stopped not far from the battlefield. Others continued to offer confused resistance in twos and threes and larger groups. Fortunately the enemy did not press his advantage, for had he done so the Muslims, without a commander, could easily have been routed. Perhaps the gallantry of the Muslim commanders and the valour with which the Muslims had fought made the enemy overcautious and discouraged him from taking bold action.

When Abdullah had fallen, the standard was picked up by Thabit bin Arqam, who raised his voice and shouted, *"O Muslims, agree upon a man from among you to be the commander."* He then spied Khalid, who stood next to him, and offered him the standard. Khalid was conscious of the fact that as a new convert he did not hold a high position among the Muslims, and Thabit bin Arqam was a Muslim of long standing. This consideration was important. He declined the offer of Thabit, saying: *"You are more deserving than I" "Not I,"* replied Thabit, *"and none but you!"* [1] This was really a windfall for the Muslims, for they knew of the personal courage and military ability of Khalid. They all agreed to his appointment, and Khalid took the standard and assumed command.

The situation now was serious and could easily have taken a turn for the worse, leading, rapidly to the total defeat of the Muslims. The commanders before Khalid had shown more valour than judgement in fighting this battle. Khalid regained control over his small army and organized it into a neatly deployed fighting force. He was faced with three choices. The first was to withdraw and save the Muslims from destruction, but this might be regarded as a defeat and he would then be blamed for having brought disgrace to Muslim arms. The second was to stay on the defensive and continue fighting; in this case the superior strength of the enemy would eventually tell and the battle end in defeat. The third was to attack and throw the enemy off balance, thus gaining more time in which to study the situation and plan the best course of action. The last choice was closest to the nature of Khalid, and this is the course that he adopted.

The Muslims attacked fiercely along the entire front. They surged forward with Khalid in the lead. The example of Khalid gave fresh courage to the Muslims, and the battle increased in violence. For some time desperate hand-to-hand fighting continued; then Qutba, commanding the Muslim right, dashed forward and killed the Christian commander, Malik, in a duel. This resulted in a setback for the enemy and led to, a certain amount of confusion. The Christian Arabs now pulled back, still fighting, with a view to gaining time for reorganization. At this moment Khalid had his tenth sword in his hand, having broken nine in fierce combat.

1. Ibn Sad: p. 638.

As the Christian Arabs stepped back, Khalid restrained the Muslims and broke contact, pulling his force back a short distance. The two armies now faced each other out of bow range, both seeking time to rest and reorganize. This last round of the battle had ended in favour of the Muslims of whom so far only 12 had been killed. There is no record of enemy casualties but they must have been considerable, for each of the Muslim commanders before Khalid was a brave and skilful fighter and the nine swords that Khalid broke were broken on the bodies of Christian Arabs. The situation, however, offered no further prospect of success. Khalid had averted a shameful and bloody defeat and saved the Muslims from disgrace and disaster; he could do no more. That night Khalid withdrew his army from Mutah and began his return journey to Madinah.

The news of the return of the expedition preceded it at Madinah, and the Prophet and those Muslims who had remained in Madinah set out to meet the returning soldiers. The Muslims were in an ugly temper, for never since the Battle of Uhud had a Muslim force broken contact with the enemy and left him in possession of the battlefield. As the army arrived among the Muslims, they began to throw dust into the faces of the soldiers.

"O you who have fled!" they cried. *"You have fled from the way of Allah."* The Prophet restrained them and said, **"They have not fled. They shall return to fight, if Allah wills it."** 1 Then the Prophet raised his voice and shouted, **"Khalid is the Sword of Allah."** 2

Later the resentment against Khalid died down, and the Muslims realised the wisdom, judgement and courage which he had shown in the Battle of Mutah. And the name stuck to Khalid. He now became known as *Saifullah*, i.e. Sword of Allah. When the Prophet gave Khalid this title, he virtually guaranteed his success in future battles.

Some historians have described the battle of Mutah as a victory for the Muslims; others have called it a defeat. As a matter of fact it was neither. It was a drawn battle; but drawn in favour of the Christians, for the Muslims withdrew from the battlefield and left it in possession of their opponents. It was not a big battle; it was not even a very important one. But it gave Khalid an opportunity to show his skill as an independent commander; and it gained him the title of the Sword of Allah.

1. Ibn Hisham: Vol. 2, p. 382.
2. Waqidi: *Maghazi*, p. 322.

Chapter 7

"When comes the Help of Allah and Victory..." [Quran 110:1]

As stated earlier, two of the tribes of Makkah had entered the Truce of Hudaibiya, the Khuza'a on the side of the Muslims and the Bani Bakr on the side of the Quraish. These two tribes had an old feud dating back to pre-Islamic days, which had lain dormant during the past few years, and it might have been expected that now that they had joined the truce, peace would prevail between them. But this was not to be. The Bani Bakr once again took up the thread of the feud. They organized a night raid on the Khuza'a in which they were secretly assisted by the Quraish, who gave them not only weapons but also a few warriors, among whom were Ikrimah and Safwan bin Umayyah. In this raid twenty of the Khuza'a were killed.

A delegation of the Khuza'a at once rode to Madinah and informed the Prophet of this flagrant breach of the truce. The visitors invoked the alliance between their tribe and the Muslims and asked for help.

Abu Sufyan had not been directly concerned with the assistance given by the Quraish in this raid. He was now seriously alarmed as he had not wished to break the truce; and fearing Muslim retaliation, he travelled to Madinah to negotiate a fresh truce. On arrival at Madinah he first went to see his daughter, Umm Habiba, wife of the Prophet, but she gave him the cold shoulder. He next went to the Prophet and spoke to him, proposing a fresh truce, but the Prophet remained silent and this did more to frighten him and disturb his peace of mind than any threat could have done.

Not knowing just where he stood with the Prophet, Abu Sufyan decided to enlist the aid of the important Companions. He went to Abu Bakr with the request that he talk with the Prophet and urge a fresh truce, but Abu Bakr refused. He next went to Umar who, as warlike as ever, replied, *"By Allah, if I had nothing more than an army of ants, I would wage war against you."* Abu Sufyan then went to the house of Ali, and there spoke first to Fatimah and then to Ali. *"Once the Messenger of Allah has made up his mind,"* explained Ali, *"nothing can dissuade him from his purpose."*

"Then what do you advise?" asked Abu Sufyan.

"You are a leader of the Quraish, O Abu Sufyan! Keep peace among men." [1]

This advice could be interpreted in many ways, but somehow it gave Abu Sufyan more satisfaction than he had got from the others. Not knowing what else to do, he returned to Makkah. He had achieved nothing.

Soon after the departure of Abu Sufyan, the Prophet ordered immediate preparations for a large-scale operation. His intention was to assemble and move with such speed, and observe such strict secrecy, that the Quraish would not get knowledge of the coming of the Muslims until the Muslims were virtually knocking at their door. Thus the Quraish would not have time to organise another alliance with neighbouring tribes to face the Muslims. While the assembly of forces was in progress, the Prophet came to know that a woman was on her way to Makkah with a letter warning the Makkans of the preparations being made against them. He sent Ali and Zubair in haste after her. These two stalwarts caught up with the woman, found the message and brought back message and messenger to Madinah.

1. Ibn Hisham: Vol. 2, pp. 396-7.

The move of the Muslim army started from Madinah on January 1, 630 (the 10th of Ramadan, 8 Hijri). Many contingents from Muslim tribes had joined the Prophet at Madinah, and other contingents fell in on the way. Thus the Muslim army soon swelled to an all-time high figure of 10,000 warriors. With this force the Prophet arrived at Marr-uz-Zahran, 10 miles north-west of Makkah, without the Quraish having any knowledge of his movement. 1 This was the fastest move the Muslim army had ever accomplished.

Abbas, uncle of the Prophet, had about now made up his mind to join the Muslims and accept the true faith. While the Muslim army was at Juhfa, it met Abbas and his family on their way to Madinah. The conversion of Abbas was received joyfully by the Prophet, with whom the relations of Abbas had always been cordial.

When the Muslims got to Marr-uz-Zahran, Abbas became deeply concerned about the fate of the Makkans. He was afraid that if the Muslims took Makkah by force, the operation would result in the destruction of the Quraish. He therefore set out on the Prophet's mule, with the Prophet's permission, to warn the Quraish of the serious consequences of resistance and persuade them to send envoys of peace to the Muslims. At about this time Abu Sufyan had come out of Makkah to carry out a personal reconnaissance and see if any Muslims were in sight. Abbas and Abu Sufyan met while the former was halfway to Makkah.

"What news do you bring, O Father of Fadhl?" asked Abu Sufyan.

"The Messenger of Allah", replied Abbas, *"comes with any army of 10,000 men."*

"Then what do you advise us to do?"

"If the Muslims capture Makkah against resistance, they will certainly cut off your head. Come with me to the Prophet, and I shall ask him to spare your life."

Abu Sufyan mounted the mule behind Abbas and, so mounted, they rode to the Muslim camp, arriving there after nightfall. It so happened that on this night Umar was the officer of the guard and was walking around the camp to see that the sentries were vigilant. He was the first to see and recognise the two visitors and exclaimed: *"Ah! Abu Sufyan, enemy of Allah! Praise be to Allah that you have come into our camp without a safe conduct."* Umar then ran to the tent of the Prophet, and Abbas, guessing Umar's purpose, urged his mule forward. The three of them arrived at the Prophet's tent simultaneously, and a heated argument arose between Umar and Abbas. Umar was asking for permission to cut off the head of Enemy Number One, while Abbas was insisting that he had given protection to Abu Sufyan, and so he could not be harmed until he had been heard. The Prophet dismissed all three of them with instructions to come again in the morning. Abbas took Abu Sufyan to his tent where he spent a sleepless night, wondering what his fate would be.

The following morning, as Abbas and Abu Sufyan were going to the Prophet's tent, the latter saw them coming and remarked, **"One comes who intends to become a Muslim but is not a Muslim at heart."** As they arrived at the tent, the Prophet asked, **"O Abu Sufyan! Do you not know that there is no Allah but Allah?"**

"I have now realised it. If other gods in whom I believed had existed, they would certainly have helped me."

"And do you not know that I am the Messenger of Allah?"

This was a terrible moment for Abu Sufyan. He was a proud leader of the Quraish, one of the nobles of the tribe, a descendant of Umayyah. He had always regarded himself as second to none, and in this he was right. He was virtually the ruler of Makkah-a man all Makkans held in respect and reverence. Now he stood like a humble supplicant before the very man whom he had persecuted and fought for years, and for whose destruction he had strained every nerve.

1. Marr-uz-Zahran is a small valley which in its lower portion becomes the Wadi Fatimah and crosses the present Jeddah-Makkah highway about 20 miles from Makkah.

"On this point", replied Abu Sufyan, *"there is some little doubt in my mind."*

Abbas now turned fiercely on Abu Sufyan. *"Woe to you, O Abu Sufyan!"* he hissed. *"Submit, or your head will be cut off!"*

"I bear witness", said Abu Sufyan hastily, *"that Muhammad is the Messenger of Allah!"*

Abbas now had a word with the Prophet out of Abu Sufyan's hearing. *"O Messenger of Allah"*, he whispered, *"Abu Sufyan is a proud man. He has dignity and self-respect. Will you not be gracious to him and give him some special token of esteem?"* [1]

At this the Prophet declared, *"Whoever enters the house of Abu Sufyan shall be safe."* The face of Abu Sufyan lit up. He had been especially honoured by Muhammad. The Prophet continued: *"Whoever locks his door shall be safe. Whoever remains in the mosque shall be safe."*

Abu Sufyan now returned to Makkah where the people had gathered, awaiting news of their fate. Abu Sufyan addressed the crowd: *"O Quraish! Muhammad has come with power that you cannot match. Submit to him and be safe. Whoever enters my house shall be safe."* This led to an uproar in the crowd. *"And how many do you think could fit into your house?"* the people asked with sarcasm. Abu Sufyan then added, *"Whoever stays in his house and locks his door shall be safe. Whoever remains in the mosque shall be safe."*

This appeased the crowd but could not appease his wife, Hind. She sprang at him like a wild cat, slapped his face and caught him by his moustaches. *"Kill the fat old fool!"* she screamed at the crowd. *"He has turned away from us."* Since Hind was no light-weight, the experience must have been a painful one for Abu Sufyan. However, he managed to shake her off and walked away to his house.'

The Muslims expected that there would be some opposition to their entry into Makkah. They could not assume that it would be an entirely peaceful operation although the Prophet hoped that blood would not be shed. With hardened anti-Muslims like Ikrimah and Safwan you could never tell. The plan of the Prophet was therefore designed to conquer Makkah as a military operation.

Makkah lies in the Valley of Ibrahim and is surrounded and dominated by black, rugged hills which rise in places to over 1,000 feet above the valley floor. The town was then approached over four routes, each one going through a pass in the hills. These routes came in from the north-west (almost north), the south-west, the south and the north-east. The Prophet divided his army into four columns, one to advance on each route. The main column, which was commanded by Abu Ubaidah and with which the Prophet travelled in person, would enter Makkah along the main Madinah route, from the north-west, via Azakhir. The second column, under Zubair, would enter from the south-west, through a pass west of the Hill of Kuda. The third column, under Ali, would enter from the south, via Kudai; and the fourth, under Khalid, would enter from the north-east, via Lait and Khandama.

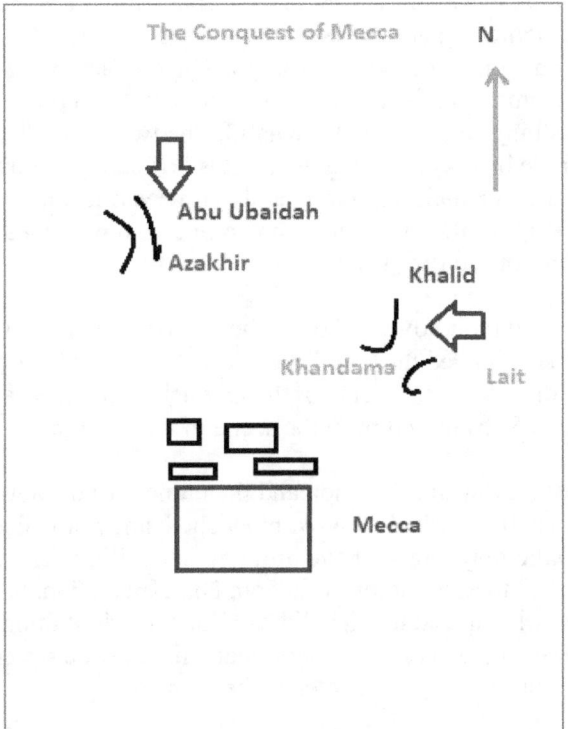

The advance consisted of convergent thrusts aimed at a single central objective which would have the effect of chopping up the enemy into small portions and also force dispersion on him, so that he would be unable to concentrate for battle on any one axis of advance. Moreover, even if the enemy succeeded in holding up the advance on some axes, the attackers would have other axes on which to break through and thus enjoy better prospects of success. All approaches were used to meet this requirement of military tactics. This was also done to prevent the escape of the Quraish; but later, when vigilance had been relaxed, some individuals did succeed in getting away.

1. Ibn Hisham: Vol. 2, pp. 402-5, Ibn Sad: p. 644; *Waqidi*: Maghazi: pp. 327-31.
2. The whole area covered by
hilly, but since the hills could not be accurately drawn without the aid of large scale topographical maps, no hills are shown on the map-just the places and the directions of the advancing columns.

The Prophet emphasised that there must be no fighting unless there was armed resistance by the Quraish. He also ordered that there would be no killing of the wounded, no pursuit of fugitives and no slaying of captives.

The entry into Makkah took place on January 11, 630 (the 20th of Ramadan, 8 Hijri). It proved a peaceful and bloodless operation except in the sector of Khalid. Ikrimah and Safwan had got together a band of dissidents from the Quraish and other tribes and decided to make the Muslims fight for victory. They met Khalid's column at Khandama, and this was a new and strange experience for Khalid. The two enemy leaders who were now opposing him in battle had been his dearest friends Ikrimah and Safwan; and the latter was also the husband of Khalid's sister, Faktah. However, Islam cancelled all relationships and friendships of the Ignorance, and no one who was not a Muslim could have a claim on a Muslim for old time's sake.

The Quraish opened up with their bows and drew their swords; and this was all that Khalid was waiting for. He charged the Quraish position, and after a short and sharp clash, the Quraish were driven back. Twelve of the Quraish were killed at a loss of only two Muslims. Ikrimah and Safwan fled from the scene of the encounter.

When the Prophet came to know of this action and the number of infidels killed, he was displeased with Khalid. He had wished to avoid bloodshed; and knowing Khalid's violent nature he feared that Khalid may himself have brought on a military engagement. Khalid was duly sent for and asked to account for his action. His explanation, however, was accepted by the Prophet, who agreed that Khalid had done the right thing. He had, after all, merely hit back. It was in the nature of Khalid that whenever he struck, he struck very hard. There was no moderation in the character of the man.

As soon as Makkah was occupied by the Muslims, the Prophet went to the Kabah and circumambulated the House of Allah seven times. This was a great moment in the life of Muhammad. It was more than seven years since he had fled as a fugitive from Makkah with the Quraish at his heels, thirsting for his blood. Muhammad was no longer the fugitive. He was no longer a voice crying in the wilderness. Muhammad had returned, and he had returned as master with Makkah at his feet. The Quraish trembled as they waited in the mosque, for they knew the savage nature of Arab vengeance.

The Prophet turned and looked at the Quraish. There was a hushed silence as the assembled populace gazed at him, wondering what their fate would be. ***"O Quraish!"*** called the Prophet. ***"How should I treat you?"***

"Kindly, O noble brother, and son of a noble brother!" the crowd replied.

"Then go! You are forgiven." [1]

The Prophet now entered the Kabah and saw the idols arranged along its walls-idols of all shapes and sizes. In and around the Kabah there were 360 idols carved of wood or hewn out of stone, including a statue of Ibrahim holding divining arrows. The Prophet had a large stick in his hand, and he set about smashing these idols to pieces. When the task was finished he felt as if a great weight had been lifted off his shoulders. The Kabah had been cleansed of the false gods; now only the true Allah would be worshipped in the House of Allah. The Prophet's joyous cry (a Quranic verse) rose above the Kabah: *"Truth has come and falsehood has vanished!"* 2

The next few days were spent in consolidation and reorganisation. Most of the people of Makkah accepted Islam and swore allegiance to the Messenger of Allah.

Before his entry into Makkah, the Prophet had announced the names of 10 persons-six men and four women-who were to be killed at sight, even if they took shelter within the Kabah. These 10 were what we would today call 'war criminals'. They were either apostates or had taken part directly or indirectly in the torture or betrayal of Muslims. At the head of the list was Ikrimah, and Hind also was one of them.

1. Ibn Hisham: Vol. 2, p. 412.
2. *Ibid*. Vol. 2, p. 417; Quran: 17:81.

When he withdrew from the engagement with Khalid, Ikrimah hid in the town, and as the Muslims relaxed their vigilance, he slipped out and fled to the Yemen with the intention of taking a boat to Abyssinia. Ikrimah's wife, however, became a Muslim and pleaded her husband's case with the Prophet, who agreed to spare his life. This woman travelled in haste to the Yemen, where she found her husband and brought him back. On arrival at Makkah, Ikrimah went straight to the Prophet and said, *"I am one who has erred and now repents. Forgive!"* 1 The Prophet accepted his submission, and Ikrimah joined the brotherhood of Islam.

Safwan bin Umayyah, though not on the war criminals' list, feared for his life and fled to Jeddah with the intention of crossing the Red Sea and seeking refuge in Abyssinia. A friend of his, however, asked the Prophet to spare his life and accept his submission. The Prophet had in any case no intention of killing Safwan and let it be known that he would gladly accept the return of Safwan. This friend then went to Jeddah and brought Safwan back. The man submitted to the Prophet, but it was a personal and political submission. As for Islam, he asked the Prophet to allow him two months in which to make up his mind. The Prophet gave him four months.

Of the war criminals actually only three men and two women were killed. The remainder were pardoned, including Hind, who became a Muslim.

Having destroyed the idols in the Kabah, the Prophet sent out small expeditions to the neighbouring settlements where other idols were known to exist in local temples. Khalid was sent to Nakhla to destroy Uzza, the most important of the goddesses. He set out with 30 horsemen. 2

It appears that there were two Uzzas, the real Uzza and a fake. Khalid first located the fake and destroyed it, then returned to the Prophet to report completion of duty. *"Did you see anything unusual?"* asked the Prophet. *"No." **"Then you have not destroyed Uzza"***, said the Prophet. ***"Go again."***

Angry at the mistake that he had made, Khalid once again rode to Nakhla, and this time he found the real Uzza. The custodian of the temple of Uzza had fled for his life, but before forsaking his goddess he had hung a sword around her neck in the hope that she might be able to defend herself. As Khalid entered the temple, he was faced by a naked black woman who stood in his way and wailed. Khalid did not stop to decide whether she was there to seduce him or to protect the idol, but drew his sword and with one powerful stroke cut the woman in two. He then smashed the idol, and returning to Makkah, gave the Prophet an account of what he had seen and done. *"Yes,"* said the Prophet, ***"that was Uzza; and never again shall she be worshipped in your land."*** 3

On or about January 20, 630 AH, after the destruction of the idols, occurred the unfortunate incident of the Bani Jazima. The Prophet sent a number of expeditions to the tribes living in the neighbourhood of Makkah to call them to Islam, and instructed the commanders not to fight those who accepted the call. Here again the Prophet's intention was to avoid bloodshed.

The expedition to the area of Tihama, south of Makkah, was commanded by Khalid. It consisted of 350 horsemen from several tribal contingents, the largest number being from the Bani Sulaim, and included a few Ansars and Emigrants. The objective of this force was Yalamlam, about 50 miles from Makkah.

When Khalid reached Al Ghumaisa, about 15 miles from Makkah on the way to Yalamlam, he met the tribe of Bani Jazima. The tribesmen saw the Muslims and took up their weapons, at the same time calling, *"We have submitted. We have established prayers and built a mosque."*

"Then why the weapons?" asked Khalid.

"We have a feud with certain Arab tribes and have to defend ourselves against them."

1. Waqidi: Maghazi, p. 332.
2. There was the Nakhla Valley, now known as Wadi-ul-Yamaniya, through which ran the main route between Makkah and Taif; and there was the Nakhla, at which was the goddess Uzza, and this lay north of the Wadi-ul-Yamaniya. It was about 4 or 5 miles south of the present Bir-ul-Batha.
3. Ibn Sad: p. 657.

"Lay down your arms!" ordered Khalid. *"All the people have become Muslims and there is no need for you to carry weapons."*

One man from the Bani Jazima now shouted to his comrades: *"This is Khalid, son of Al Waleed. Beware of him! After the laying down of arms there will be a binding of hands, and after the binding of hands there will be a severing of heads!"* 1

There was an old feud between the clan of Khalid and the Bani Jazima. In pre?Islamic days a small Quraish caravan was returning from the Yemen when it was set upon by the Bani Jazima, who looted the caravan and killed two important individuals-Auf, father of

Abdur-Rahman bin Auf, and Fakiha, son of Al Mugheerah, an uncle of Khalid. Abdur-Rahman had later killed the murderer of his father and thus avenged his father's blood, but the death of Fakiha had not been avenged. All this, however, happened during the Ignorance.

The people of the Bani Jazima now began to dispute with the man who was warning them against Khalid. *"Do you want to have us slaughtered?"* they asked him. *"All the tribes have laid down their arms and have become Muslims. The war is over."* 2 After a brief argument the tribe laid down its arms.

The cause of what happened next is not clear. Perhaps Khalid reverted momentarily to the tribal vindictiveness of the Ignorance. (He had been a Muslim for only a few months.) On the other hand, perhaps there was an excess of Islamic zeal in the heart of Khalid and he doubted the truth of the declaration of faith by the tribe. As the tribesmen laid down their arms, Khalid ordered his men to tie their hands behind them. He then ordered that all the captives be put to the sword. Luckily only the Bani Sulaim obeyed the order and killed the captives in their hands, whose number is not known. Other tribal contingents refused to carry out the order. There was a strong protest from Abdullah, son of Umar, and Abu Qatadah, but Khalid rejected the protest. Abu Qatadah immediately rode to Makkah and informed the Prophet of what Khalid had done.

The Prophet was horrified. He raised his hands towards heaven and exclaimed: **"O Lord! I am not responsible for what Khalid has done."** 3 He then sent Ali with a good deal of money to soothe the feelings of the Bani Jazima and pay indemnity for the blood that had been shed. Ali carried out the mission with generosity and did not return until the tribe was fully satisfied.

Khalid was now sent for by the Prophet who demanded an explanation for what he had done. Khalid said that he did not believe that they really were Muslims that he had the impression that they were deceiving him, and that he believed that he was killing in the way of Allah.

Present with the Prophet was Abdur-Rahman bin Auf. When he heard the explanation of Khalid, he said, *"You have committed an act of Ignorance in the days of Islam."*

Khalid now thought that he saw a way out of this delicate predicament, and he replied, *"But I took revenge for the killing of your father."* *"You lie!"* snapped Abdur-Rahman. *"I killed the murderer of my father a long time ago and vindicated the honour of my family. You ordered the slaughter of the Bani Jazima in revenge for the death of your uncle, Fakiha."*

This led to a heated argument between the two. And this was a mistake on the part of Khalid, for Abdur-Rahman was one of the Blessed Ten and thus had a position which few could challenge. Before the argument could get out of hand, however, the Prophet intervened and said sternly, **"Leave my Companions alone, O Khalid! If you possessed a mountain of gold and spent it in the way of Allah, you would not achieve the status of my Companions."** 4 He was referring, of course, to his early Companions, for Khalid too was a Companion.

Thus was Khalid put in his place. He was pardoned; but he learnt the important lesson that he, as a later convert, did not have the same status as the early Companions, especially the Blessed Ten. He was to keep this lesson in mind on many future occasions.

1. Ibn Sad: pp. 659-60; Ibn Hisham: Vol. 2, p. 429.
2. *Ibid*
3. *Ibid*
4. Ibn Sad: Vol. 2, p. 431.

Chapter 8

"Assuredly Allah did help you in many battlefields, and on the Day of Hunayn: behold! Your great numbers elated you, but they availed you nothing. The land, for all that it is wide, did constrain you, and you turned back in retreat.
But Allah did pour His calm on the Messenger and on the Believers, and sent down forces which you saw not, and He punished the Unbelievers: Thus does He reward those without Faith. Again will Allah, after this, turn in mercy to whom He will, for Allah is Oft-Forgiving, Most Merciful."
[Quran 9: 25-27]

Hardly had the people of Makkah sworn allegiance to the Prophet and life returned to normal in the town, when hostile winds began to blow from the east. The powerful tribes of the Hawazin and the Thaqeef were on the war-path.

The Hawazin lived in the region north-east of Makkah and the Thaqeef in the area of Taif. They were neighbouring tribes, who now feared that the Muslims, having conquered Makkah, would attack and catch them dispersed in their tribal settlements. To avoid being taken at a disadvantage, they decided to mount an offensive themselves, hoping to benefit from their initiative. The two tribes concentrated at Autas, near Hunain, where they were joined by contingents from several other tribes. This again was a coalition like the one which had assembled for the Battle of the Ditch. The total strength of the assembled tribes was 12,000 men, and the over-all commander was the fiery, 30-year-old Malik bin Auf. This young general decided to make his men fight in a situation of such serious danger that they would fight with the courage of desperation. He ordered the families and the flocks of the tribes to join the men.

Another leader in the coalition was the venerable Duraid bin As-Simma. Hoary with age, this man had lost the strength and vitality to lead men in battle, but he was a sage with a clear mind who accompanied his men wherever they marched; and since he was an experienced veteran, his advice on matters of war was widely sought. His military wisdom was unchallenged.

At Autas the aged Duraid heard the noises which usually, arise wherever families and animals are gathered. He sent for young Malik and asked, *"Why do I hear the call of camels, the braying of donkeys, the bleating of goats, the shouting of women and the crying of children?"* Malik replied, *"I have ordered the families and the flocks to muster with the army. Every man, will fight with his family and his property behind him and thus fight with greater courage."*

"Men fight with swords and spears, not with women and children", said Duraid. *"Put the families and the flocks at a safe distance from the field of battle. If we win, they can join us. If we lose, at least they shall be safe."*

Malik took this as a challenge to his judgement and his, ability to command the army. *"I shall not send them away"*, he bristled. *"You have grown senile and your brain is weak."* At this Duraid withdrew from the argument and decided to let Malik have his way. Malik then returned to his officers and, said, *"When you attack, attack as one man. As our attack begins, let all scabbards be broken."* [1] This breaking of scabbards was practised by the Arabs to signify an attitude of suicidal desperation.

As it happened only the Hawazin brought their families and their flocks to the camp. Other tribes did not do so.

The Prophet did not want any more bloodshed, but had, no choice except to set out to face this new enemy. He had no intention of waiting for another coalition to form against him and attack him as had happened three years before at the Battle of the Ditch. Moreover, if he waited on the defensive in Makkah and the enemy remained poised at Autas, the situation would lead to a stalemate which could last for months; and the Prophet could not afford to waste all that time. He had to attend to organisational matters and set about the conversion of the tribes of Arabia while the psychological impact of the fall of Makkah was still fresh in the minds of the Arabs. With a large hostile concentration at Autas, he would not be able to carry out these tasks. In any case, a strong enemy challenge to his authority at this stage would reduce the impact the Muslim conquest of Makkah had made on the Arab mind. This challenge had to be met. This opposition had to be crushed. The Prophet's decision to advance from Makkah created the unusual situation of both sides moving forward to fight an offensive battle.

On January 27, 630 (the 6th of Shawal, 8 Hijri), the Muslims set out from Makkah. The army consisted of the original 10,000 men who had conquered Makkah plus 2,000 new converts from among the Makkans. These new Muslims were of doubtful value as Islam had not really entered their hearts; they had come because they supposed that this was the right thing to do. Among them were Abu Sufyan and Safwan bin Umayyah. The latter had been given four months in which to make up his mind about the new faith, but was now favourably inclined towards the Prophet and had gone so far as to lend the Muslims 100 coats of mail for the forthcoming battle.

1. Ibn Hisham: Vol. 2, pp. 438-9.

The Muslim advance from Makkah was led by a contingent of 700 men from the Bani Sulaim, operating under the command of Khalid. During the evening of January 31, the Muslims arrived in the Valley of Hunain and established their camp.

Hunain is a valley running from Shara'i-ul-Mujahid (new), which is 11 miles east-north-east of Makkah, to Shara'i Nakhla (old) which is 7 miles further east. The valley continues eastwards for another 7 miles and then turns north towards Zaima. (None of these places were then in existence.) Between the Shara'i's the valley is quite wide, about 2 miles in most places, but beyond the old Shara'i it narrows down to between a quarter and a half-mile, and as it approaches Zaima it gets narrower still. It is this second portion of the Hunain Valley which is a defile, and the defile is narrowest near Zaima. Beyond Zaima the Taif route winds into the Wadi Nakhlat-ul-Yamaniya.

While the Muslims were moving towards Hunain, each side had sent out agents to get information about the other side. Both sides were well informed of opposing strengths, locations and movements. An agent sent by the Prophet mixed with the Hawazin at Autas, got to know the exact strength of the coalition and slipped out unseen to give this information to the Prophet. When he gave his report, Umar was also present, and for some reason did not believe the intelligence conveyed by the agent. He called the agent a liar, whereupon the agent replied, *"If you call me a liar, you call the truth falsehood. And you had called a liar one who is better than me."* The man was alluding to the time when Umar, before his conversion, was a violent enemy of the Prophet.

Umar suddenly turned to the Prophet and said, *"Did you hear that?"* **"Steady, O Umar!"** replied the Prophet. **"You were once misguided, and Allah showed you the way."** 1 Umar said no more.

As the Muslims arrived at their new camp in the Hunain Valley, news of their arrival was conveyed to Malik bin Auf by his agents. He guessed that the Muslims would know that his army was at Autas, and would expect to fight him at or near Autas. And he put into effect his plan to outwit the Muslims.

Before dawn on February 1, 630 (the 11th of Shawal, 8 Hijri) the Muslims formed up in marching order to advance to Autas where they expected to engage the enemy. It was their intention to get through the defile of Hunain before the enemy came to know of their movement. The advance guard again consisted of the Bani Sulaim under Khalid, and behind it marched various Muslim units, including the group of 2,000 Makkans. The camp was left standing as the base of the operation.

As the first glow of dawn appeared in the eastern sky, the advance guard entered the defile (about 2 miles short of Zaima.) Eagerly anticipating a lively battle with a surprised enemy at Autas, Khalid increased his pace. And then the storm broke!

Khalid was the first to receive the shock of the ambush. The quiet of the dawn was shattered by a thousand piercing yells, and the arrows came not in tens or twenties but in hundreds. They came like hailstones, whistling and hissing, striking horse and man. The Bani Sulaim did not stop to act against the enemy. They did not stop to think or take cover. They turned as one man and bolted. Khalid's shouts to his men to stand fast were lost in the noise and confusion. He himself was badly wounded and was carried away with the tide of fleeing men and horses; but after riding a short distance he fell off his horse and lay still, unable to move because of his wounds.

As the Bani Sulaim turned in panic and fled, they ran into other units which occupied the narrow track, who now became aware that something terrible had happened. The half-hearted Makkans turned and joined the flight, followed by several other Muslim units. Some of the Muslims fled to the camp, but the majority of them merely dispersed and took cover some distance behind the scene of the ambush on the other side of the track. No one knew quite what happened. The confusion increased as camel mounted camel and horses and men ran into each other in a blind urge to get away.

1. Ibn Hisham: Vol.2, p. 440.

Malik bin Auf had surprised his would-be surprisers. During the night he had moved his army into the defile of Hunain which allowed no room for manoeuvre. His men moved into position on both sides of the track and hid behind boulders and in broken ground which afforded excellent cover. In front were the Hawazin, with a few groups of Thaqeef. Then came the Thaqeef and behind them were other tribal contingents. Malik had devised a masterly plan. He had delayed his move till after dusk, so that the Muslims would continue to believe that his army was at Autas, and then placed it in ambush in the defile of Hunain with the intention of annihilating the Muslims or driving them back in panic to Makkah and beyond. Behind the site of the ambush was a narrow pass [1] to which Malik could withdraw in case the battle did not go according to his plan. As long as this pass was secure, the Muslims would not be able to advance to Autas, Malik's base.

Most of the new Meccan converts were delighted at this setback to the Muslims. Abu Sufyan remarked, *"This retreat will not stop until they get to the sea!"* Present with Safwan bin Umayyah was his half-brother, who said, *"Now the sorcery of Muhammad will be exposed." "Silence!"* Safwan snapped at him. *"May Allah break your mouth! I would rather see a man of the Quraish ruling over us than a man of the Hawazin!"* [2]

The Prophet was left standing on the track with nine of his Companions, including Ali, Abu Bakr, Umar, and Abbas. As the Muslims ran past, he shouted to them, **"O Muslims! I am here! I, the Messenger of Allah! I, Muhammad, son of Abdullah!"** [3] But his cries were of no avail. The leading elements of the Hawazin got to the place where the Prophet stood, and here Ali brought down the first infidel to fall at Hunain-a man mounted on a red camel, carrying a long lance at the end of which flew a black pennant. This man was chasing the Muslims as they fled. Ali pursued the man, along with a fellow Muslim, and catching up with him cut the tendons of the camel's hind legs with his sword. The infidel fell with the camel, and the other Muslim cut off his head.

The Prophet now moved towards the right with his group and took shelter on a rocky spur. A few men of the Thaqeef came towards the Prophet's group, but were driven back by the Companions.

Malik bin Auf had done to the Muslims what no one had ever done before. For the Muslims this was the first, and bitter, experience of being ambushed, and many of them lost their heads and fled from the scene of action.

Malik had struck brilliantly; but unfortunately for him, his men had not performed as expertly as he had hoped. They had not waited until the main body of the Muslims had entered the trap, but had opened up when just the advance guard was in their field of fire. And Malik now made the mistake of being satisfied with what he had achieved so far; beyond advancing a few hundred yards he made no attempt to pursue the Muslims. If he had done so, the story of this battle might have read differently. Moreover, the archery of the Hawazin was extremely poor. While several Muslims and their mounts were wounded, none were killed in the ambush.

The Holy Prophet surveyed the scene before him, and the scene was anything but promising. He decided not to let Malik get away with such an easy victory. He turned to Abbas and ordered him to call the Muslims to rally around him. Abbas was a large man with a powerful voice which, according to some accounts, could be heard miles away. Now he yelled at the top of his voice: *"O Muslims! Come to the Messenger of Allah! O Ansar...O Companions...O ..."* He called each tribe in turn to report to the Prophet.

The call was heard by most of the Muslims and they at once began to move to where the Prophet stood. As soon as the first 100 men had gathered beside the Prophet, he ordered a counter-attack. These men assailed those of the Hawazin who were nearest to the Prophet and drove them back. Soon the assembling Muslims increased in number until thousands of them had rejoined the Prophet. When the Prophet felt that sufficient strength had been gathered around him, he ordered a general attack against the Hawazin.

1. I have not been able to place this pass. It was probably at or near Zaima.
2. Ibn Hisham: Vol.2 pp. 443-5.
3. *Ibid*

This time it was Malik who was surprised. Having been certain that victory was his, he now found that his own army was under attack. The hand-to-hand fighting became more desperate, and this is just what the Muslims wanted, for in this sort of violent fighting their superiority in swordsmanship put the odds in their favour. In close-quarter battle the Muslims had no equal. Gradually the Hawazin were pressed back and as the Prophet saw their men fall before the onslaught of the Muslims, he affirmed

"In truth, I am the Prophet,
I the son of Abdul Muttalib."

He then turned to those who stood next to him and remarked, *"Now the oven heats up!"* [1]
Malik decided that he was getting the worst of the fighting and put his withdrawal plan in action. The Thaqeef were already in position a short distance behind the Hawazin. Leaving the Thaqeef to act as a rearguard, he pulled the Hawazin back to safety. The Muslims moved forward and made contact with the Thaqeef, who now began to receive heavy punishment from the Believers. Soon after this contact, the Thaqeef turned and took to their heels, followed by other tribal contingents, some of which had taken no part in the fighting. In the meantime Malik had got the Hawazin safely to the pass, and here he deployed them to fight a defensive battle while waiting for stragglers to catch up. As long as he held this pass, the families and the flocks of the Hawazin were safe.

The Muslims had not only recovered from the shock of the ambush but had counter-attacked, regained their position and driven the enemy from the battlefield. This was a tactical victory, but more was to come.

While the Muslims were stripping the Thaqeef dead of their weapons and clothing, an interesting incident involving two Muslims took place. One was an Ansar from Madinah and the other a man by the name of Mugheerah bin Shu'ba, who belonged to the tribe of Thaqeef. Among the Thaqeef dead was a Christian slave who had died beside his master. As the Ansar stripped this slave, he noticed that the dead man was not circumcised. Amazed at this discovery, for circumcision was a universal practice among the Arabs, he called aloud to those who stood around him: *"O Arabs! Did you know that the Thaqeef are not circumcised?"* Mugheerah, who stood next to the Ansar, was horrified to hear this, as the spread of such a report would mean disgrace for the Thaqeef. He knew the dead slave and could understand how the misunderstanding had arisen. *"Don't say that!"* he hissed at the Ansar. *"This man was a Christian slave."*

"No, he was not", insisted the Ansar. *"I am sure that he is one of the Thaqeef."* And he remained unconvinced until Mugheerah had undressed several bodies of the Thaqeef and pointed out familiar signs! [2]

The Muslim army having fully re-assembled, except for a few who had fled, the Prophet decided to press his advantage. He organised a strong cavalry group and sent it forward to clear the valley before the Hawazin had time to recover and reorganise. This group was formed of several contingents, including the Bani Sulaim, over whom Khalid had regained control. Khalid had missed the Muslim counter-attack. He had lain where he fell in the flight of the Bani Sulaim until the counter-attack was over. Then the Holy Prophet came to him and blew upon his wounds, whereupon Khalid arose, feeling strengthened and fit for battle again. 3 He quickly got the Bani Sulaim together.

The entire group was placed under command of Zubair bin Al Awwam, who now advanced along the valley and contacted Malik at the pass. After a short, brisk engagement, Malik was driven off the pass. The whole valley was now in Muslim hands. The Prophet left Zubair's mounted group at the pass, to hold it as a firm base and guard it against a possible return of the Hawazin, and sent another group under Abu Amir to Autas. This was the camp of the Hawazin, who on being driven off the pass had taken up positions around the camp to defend their families and flocks. On the arrival of the Muslims, a fierce clash took place at Autas. Abu Amir killed nine men in personal combat and was killed by his tenth adversary, whereupon the command of the Muslim group was taken over by his cousin, Abu Musa, who continued the attack on Autas until the Hawazin broke and fled. The camp of the Hawazin fell into Muslim hands, and here this Muslim group was joined by the cavalry group of Zubair, with Khalid in the lead.

1. Ibn Sad: p. 665.
2. Ibn Hisham: Vol. 2, p. 450.
3. Isfahani: Vol. 15, p. 11.

The enemy coalition had now completely disintegrated. The Hawazin and other tribes dispersed to their various settlements while the Thaqeef, led by Malik, hastened to Taif where they decided to resist till the bitter end. The Battle of Hunain was over. Muslim casualties in this battle were surprisingly few, thanks to the indifferent archery of the Hawazin. While many Muslims had been wounded, only four lost their lives. The reason for this lay in the superior skill and courage of the Muslims, which enabled their champions to take on three or four opponents at a time, killing them one by one. Seventy of the unbelievers were killed in the valley, at the pass and at Autas, and these included the sage, Duraid, who had given such sound advice but in vain. In the enemy camp at Autas, the Muslims captured 6,000 women, children and slaves and thousands of camels, goats and sheep. 1

This was the first time that the Muslims had been ambushed in a large-scale operation by their enemies. This was the second instance in history of the ambush of an entire army by an entire army (the first being the ambush of the Romans by Hannibal at Lake Trasimene in 217 B.C.). Malik had made a brilliant and flawless plan to annihilate the Muslims, but because of the poor performance of his men could not achieve the mission that he had set himself. In spite of this poor performance, however, he would have won a resounding victory had his enemy not been the Muslims. It was the determination of the Prophet not to accept defeat, and the faith of the Muslims in their leader, which turned defeat into victory for them. Unlike Malik, the Prophet was not content with a limited gain and pressed his advantage to rout the enemy and capture the entire enemy camp with all its booty.

This was the first time that Khalid had been taken by surprise. He had always known the value of surprise, but this time he had been at the receiving end of it. He saw how his otherwise brave men had panicked at the sudden appearance of the enemy at an unexpected time and an unexpected place. He made up his mind never again to be caught unawares. And he never was.

1. No one today knows the location of Autas, but it must have been in the valley proper, as a camp with 6,000 people (excluding soldiers) and thousands of camels, goats and sheep could not be established on a hillside or in some little wadi. I have placed it a little beyond Zaima, but it could have been elsewhere.

Chapter 9

It was said, "O Messenger of Allah! Pray to Allah against the tribe of Thaqif [of Ta'if]." He said, "O Allah! Guide Thaqif, and bring them (to us)." 1

The Prophet had routed the enemy at Hunain and driven him from Autas. He now decided to give Malik bin Auf no time to recover his breath and organise further resistance. Consequently, he sent the captives and the flocks taken at Autas with an escort to Jirana, to be kept under guard until the return of the army, and the very next day he set out for Taif, where major resistance was to be encountered. But he moved cautiously, for after the unpleasant experience of the ambush at Hunain, he had no intention of letting the army walk into another trap. The country now was hilly, consisting of steep ridges rising up to the plateau on which stood Taif; and in this terrain a wily commander like Malik could lay an ambush almost anywhere.

Leaving Autas, the Prophet marched through the Nakhla Valley and then turned south into the Wadi-ul-Muleih. From this valley he crossed into the Wadi-ul-Qarn, and following this *wadi*, reached the plateau 7 miles north-west of Taif. So far the Muslims had encountered no opposition and scouts had reported no sign of the Thaqeef outside Taif; but hoping to surprise Malik, the Prophet shifted his axis. Cutting across the difficult terrain north of Taif, he got to the less hilly region lying east of the town, between Nikhb and Sadaira. 2 From here he marched to Taif, coming in from the rear. Throughout this march, Khalid again led the army with the Bani Sulaim as advance guard.

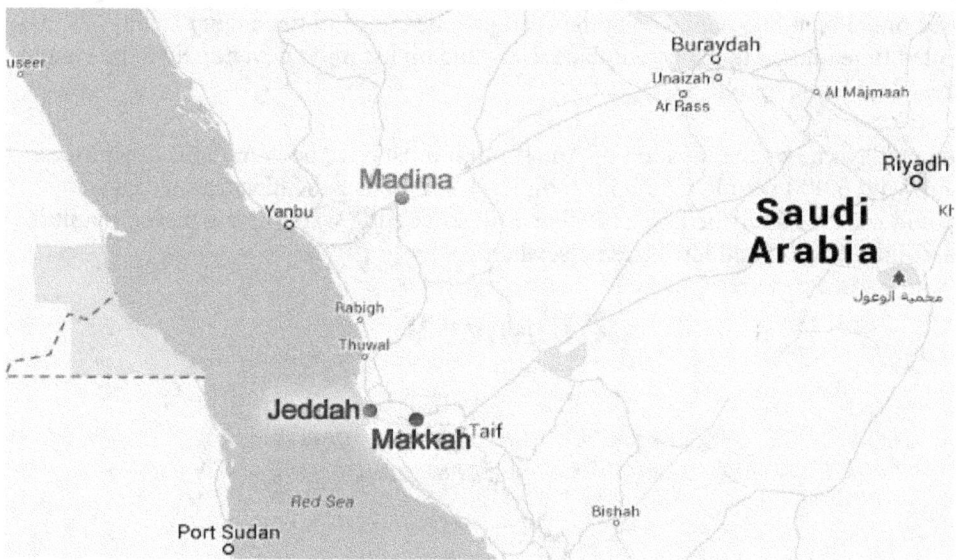

But Malik bin Auf, in spite of his lack of years, was not a man to be caught unawares. Having suffered grievously in his clash with the Muslims at Hunain and Autas, he was determined not to accept battle with the Muslims again in the open: he would fight them on his own terms. Consequently, he kept his army within the walled city of Taif and speedily stocked it with sufficient provisions to withstand a long siege. Here the Thaqeef, under their brave young general, awaited the arrival of the Muslims.

The Muslims got to Taif on February 5, 630 (the 15th of Shawal, 8 Hijri), and started a siege which was to last 18 days. On arrival at Taif, the camp was set up too close to the wall of the town and this mistake was punished by the Thaqeef archers, who showered the camp with arrows. A few Muslims were killed before the camp was moved away and established in the area where the mosque of Ibn Abbas stands today. Groups of Muslims were now deployed around the fort to prevent entry and escape; and Abu Bakr was made responsible for the siege operations.
Most of the time, fighting between the two armies consisted of exchanges of archery. The Muslims would close up to the town and try to pick off the Thaqeef archers on the wall, but the Thaqeef had the odds in their favour as they had some cover in the open. So the Muslims got the worst of these engagements and many of them were wounded, including Abdullah, son of Abu Bakr, who later died of his wounds.

Thus some days passed. After the fall of Makkah, the Prophet had sent two Muslims to Jurash, in the Yemen, to learn all about siege warfare. These two men did not, however, return till after the Siege of Taif and thus could play no part in the siege. But Salman the Persian again came to the help of the Muslims as he had done in the Battle of the Ditch. As a Persian he knew something about more sophisticated forms of warfare. Under his instructions, the Muslims constructed a catapult and used it to hurl stones into the town; but the Muslims were amateurs at this business and the catapult produced no significant effect.

Salman next decided to use a testudo. (A testudo was a large shield, usually made of wood or leather, under which a group of assailants could advance to the gate of the fort, safe from enemy missiles, and either crash through the gate with a battering ram or set fire to it.) Under the instructions of Salman, the Muslims constructed a testudo of cowhide, and a group of them advanced under its protection to set fire to the wooden gate of Taif. As they got to the gate, however, Malik and his men poured red-hot scraps of iron onto the testudo. These pieces burnt the testudo and terrified those under it, so that they hurriedly dropped the unfamiliar equipment and ran back. As they ran, the Thaqeef fired a volley of arrows at them and killed one of them.

1. Mukhtasar Sirat Al-Rasul sall-Allahu 'alayhi wa sallam, of Sheikh Muhammad bin Abdul Wahhab.
2. The Wadi-ul-Muleih runs between the present Taif Airport and Seil-ul-Kabeer. The Wadi-ul-Qarn, in its upper reaches, crosses the present Taif-Makkah Highway 7 miles from Taif. Sadaira is 25 miles east of Taif on the Turaba road, and Nikhb lies just 3 miles east-south-east of Taif. The Wadi-un-Nikhb was known in ancient times, according to local tradition, as the Wadi-un-Naml-the Valley of Ants-through which Solomon marched towards the Yemen for his encounter with the Queen of Sheba. The story of Solomon is narrated in the Quran (27: 16-44).

Two weeks passed and the end was not in sight. The Thaqeef would not come out to fight; the Muslims could not get in to fight. Every time they approached the town they were driven back with, arrows. One day Abu Sufyan also took part in a sally towards the town and stopped an arrow with his eye. He lived thereafter as a one-eyed man. [1]

February can be very cold in the region of Taif, and the weather during the siege was unpleasant. The Muslims tried to force the Thaqeef out to give battle by destroying some vineyards near Taif; but the Thaqeef refused to leave the security of their fort. Malik was much too clever a general to risk a battle under conditions which would favour his opponent. Finally the Holy Prophet called a council of war and sought the advice of his officers. One of them said, *"When you corner a fox in its hole, if you stay long enough you catch the fox. But if you leave the fox in its hole it does you no harm."* [2] Abu Bakr advised a return to Makkah, and Umar concurred with him.

The Prophet could not wait indefinitely for the fall of Taif as he had more important matters to attend to. He proposed that the siege be raised and the army return to Makkah; but some Muslim hot-heads protested against this and insisted that they fight on until victory was gained. ***"Then you can attack tomorrow."*** [3] said the Prophet.

The next day a few of these battle-hungry Muslims again approached the fort with a view to capturing it, but were severely punished by the Thaqeef archers. They returned in a more philosophical mood and agreed with the Prophet that it might be best to leave the fox in its hole.

On February 23, 630 (the 4th of Dhul Qad, 8 Hijri) the siege was raised. The Muslims had lost 12 men and a large number had been wounded. The Thaqeef remained defiant. Ten months later, however, this tribe was to accept Islam and prove staunch in its faith.

The Muslims arrived at Jirana on February 26, and here the Prophet distributed the spoils taken at Autas. To show the newly converted Makkans that there was no discrimination against them for having delayed their acceptance of the new faith, the Prophet also gave them a share of the spoils. But hardly had the women, children and animals been distributed among the Muslims, when a delegation of the Hawazin came to the Prophet

and declared that the tribe had accepted Islam. *"Will you not return to us what you captured from us in battle?"* the delegates pleaded. Actually they had no right to demand a return of what they had lost, because they had lost it as infidels and not as Muslims; but the Prophet was generous. **"Are your women and children dearer to you or your property?"** he asked them. *"Return to us our women and children and you can keep the rest"*, they replied. 4

The Prophet now appealed to his army to return the women and children of the Hawazin. Every soldier responded to the Prophet's appeal and returned the captives in his hands, with the exception of Safwan bin Umayyah, who refused to part with a girl who had been given to him as his share of the spoils. She must have been very beautiful!
A few days later Malik slipped out of Taif and came to the Muslim camp. He became a Muslim and was amply rewarded by the Prophet. It is a pity that this brilliant young soldier was given no important role in later Muslim campaigns, for he had the makings of a superb general.

The Holy Prophet and the army of Islam now returned to Madinah, arriving there in the latter part of March 630. Thus ended the eighth year of the Hijra. The year that followed was to become known as the Year of Delegations, for during this year most of the tribes of Arabia sent delegations to Madinah and submitted to the Prophet. Not all the delegates, or the tribal chiefs who sent them, were motivated by a desire for the true religion, as we shall see later. While some were sincere seekers of the truth, others came for political reasons. Some came out of sheer curiosity, and a few were downright scoundrels.

1. According to some sources, Abu Sufyan lost his eye at Yarmuk and not at Taif.
2. Ibn Sad: p. 675.
3. *Ibid*
4. Ibn Hisham: Vol. 2, p. 489.

Chapter 10

"... And you see the people enter Allah's Religion in crowds, then celebrate the Praises of your Lord, and pray for forgiveness from Him: for He is Oft-Returning (in Grace and Mercy)."
[Quran 110:2-3]

In the ninth year of the Hijra only one major operation was carried out by the Muslims-the expedition to Tabuk, led by the Holy Prophet in person. It turned out to be a peaceful operation; but no matter how peacefully other people went about their tasks, Khalid always managed to find adventure and violence.

During the long, hot summer of 630, reports arrived at Madinah that the Romans had concentrated large forces in Syria, and had pushed their forward elements into Jordan. Heraclius, the Byzantine Emperor, was himself in Emessa.

In the middle of October 630, the Prophet ordered the Muslims to prepare for battle with the Romans. The purpose of the expedition was not just to fight the Romans, for that could have been done later when the weather had improved. The Prophet also wanted to put the faith of the Muslims to test by making them march out in the fierce heat of summer. Under these conditions only true Believers would respond.

And the true Believers did. The vast majority of the Muslims answered the call cheerfully and began preparations for the expedition; but some did take unkindly to the call to arms. The October of this year was an unusually hot month, and the cool shade of the date orchards proved too tempting for these Muslims. Men wanted nothing more than to rest in the shade until the worst of the heat was over. The Hypocrites, as usual, went about dissuading the Muslims from joining the expedition and gave trouble enough; but on this occasion even a few proven Muslims faltered.

In late October 630 (mid-Rajab, 9 Hijri) the Muslims set out for Tabuk. This was the largest army that had ever assembled under the standard of the Prophet. It consisted of men from Madinah, from Makkah and from most of the tribes which had accepted Islam. One source had placed the strength of this army at 30,000 warriors, including 10,000 cavalry, but this is probably an exaggeration.

On arrival at Tabuk the Muslims came to know that the Roman elements in Jordan had withdrawn to Damascus. There was no need to go further. But the Prophet decided to subdue the tribes living in this region and bring them under the political control of Islam. The important places in the region were Eila (near the present-day Aqaba), Jarba, Azruh and Maqna-all lying along the Gulf of Aqaba. Pacts were made with these tribes and they all agreed to pay the *Jizya*. 1

One important region which the Prophet wished to subdue was a little farther away from Tabuk. This was Daumat-ul-Jandal (the present-day Al Jauf), ruled by Ukaidar bin Abdul Malik, a Christian prince from the tribe of Kinda who was famous for his love of hunting. To subdue this region, the Prophet sent Khalid with 400 horsemen and instructions to capture Ukaidar. ***"You will probably find him hunting the wild bull"***, said the Prophet. 2

Khalid arrived at the walled town of Daumat-ul-Jandal on a bright, moonlit night in late November, 630 (mid-Shaban, 9 Hijri). Hardly had he deployed his force near the town when the gates opened and out came Ukaidar with a few friends mounted on horses and armed with hunting weapons. Perhaps owing to the heat of the day Ukaidar had decided to hunt in the cool of the night, and the bright moonlight promised good hunting. Khalid took a few of his men and rushed at the hunting party. While Khalid himself pounced on Ukaidar and brought him-down from his horse, his men assailed the other members of the party. Ukaidar's brother, Hassaan, resisted capture and was killed; but the rest galloped back to the fort and, once inside, locked the gate.

Khalid now returned to Tabuk with his distinguished prisoner. Ukaidar entered into a pact with the Prophet, paid a heavy ransom for himself and agreed to the *Jizya*.

1. A tax levied on non-Muslims. In return they were exempt from military service and their safety was guaranteed by the Muslim State.
2. Ibn Hisham: Vol. 2, p. 526.

Soon after this incident the Muslim army left Tabuk to return to Madinah. It arrived home in the middle of December 630, by which time the weather had become very pleasant.

After Tabuk there was no major military activity during the lifetime of the Prophet. Delegations came from all the tribes of Arabia, swore allegiance to the Prophet, accepted Islam and agreed to pay certain taxes. For each tribe the Prophet appointed a leader from the converted members of the tribe. The Prophet thus remained busy with affairs of state, consolidating the gains of Islam and raising the edifice of the new state. Several small expeditions were sent by him to various places in Arabia. The mission given to them was to call the tribes to accept Islam, but in case of armed opposition the tribe concerned was to be fought and subdued.

In July 631 (Rabi-ul-Akhir, 10 Hijri), the Prophet sent a military expedition under the command of Khalid to the tribe of Bani Harithah bin Kab in Najran, which lies to the north of the Yemen. The instructions to Khalid were: **"Call the tribe thrice to accept Islam. If they respond favourably, do them no harm. If they refuse, fight them."** 1 With Khalid went 400 mounted warriors.

Khalid arrived at Najran and made contact with the Bani Harithah bin Kab. He called upon them to submit to Islam, and they accepted his call. No blood was shed. Khalid remained with the tribe for several months, teaching them the ways of Islam; and when he was satisfied that they had become good Muslims, he wrote to the Prophet and informed him of the progress of his mission. The Prophet sent Khalid an appreciative letter in reply and instructed him to return to Madinah and bring a delegation of the Bani Harithah bin Kab with him. Khalid returned with the delegation in January 632 (Shawal, 10 Hijri). The Prophet received the delegation with the usual courtesy shown to all delegations. The terms of submission were explained to the delegates, a leader was appointed for the tribe and the delegation then returned to Najran.

This was the last mission carried out by Khalid in the time of the Prophet. 2

1. Ibn Hisham: Vol. 2, p. 592.
2. For opinions regarding other missions supposed to have been carried out by Khalid, see Note 2 in Appendix B.

Part II: The Campaign of the Apostasy

Chapter 11

(The Gathering Storm)

"The desert Arabs say, 'We believe.' Say, 'You have not believed, but say: 'We have submitted,' for Faith has not yet entered your hearts.
But if you obey Allah and His Messenger, He will not belittle anything of your deeds: For Allah is Oft-Forgiving, Most Merciful.'
Only those are Believers who have believed in Allah and His Messenger, and have never since doubted, but have striven with their belongings and their persons in the Cause of Allah: such are the sincere ones."
[Quran 49:14-15]

Apostasy had actually begun in the lifetime of the Prophet, and the first major action of the apostasy was fought and satisfactorily concluded while the Prophet still lived. But the real and most serious danger of apostasy arose after the Prophet's death, when a wild wave of disbelief-after-belief moved across the length and breadth of Arabia and had to be tackled by Abu Bakr. Hence the Campaign of the Apostasy is here taken up as a whole, although chronologically the first of these events belongs to Part I of this history.

The first major event of the apostasy occurred in the Yemen and is known as the Incident of Aswad Al Ansi. Aswad was a chief of the Ans-a large tribe inhabiting the western part of the Yemen. His actual name was Abhala bin Kab, but because of his very dark colour he was called *Aswad*, i.e. the Black One. A man of many qualities, few of them enviable, he was, before the apostasy, known mainly as a tribal chieftain and a soothsayer.

During the tenth year of the Hijra, the people of the southern and south-eastern regions of the Arabian peninsula had been converted to Islam. The Prophet had sent envoys, teachers and missions to various places to accomplish this task and the task had been duly completed. But the majority of the inhabitants of these regions had not become true Muslims, their conversion being more a matter of form than a sincere change of heart.

Before this conversion the Yemen was governed, on behalf of the Persian Emperor, by a noble-born Persian named Bazan. [1] This man became a Muslim and was confirmed in his appointment as governor of the Yemen by the Prophet. As he was a wise and virtuous officer, the province prospered under his rules; but shortly before the last pilgrimage of the Prophet, Bazan died, and the Prophet appointed Bazan's son, Shahr, as governor at San'a. Peace continued to prevail in the Yemen, and no clouds darkened the southern skies.

Then, at about the time of the Prophet's last pilgrimage, Aswad decided that he would become a prophet. He gathered his tribe, recited some of his verses, claiming that they were verses of the Quran revealed to him, and announced that he was a messenger of Allah.

Aswad had a donkey which he had trained to obey certain commands, and he used this donkey to demonstrate his powers. He would give the order, *"Bow before your lord"*, and the donkey would bow its head before Aswad. He would then command, *"Kneel before your lord!"*, [2] and the donkey would kneel. Because of this, Aswad became known in the region as Dhul Himar-the One of the Donkey, or 'Donkey-Walah'. Some chroniclers, however, maintain that he was known not as Dhul Himar, but as Dhul Khumar, i.e. the Drunk. [3] This could be true because he was heavily addicted to alcohol and often in a drunken stupor. Nevertheless, his tribe followed him, believing him to be a genuine prophet; and in this error they were joined by some of the lesser tribes of the Yemen.

Aswad organised a column of 700 horsemen and rode to Najran. He captured the town with no difficulty and drove out its Muslim administrator. Elated by this easy victory, he left his own man to govern Najran and moved on San'a. Shahr, the newly appointed Muslim governor of the Yemen, heard of the fall of Najran, came to know of the intentions of Aswad and decided to tackle Aswad before he could reach San'a. Organising a small armed force (he did not have many warriors), he marched out to meet his adversary, and the two forces met some distance north of San'a. The short, brisk engagement that followed ended in Aswad's favour. The Muslims suffered a defeat and Shahr was killed in battle, leaving behind a beautiful young widow named Azad. Five days later Aswad entered Sana' as a conqueror. He had worked fast for his unholy mission, for it was now only 25 days since he had first gathered his tribe and proclaimed his prophethood.

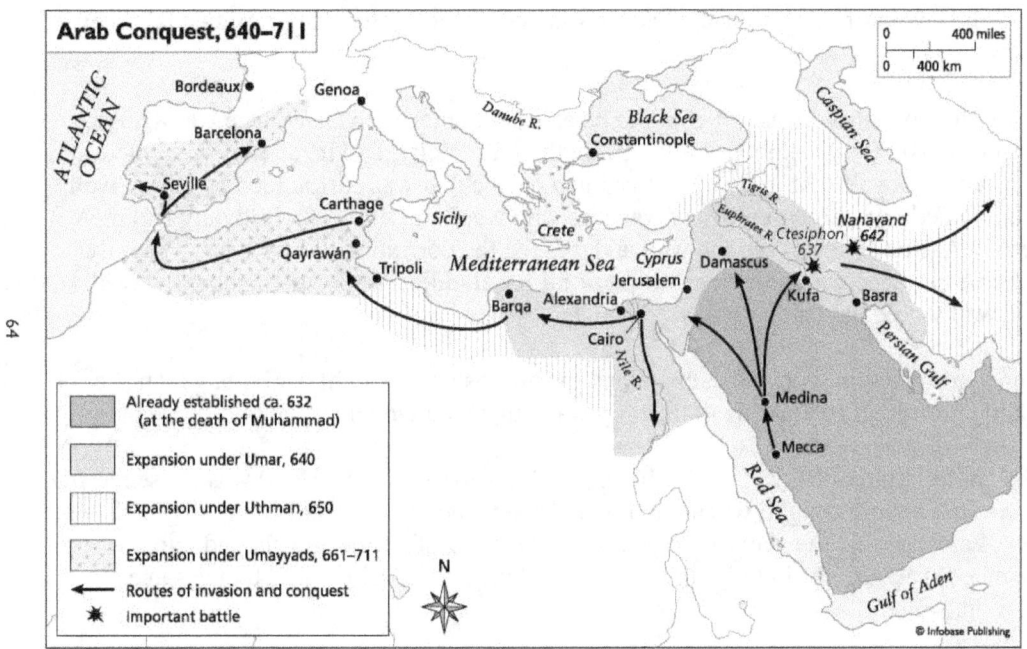

1. Called Bazam by some historians.
2. Balazuri: p. 113.
3. *Ibid*

Most of the Yemen was now his. And in order to get the maximum pleasure from his military and political success, Aswad forcibly married the lovely Azad. The poor widow had no choice but to submit to the drunken embraces of the loathsome Donkey-Walah.

Having occupied Najran and San'a, Aswad consolidated his gains and extended his sway over all Yemen, many tribes of which acknowledged him as ruler and prophet. As his authority grew, he began to feel discontented with the title of prophet and proclaimed himself Rahman of Yemen. [1] The word 'Rahman' means the Merciful One, and is one of the titles by which Muslims know Allah. Thus Aswad attempted to enter the divine province to which no man has laid claim without suffering disastrous consequences. Anyway, to his followers he became known as the Rahman of Yemen. His drunken orgies continued, as did his enjoyment of the ill-starred Azad, whose loathing for him grew so intense that she confided to a friend: *"To me no man is more hateful than he."* [2] In his viciousness Aswad also turned against the family of the Persian Bazan and heaped every manner of indignity and insult upon the surviving members. By doing so he earned the bitter hostility of a stalwart and true Muslim by the name of Fairoz Al Deilami-a member of this Persian family and a cousin of Azad.

Unknown to the false prophet, the real Prophet at Madinah had already initiated measures to deal with him. Having received full reports of Aswad's mischief, the Holy Prophet sent Qais bin Hubaira to organise the liquidation of Aswad. Qais got to Sana' undetected, laid the foundations of an underground movement against the impostor and made contact with the Persian Fairoz. Qais and Fairoz became the brains of the organisation that was to draw the sword of vengeance against Aswad and his apostates. In secret they laid their plans.

The killing of Aswad was not going to be an easy matter. The Black One was a huge, powerfully-built man, known for his strength and ferocity, and he already suspected Fairoz of disloyalty. Moreover, he lived in a palace that was surrounded by a high wall and guarded by a large number of warriors who were chosen for their loyalty and their faith in Aswad. They paced the wall and treaded the corridors of the palace. The only possible entrance was over a certain part of the wall adjacent to the chamber of Azad. The wall would have to be scaled.

Fairoz got in touch with Azad, explained his purpose and sought her help, which she readily promised, seeing this as the only way out of the wretched life that she led.

The fateful night of May 30, 632 (the 6th of Rabi-ul-Awwal, 11 Hijri) was chosen as *the night*. Just after midnight, when the moon had set, and at a moment when no guards were near, Fairoz scaled the wall of the palace with the aid of a rope and slipped into the chamber of Azad. She hid him in the room and the two cousins, fired by the same mission, waited.

Shortly before dawn Azad stole out of her room and walked to Aswad's chamber, which was next to hers. She knew that there was a sentry on duty nearby, though not in sight. She opened the door, looked in, and then returned to Fairoz. The fire of vengeance burnt in her eyes as she whispered, *"Now! He is lying drunk!"*

Fairoz, followed by Azad, tiptoed out of her chamber and to the door of Aswad's room. The woman stationed herself at the door while Fairoz entered with drawn sword. Suddenly Aswad sat up in bed and stared in horror at Fairoz, whose appearance left no doubt as to his purpose. In the face of this danger the drunkenness of the Black one vanished; but before he could get off the bed, Fairoz sprang forward and struck him on the head with his sword. Aswad fell back on his pillow. According to the chroniclers, *"He began to bellow like a bull"* [3]

His cries attracted the attention of the sentry who rushed to Aswad's chamber. He saw Azad standing by the door and asked, *"What is the matter with the Rahman of Yemen?"* The plucky girl raised her finger to her lips. *"Shush!"* she whispered. *"He is receiving a revelation from Allah!"* [4] The sentry nodded knowingly, and disregarding the shouts of his master, walked away.

1. Balazuri: pp. 113 - 125.
2. Tabari: Vol. 2, p. 467.
3. Balazuri: p. 114.
4. *Ibid*

Azad waited until the sentry had turned the corner of the corridor, then rushed into the room. She saw Fairoz standing beside the bed, waiting for a chance to strike again, while the impostor writhed in his bed, waving his arms about. The two now worked together. The woman hastened to the head of the bed, caught the hair of Aswad in both her hands and held his head down. Fairoz drew his dagger and with a few deft strokes severed the black head from the enormous body. Thus ended the career of the false prophet, Abhala bin Kab, alias the Black one, alias the Donkey-Walah, alias the Drunk. His mischief lasted three months and ended with his death, six days before Prophet Muhammad passed away.

With Aswad's death his movement collapsed. The Muslim resistance organised by Qais in San'a turned in violent vengeance against the followers of Aswad, many of whom were killed. But many escaped to create trouble for Muslim rulers at a later stage. Many became Muslims again, and of these some again apostatised. Fairoz was appointed governor of San'a.

The messenger who carried the good news to Madinah arrived there shortly after the death of the Holy Prophet. The report of the destruction of the mischief of Aswad Al Ansi brought some solace to the heart-broken Muslims.

Madinah was now going through a crisis which was at once emotional, spiritual and political. The loss of the beloved Muhammad had left the Muslims devastated. For the past 10 years the Prophet had been everything to them-commander, ruler, judge, teacher, guide, friend. There was no aspect of life in which he had not participated. They had taken all their problems to him, and he had settled, decided, directed, comforted. In the warm light of his presence they had felt safe from trouble and misfortune. Now that light had gone out. The Muslims felt alone and frightened-in the words of the chroniclers: *"like sheep on a cold, rainy night."* [1]

The crisis deepened as reports of the revolt spreading over Arabia began to arrive. All the tribes of Arabia, with the exception of those in Makkah and Madinah and the Thaqeef in Taif, revolted against the political and religious authority of Madinah and broke their oaths of allegiance. False prophets arose in the land and claimed a share in Muhammad's prophethood. These impostors, having seen the affection and reverence in which the Holy Prophet was held, and unmindful of the trials and sufferings which he had experienced before his efforts bore fruit, decided that prophethood was a good thing and that they too should get the benefit of it. Apart from Aswad, there were two impostors (possibly three) and one impostress. There were others-chieftains and elders, who did not claim prophethood, but united with the false prophets in their perfidious designs to extinguish the flame of Islam and return to the tribal independence of the Ignorance. The flames of the apostasy raced like wild fire across all Arabia, threatening to engulf Makkah and Madinah-the spiritual and political centres of the infant state of Islam.

The chief cause of the apostasy was lack of true faith. Most of the tribes, converted in the ninth and tenth years of the Hijra, had taken to Islam for political reasons. They had found it expedient. They saw Muhammad as a powerful political boss rather than a prophet with a new message. The true Muslims were the Muslims of Makkah and Madinah, especially the latter who had been in contact with the Holy Prophet for many years and had drunk deep at the fountain of truth which the Prophet had revealed. The outlying tribes had not enjoyed this spiritual experience. In many cases, when a chief became a Muslim the tribe followed his example out of tribal loyalty rather than religious conviction. With the death of the Prophet the tribes felt free to renounce their allegiance, which, as they saw it, had been made to a person and not to Madinah or to Islam. Muhammad was dead; and now they could throw off the yoke of discipline which the new faith had imposed in limiting the number of wives a man could marry, in collecting taxes for the benefit of the community, in enforcing prayers and fasting. The strong leaders who led the revolt preferred to be free to exploit the weak to their own advantage, unhampered by the restrictions which Islam placed upon them.

The fears of the Muslims deepened when Abu Bakr became caliph-the first caliph in Islam. Abu Bakr had never been known for any great quality of leadership, let alone the ability to steer the ship of state through the storm that gathered on every side and threatened the very existence of Islam. What was needed at this critical juncture was a strong, robust and capable leader. And what was the image of Abu Bakr? A small, slender, pale man, he had deep-set eyes under thin, delicate eyebrows. By now he had a pronounced stoop which heightened the impression of age and senility, in spite of the fact that he dyed his beard. A mild, gentle and tender-hearted individual, he was easily moved to tears.

1. Tabari: Vol. 2, p. 461.

As the Muslims gathered to take the oath of allegiance, Abu Bakr made the first speech of his caliphate-a speech that further emphasised his modesty and humility and gave no promise of strength. He said:

"Praise be to Allah! I am now in authority over you, but I am not the best among you. If I act virtuously, help me. If I act wrongfully, correct me. Truth is honesty, falsehood is treachery.
The weak among you is strong in my sight, until I give him what is due to him, if Allah wills it. And the strong among you is weak in my sight, until I take what is due from him, if Allah wills it.
Let none among you abjure the holy war in the way of Allah, for no people do so but Allah strikes them with disgrace. And among no people does vice become general but Allah inflicts upon them terrible punishment.
Obey me while I obey Allah and His Messenger; and if I disobey Allah and His Messenger, you are not obliged to follow me.
Forget not your prayers. May Allah have mercy upon you!" [1]

Abu Bakr's virtues and outstanding services to Islam were well known. His personal courage, his devotion to the Prophet, who had given him the title of the Truthful One, his high moral principles and his faith as one of the staunchest of believers were unquestioned. As the third male to embrace Islam his position among the Blessed Ten was high indeed. [2] But did such virtues make for leadership in troubled times? And then there was the departure of the Army of Usama, which further imperilled Madinah and increased the alarm of the Muslims.

About the middle of May 632, the Holy Prophet, now ailing, had ordered a large expedition to be prepared for the invasion of Jordan. Every body was to join it. As commander of the expedition, he appointed Usama-a young man of twenty-two. Usama was the son of Zaid bin Harithah, the Prophet's freedman, who had been the first of the Muslim commanders to fall at the Battle of Mutah. Although Usama was common-born and enjoyed no family standing among the Quraish, the Prophet put him in command over all the older and more distinguished warriors from the best clans. The warriors gathered at a campsite just west of Uhud, and the force thus concentrated became known as the Army of Usama. This was the last expedition ordered by the Prophet; and it could mean war with the Romans.

Usama was given the Jordanian area of Mutah as his geographical objective. *"Go to the place where your father was killed"*, ordered the Holy Prophet. *"Raid those territories. Go fast; take guides with you and send your scouts and agents ahead of you."* [3] Shortly before his death the Prophet remarked, *"Remember to dispatch the Army of Usama!"* [4] The army was still in camp when, on Monday, June 5, 632, (the 12th of Rabi-ul-Awwal, 11 Hijri) the Holy Prophet passed away. On the same day Abu Bakr, son of Abu Quhafa, became caliph.

The following day Caliph Abu Bakr issued instructions for the Army of Usama to prepare for the march. All the distinguished Companions who were available for war were sent to join the Army in its camp and serve under the command of the youthful Usama. Even Umar, one of Abu Bakr's closest friends, was sent to the camp.

For the next few days the preparations continued even as reports of the rapid spread of the apostasy arrived. Then a group of prominent Muslims came to the Caliph. *"Will you send away the Army of Usama when most Arabs have revolted, and disruption raises its head everywhere?"* they protested. *"The Muslims are few. The unbelievers are many. The army must **not** be sent away!"*

Abu Bakr was adamant. *"Even if wild dogs rove around the feet of the wives of the Messenger of Allah (SAWS),"* he replied, *"I would still dispatch the Army of Usama as ordered by the Prophet."* [5]

1. Tabari: Vol. 2, p. 450.
2. The first was Ali, the second Zaid bin Harithah.
3. Ibn Sad: p. 707.
4. *Ibid*: p. 709.
5. Tabari: Vol. 2, p. 461.

A few more days passed. Reports from the countryside became more alarming. Then one day Usama, who feared for Madinah and for Islam no less than the others, spoke to Umar. *"Go to the Caliph"*, he said. *"Ask him to permit the army to remain at Madinah. All the leaders of the community are with me. If we go, none will be left to prevent the infidels from tearing Madinah to pieces."*

Umar agreed to speak to the Caliph. As he was leaving the camp, he was met by a group of leaders who made the same suggestion and added: *"If he does not agree to our remaining in Madinah and we have to go, ask him at least to place an older man than Usama in command of the army."* [1] Umar agreed to put this across also.

In Madinah Abu Bakr sat on the floor of his house, getting used to the tremendous burden which the assumption of the caliphate in these stormy days had placed upon his shoulders. The strain would have shattered his nerves but for his limitless faith. Umar entered. Umar was calm and confident, for he was used to speaking to Abu Bakr as a strong, vigorous man would address a mild and submissive, albeit beloved comrade.

Abu Bakr waited until Umar had delivered the message and also expressed his own opinion regarding the proposed change of command. Then he leapt to his feet and shouted at Umar, *"O Son of Al Khattab! It was the Messenger of Allah who appointed Usama as the commander. And you want **me** to remove him from command."* [2]

Umar hastily backed out of Abu Bakr's house. He returned to the camp where the elders waited to see what news he would bring. Umar abused them roundly! [3]

On June 24, 632 (the 1st of Rabi-ul-Akhir, 11 Hijri), the Army of Usama broke camp and moved out. Abu Bakr walked some distance beside the mounted Usama and refused to let the young commander dismount from his horse. *"Every step that a Muslim warrior takes in the way of Allah"*, he explained to Usama, *"earns him the merit of 700 good deeds and the forgiveness of 700 sins."* [4]

Abu Bakr asked if he could retain Umar with him as adviser, to which Usama readily agreed. Then he gave his parting instructions to the Army Commander: *"Carry out your task. Start the operation with raids against the Quza'a. Let nothing deter you from accomplishing the mission given you by the Messenger of Allah."* [5] And the Army of Usama marched away.

The dispatch of the Army of Usama was a mistake in the circumstances which had arisen since the Prophet's death.[6] Some Muslim writers have stated that it was a wise move on the part of Abu Bakr, as it gave a show of strength to the rebels and thus deterred them from greater violence. Actually, this was not the case. Although Usama carried out his mission with efficiency and speed, his operation had no bearing whatever on the major actions of the apostasy which were fought in North-Central Arabia. The dispatch of the Army of Usama was an act of faith displaying complete submission to the will of the departed Prophet, but as a manoeuvre of military and political strategy, it was anything but sound. This is also proven by the fact that all the Muslim leaders were opposed to the move-leaders who produced, in this and the following decades, some of the finest generals of history.

Abu Bakr was moved to this decision by nothing other than his desire to carry out the last military wish of the Prophet. It was not lack of strategical judgement which led him to send of the Army of Usama, for Abu Bakr had ample military ability, as he was to prove soon after in his direction and conduct of the war against the apostates and the invasions of Iraq and Syria.

The Army of Usama was gone. Reports of ever-spreading revolt and of the concentration of hostile tribes became more serious day by day. The apprehensions of the Muslims increased. In contrast, the apostates rejoiced at the news of Abu Bakr's assumption of the caliphate and the departure of the army. With Abu Bakr at the helm of Muslim affairs, they thought, their objective of crushing the new Muslim State would be more easily achieved. The rebels were relieved that they did not have to deal with the fiery Umar or the peerless Ali. They would only have to deal with a nice old man!

But the Muslims were in for some pleasant surprises and the apostates for some rude shocks at the hands of 'the nice old, man'-such shocks that one rebel chieftain, fleeing from the columns of Abu Bakr, would cry in terror: *"Woe to the Arabs from the son of Abu Quhafa!"* [7]

1. Tabari: Vol. 2, p. 462.
2. *Ibid*: Vol. 2, P. 462.
3. *Ibid*.
4. *Ibid*.
5. *Ibid*: Vol. 2, p. 463.
6. Only a purely material perspective could regard this as a mistake, and even then this statement is difficult to defend categorically. It could easily be argued that the astonishing victories against the apostates, Persians and Romans during Abu Bakr's short caliphate were partly from the blessing of this decision to continue a matter begun by the Prophet (SAWS), one of the first major decisions made by Abu Bakr as caliph.
7. Balazuri: p. 104.

Chapter 12

(Abu Bakr Strikes)

"If the faith of Abu Bakr were to be weighed against the faith of all the people of the earth, his faith would outweigh theirs." [Umar bin Al-Khattab]

"Abu Bakr was the bravest of the people." [Ali ibn Abi Talib][1]

The apostasy had become so general that it affected every tribe in Arabia with the exception of the people in Makkah and Madinah and the tribe of Thaqeef at Taif. In some cases the entire tribe apostatised. In other cases part of the tribe apostatised while part continued to follow the true faith; and among those who remained Muslims, many had to pay with their lives for their faith. The flames of disbelief were fanned by two false prophets, Tulaiha bin Khuwailid and Musailima bin Habib, and a false prophetess by the name of Sajah bint Al Harith. Musailima had been an impostor for some time, while Tulaiha made his claim to prophethood during the illness of the Holy Prophet. The most immediate threat to Madinah was posed by Tulaiha and the tribes of West-Central and North-Central Arabia that followed him. These tribes were the Ghatfan, the Tayy, the Hawazin, the Bani Asad and the Bani Sulaim.

The concentrations of apostates nearest Madinah were located in two areas: Abraq, 70 miles north-east of Madinah, and Zhu Qissa, 24 miles east of Madinah. These concentrations consisted of the Ghatfan, the Hawazin and the Tayy. A week or two after the departure of the Army of Usama, the apostates at Zhu Qissa sent a delegation to Abu Bakr. *"We shall continue the prayers"*, said the delegates, *"but we shall not pay any taxes."* Abu Bakr would have none of it. *"By Allah"*, he replied, *"if you withhold a single ounce of what is due from you, I shall fight you. I allow you one day in which to give your reply."* [3]

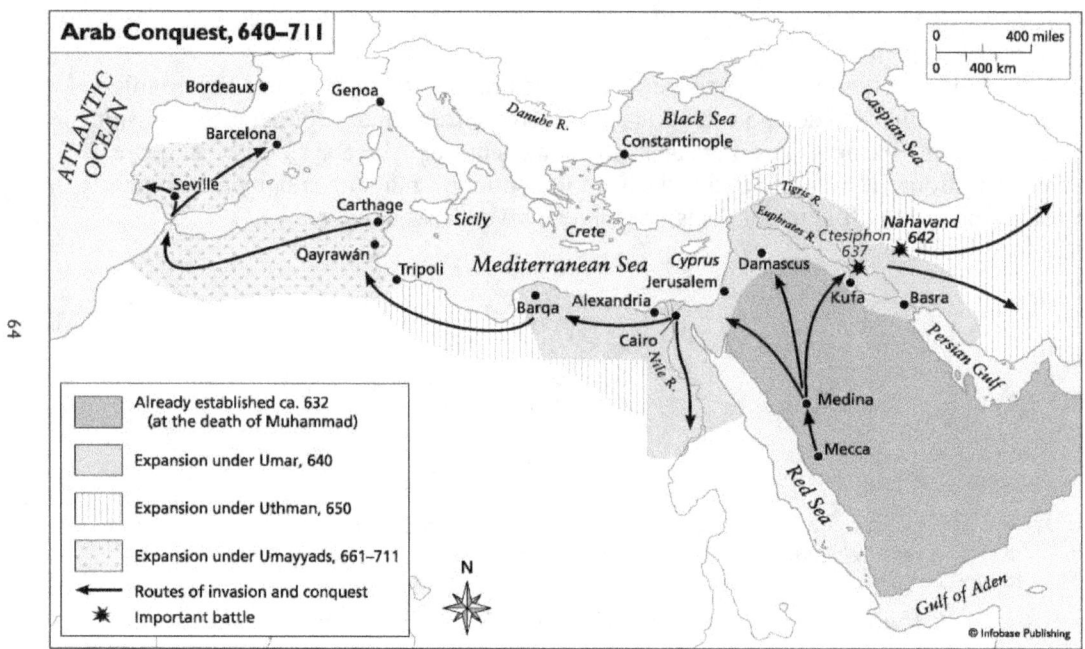

The envoys were taken aback by the determination and confidence of the new Caliph who seemed to be entirely unaware of the weakness of his position. And he had given them one day! The following morning, before the single day's ultimatum had expired, the envoys slipped out of Madinah, which meant a rejection of Abu Bakr's demands. Soon after their departure, Abu Bakr sent his own envoys to all the apostate tribes, calling upon them to remain loyal to Islam and continue to pay their taxes.

But the apostate envoys from Zhu Qissa, before leaving Madinah, had had a good look at the place, and their keen eyes had noticed the absence of warriors. On returning to Zhu Qissa they told their comrades about their conversation with Abu Bakr and the very vulnerable state of Madinah. Meanwhile Tulaiha, who was now at Samira, had reinforced the apostates at Zhu Qissa with a contingent under his brother, Hibal-a wily and resourceful general. When the apostates heard the reports of the envoys, the temptation proved too much for them; they decided to have a crack at Madinah while it was still defenceless. Consequently, the force at Zhu Qissa moved forward from Zhu Hussa 4, from where, after forming a base, part of the force advanced still nearer Madinah and went into camp, preparatory to attacking the town. It was now the third week of July 632 (late Rabi-ul-Akhir, 11 Hijri).

Abu Bakr received intelligence of this move and at once undertook the organisation of the defences of Madinah. The main army was out under Usama, but Madinah was not as defenceless as the rebels had imagined. Quite a few warriors were still there, especially from the clan of Bani Hashim (the Prophet's own clan) who had remained behind to mourn their departed kinsman. From these remnants Abu Bakr scraped together a fighting force. The confidence of Abu Bakr, never shaken, was strengthened by the thought that he had such stalwarts with him as Ali, Zubair bin Al Awam and Talha bin Ubaidullah. Each of these was appointed to command one-third of the newly created force.

For three days nothing happened. The apostates, uncertain of how they should set about their task, remained inactive. Then, on orders from Abu Bakr, the Muslims sallied out of Madinah. They launched a quick attack on the forward camp of the apostates and drove them back. The apostates withdrew to Zhu Hussa. The Muslims informed Abu Bakr of their success, and the Caliph ordered them to stay where they were and await his instructions.

1. *Tarikh Al-Khulafaa* of As-Suyuti.
2. Abraq is now just a stony plain (the word means a spur or bluff) 5 miles north of Hanakiya. Zhu Qissa does not exist; its location is known only in terms of its distance from Madinah (Ibn Sad: p. 590), and it was on the road to Rabaza, which is 20 miles north-east of Hanakiya. The latter is the old Batn Nakhl.
3. Tabari: Vol. 2, p. 487; Balazuri: p. 103.
4. The location of Zhu Hussa is not known.

The following day Abu Bakr set out from Madinah with a long string of pack camels, for the riding camels had all gone with Usama and these inferior camels were the best that Abu Bakr could muster in the way of transportation. As the convoy got to the abandoned apostate camp, the Muslims who had driven the apostates away mounted these camels and the force advanced towards Zhu Hussa-the apostate base.

Here the enemy waited, and Hibal, the brother of Tulaiha, showed his military cunning. He kept his men behind the crest of a slope, some distance ahead of the base towards which the Muslims were advancing.

The Muslims, mounted on their pack camels, rode up the slope unaware of the enemy who waited just beyond the crest. When the unsuspecting Muslims got near the crest, the apostates stood up and hurled upon the forward slope a countless number of goatskins filled with water. As these goatskins rolled down the crest towards the Muslims, a wild din arose from the apostate ranks as they hammered on drums and screamed at the top of their voices. The pack camels, untrained for battle and not used to sudden loud noises or the sight of unfamiliar objects rolling towards them in large numbers, turned and bolted. The Muslims did their utmost to control their panic-stricken mounts but failed, and very soon the entire Muslim force was home again!

Hibal had reason to feel pleased with himself. He had pulled a fast one on the Muslims and driven them back to Madinah without, so to speak, firing a shot. In view of this clever trick which Hibal pulled off, it is possible that the preceding apostate withdrawal had been a feint, planned by Hibal, to draw the Muslims out of the security of their town towards Zhu Hussa. We do not know. But Hibal now made the mistake of assuming that the Muslims were frightened, and that their hasty move back to Madinah was a sign of weakness. He did not know that the Muslims were mounted on pack camels, and that it was these animals that had panicked and not the men who rode them. The part of his force that had remained at Zhu Qissa was informed of this success and called forward. The same evening the full force of the apostates advanced and re-established the camp near Madinah, from which they had withdrawn only the day before. The spirits of the apostates were high.

The Muslims, on the other hand, were very angry, and every man was determined to set the record straight in a return engagement. Abu Bakr knew that the apostates had returned to their camp near Madinah, and decided to assail them before they could complete their preparations for battle. Under his instructions, the Muslims spent most of the night reorganising their small army and preparing for battle.

During the latter part of the night Abu Bakr led his army out of Madinah and formed up for the assault. He deployed the army with a centre, two wings and a rear guard. Keeping the centre under his direct command, he placed the right wing under Numan, the left wing under Abdullah and the rearguard under Suwaid-all three of whom were sons of Muqaran. Before dawn the army was set in motion towards the enemy camp where the apostates, confident of an easy victory on the morrow, slept soundly.

This time it was Hibal who was surprised. The first glow of dawn had not yet appeared when a furious, screaming mass of Muslims fell upon the camp with drawn swords. The apostates did not stand upon the order of their going. Many were killed, but most of them found safety in flight, and did not stop until they had got to Zhu Qissa, where they paused to rest and reorganise. Their spirits were no longer so high.

This round had been won by Abu Bakr, and his was no empty success. It was a bloody tactical action in which the enemy had been driven back by the sword and not by deception alone. Abu Bakr had decided to catch the enemy unawares and thus get the benefit of surprise to offset his numerical inferiority, and in this he had succeeded. He needed quick tactical victory and he had got it. As a matter of interest it may be noted that this is the first instance in Muslim history of a night attack-a tactical method which did not achieve popularity until the First World War.

Having won this round, Abu Bakr decided to give no respite to his opponents. He would catch them before the effect of the shock wore off and while alarm and confusion kept them disorganised. As the sun rose, he marched to Zhu Qissa.

On arrival at Zhu Qissa, he formed up for battle as he had done the night before, and then launched his attack. The apostates put up a fight, but their morale was low and after some resistance they broke contact and retreated to Abraq where more clansmen of the Ghatfan, the Hawazin and the Tayy were gathered. Abu Bakr, on capturing Zhu Qissa, sent a small force under Talha bin Ubaidullah to pursue the enemy. Talha advanced a short distance and killed some stragglers, but the small size of his force prevented him from doing any great damage to the retreating apostates.

The capture of Zhu Qissa took place on or about July 30, 632 (the 8th of Jamadi-ul-Awwal, 11 Hijri). Abu Bakr left Numan bin Muqaran with a detachment to hold Zhu Qissa, and with the rest of his force rode back to Madinah. On August 2, the Army of Usama returned to Madinah; the capital of Islam was no longer in danger.

On leaving Madinah, Usama had marched to Tabuk. Most of the tribes in this region opposed him fiercely; but Usama, with the zeal and vigour of youth, swept across the land with fire and sword. He raided far and wide in the region of Northern Arabia, starting with the Quza'a, who scattered under the blows of his columns and then made their way to Daumat-ul-Jandal (where Khalid had captured Ukaidar two years before). Usama killed all those who fought him and burnt orchards and villages, leaving in his wake 'a hurricane of smoke.' 1

As a result of his operations several tribes resubmitted to Madinah and re-embraced Islam. But the Quza'a remained rebellious and unrepentant, and had to be dealt with again a short while later by Amr bin Al Aas.

Usama next marched to Mutah, fought the Christian Arabs of the tribes of Kalb and Ghasan and avenged the death of his father. There was, however, no major battle. Then he returned to Madinah, bringing with him a large number of captives and a considerable amount of wealth, part of which comprised the spoils of war and part the taxes paid by the repentant tribes. The Army of Usama was warmly welcomed by Abu Bakr and the people of Madinah, to whom its return brought comfort and assurance. It had been away for 40 days.

After the defeat of the apostates at Zhu Qissa, several apostate clans turned viciously upon those of their members who remained Muslims and slaughtered them. The killing was done mercilessly, some Muslims being burnt alive and others thrown from the tops of cliffs. Abu Bakr heard the news of these atrocities with cold anger, and swore that he would kill every infidel who had murdered a Muslim and carry fire and sword to every apostate clan.

Things were now looking up for the Muslims. The recent victories of Abu Bakr, though not decisive, had raised spirits. Some of the apostate tribes living near Madinah had repented, rejoined the faith and paid their taxes and more. The Army of Usama was back with captives and wealth. The coffers of the Muslim State were full again, providing a sound financial base for all-out war against the enemies of Islam.

But Abu Bakr decided that he needed more time before launching a general offensive, in order to rest and re-equip the Army of Usama. He consequently ordered Usama to rest his men at Madinah and while doing so also ensure the safety of the capital. His own hastily scraped together force had now begun to feel like an army; and he decided to use this army, while the Army of Usama rested and re-equipped, to fight another offensive battle against the apostates gathered at Abraq. Now Abu Bakr really prepared for war, not only to punish the tribes for the heinous crime of apostasy, but also to avenge the innocent blood of the faithful Muslims who had been murdered by the apostates.

When Abu Bakr announced his intention of leading his army to Abraq, Muslim elders tried to restrain him. *"May Allah bless you, O Caliph of the Messenger of Allah!"* they said. *"Do not endanger yourself by leading the army in person. If you should be killed, it would upset the order of things. Your very existence is a source of trouble to the unbelievers. Appoint another to command the army. Then, if he is killed, you can appoint yet another."*

Abu Bakr was shortly going to place an immense burden on the shoulders of the Muslims, both commanders and troops. He was going to ask them to strive as they had never done before and to face dangers which would shock most warriors.

He could think of no better way of making them come up to his expectations than setting the pace himself.

1. Ibn Sad: p. 709.

"No, by Allah!" he replied. *"I shall not do that. I shall not trouble others with my burden."* 1

And it was under Abu Bakr that the small army marched out to Zhu Qissa, where Numan awaited him. (This Numan was later to achieve everlasting fame as the victor of Nihawand in Persia.) Here Abu Bakr placed Numan and his brothers in command of the wings and the rear guard, as he had done for his night attack, and set out for Abraq. It was now the second week of August (third week of Jamadi-ul-Awwal).

When the Muslims got to Abraq they found that the enemy was already formed up in battle array. Without delay, Abu Bakr deployed his army and attacked the apostates.

The apostate spirits now were not as high as they had been a fortnight before. The defeated elements, which had escaped from Zhu Qissa, had joined the apostates at Abraq, and as is usual in such cases their arrival had had a depressing effect on others. For some time the apostates, who were numerically superior, resisted the Muslim attack, then they broke and fled. Abu Bakr had won another victory.

The remnants of the apostates fleeing from Abraq, and certain other clans from this region, travelled to Buzakha, whither Tulaiha the Impostor had moved from Samira. But other clans living in this area submitted to the columns that Abu Bakr sent out after the capture of Abraq to subdue the countryside. Now more taxes were gathered, to which the repentant clans gladly added gifts that were as gladly accepted.

The following day the Caliph left Abraq for Madinah. On arrival at Madinah he spent a few days in dealing with matters of state; then he moved to Zhu Qissa with the Army of Usama. But it had now ceased to be the Army of Usama, for Usama had completed his work and his army was now the Army of Islam-to be used by the Caliph as required. Usama's tenure of command was over.

At Zhu Qissa, Abu Bakr organised the Army of Islam into several corps to deal with the various enemies who occupied the entire land of Arabia except for the small area in the possession of the Muslims. This was the first time that the Muslim Army was organised into separate corps, each with its own commander, for independent missions under the general strategical guidance of the Caliph. Muslim commanders, until now essentially tacticians, would henceforth enter the higher realms of strategy and master those realms with a sure-footedness and ease that would astonish the world.

At Zhu Qissa, in the fourth week of August 632 (early Jamadi-ul-Akhir, 11 Hijri) Abu Bakr planned the strategy of the Campaign of the Apostasy. The battles which he had fought recently against the apostate concentrations at Zhu Qissa and Abraq were in the nature of immediate preventive action to save Madinah and discourage further offensives by the enemy, thus gaining time for the preparation and launching of his main forces. These actions could be described as spoiling attacks; they had enabled Abu Bakr to secure a base from which he could fight the major campaign that lay ahead.

Abu Bakr had no illusions about the task that faced him. He had to fight not one but several enemies-Tulaiha the Impostor at Buzakha, Malik bin Nuwaira at Butah, Musailima the Liar at Yamamah. He had to deal with widespread apostasy on the eastern and southern coasts of Arabia-in Bahrain, in Oman, in Mahra, in Hadhramaut, in the Yemen. There was apostasy in the region south and east of Makkah, and in Northern Arabia the Quza'a had staged a comeback after the return of the Army of Usama.

The situation of the Muslims can be compared with a small island of belief in an ocean of disbelief, a lamp shining in the darkness which held every manner of danger for the Faithful. Abu Bakr had not only to keep the flame alive, but also to dispel the darkness and crush the forces of evil that gathered threateningly on all sides. In numerical strength the apostates vastly outnumbered the Muslims, though they were not united. Abu Bakr's military strength lay in his having, among the Muslims, the finest fighting men of the time. And he had a tremendous weapon-Khalid bin Al Waleed: the Sword of Allah.

1. Tabari: Vol. 2, p. 476.

Abu Bakr planned his strategy accordingly. He formed the army into several corps. The strongest corps, and this was the main punch of the Muslims, was the corps of Khalid. This was used to fight the most powerful of the rebel forces, to crack the toughest nuts. Other corps were given areas of secondary importance in which to bring the less dangerous apostate tribes to their senses, after the main enemy opposition was crushed. Two corps were kept as reserves to reinforce the corps of Khalid or any other corps that might need assistance. The first corps to go into action was that of Khalid, and the timing of the despatch of other corps hinged on the operations of Khalid, who was given the task of fighting the strongest enemy forces one after the other. Abu Bakr's plan was first to clear the area of West Central Arabia (the area nearest Madinah), then tackle Malik bin Nuwaira, and finally concentrate against the most dangerous enemy of the lot-Musailima the Liar. Thus Abu Bakr would achieve concentration of force, by dealing with the main enemy armies separately and in turn, progressing step by step from nearer to farther regions.

The Caliph formed 11 corps, each under its own commander. [1] A standard was given to each corps. The available manpower was distributed among these corps and while some commanders were given immediate missions, others were given missions for which they would be launched later. The commanders were also instructed to pick up brave men on the way as they marched to their objectives. The 11 corps commanders and their assigned objectives were as follows:

1. *Khalid*: First Tulaiha at Buzakha, then Malik bin Nuwaira, at Butah.
2. *Ikrimah bin Abi Jahl*: Contact Musailima at Yamamah but not to get involved until more forces were built up.
3. *Amr bin Al Aas*: The apostate tribes of Quza'a and Wadi'a in the area of Tabuk and Daumat?ul-Jandal.
4. *Shurahbil bin Hasanah*: Follow Ikrimah and await the Caliph's instructions.
5. *Khalid bin Saeed*: Certain apostate tribes on the Syrian frontier.
6. *Turaifa bin Hajiz*: The apostate tribes of Hawazin and Bani Sulaim in the area east of Madinah and Makkah.
7. *Ala bin Al Hadhrami*: The apostates in Bahrain.
8. *Hudhaifa bin Mihsan*: The apostates in Oman.

9. *Arfaja bin Harsama*: The apostates in Mahra.
10. *Muhajir bin Abi Umayyah*: The apostates in the Yemen, then the Kinda in Hadhramaut.
11. *Suwaid bin Muqaran*: The apostates in the coastal area north of the Yemen.

As soon as the organisation of the corps was complete, Khalid marched off, to be followed a little later by Ikrimah and Amr bin Al Aas. The other corps were held back by the Caliph and despatched weeks and even months later. Their despatch was conditioned by the progress of Khalid's operations against the hard core of enemy opposition.

Before the various corps left Zhu Qissa, however, envoys were sent by Abu Bakr to all apostate tribes in a final attempt to induce them to see reason. These envoys were given identical instructions: they were to call upon the tribes to return to Islam and render full submission, for those tribes which submitted there would be forgiveness and peace, those tribes that resisted would be fought until no opposition remained and their women and children would be enslaved: before the attack, against any tribe, the Muslim forces would call the *Adhan* (the Muslim call to prayer), and if the tribe responded with the *Adhan* it would be assumed that it had submitted.

To the corps commanders, too, the Caliph gave identical general instructions, apart from their specific objectives. These instructions were as follows:

a. Seek the tribes which are your objectives
b. Call the *Adhan*.
c. If the tribe answers with the *Adhan*, do not attack. After the *Adhan*, ask the tribe to confirm its submission, including the payment of taxes. If confirmed, do not attack.
d. Those who submit will not be molested.
e. Those who do not answer with the *Adhan*, or after the *Adhan* do not confirm full submission, will be dealt with by fire and sword.
f. All apostates who have killed Muslims will be killed, those who have burnt Muslims alive will be burnt alive. 2

With these instructions Abu Bakr, no longer the meek, submissive Companion, launched the forces of Islam against the apostates.

1. The word 'corps' has been used in a loose sense to indicate an independent tactical command. These corps had no organisational resemblance with the modem army corps of about three divisions.
2. Tabari: Vol. 2, p. 482.

Chapter 13

(Tulaiha The Imposter)

Ibn Abbas related that the Prophet (SAWS) said, "The nations were presented to me. I saw the prophet having a party of people with him, the prophet having one or two men with him, the prophet having no one with him. Then a great mass of people was shown to me, so I thought that they were my nation, but it was said to me, 'This is Musa and his people, but look at the horizon.' Behold! A great mass of people! It was said to me, 'Look at the other horizon.' Behold! A great mass of people! It was said to me, 'This is your nation, and among them are seventy thousand who will enter the Garden without reckoning or punishment'."

He then got up and entered his house, and the people began speculating about those special believers. Some of them said, "Perhaps they are those who accompanied the Messenger of Allah, may Allah bless him and grant him peace." Others said, "Perhaps they are those who were born in Islam and so never associated any partners with Allah at all." Others said various things.

Then the Messenger of Allah, may Allah bless him and grant him peace, came out to them and they informed him of their discussion, so he said, "They are those who do not seek spiritual cures for physical ailments, who do not practice cauterisation, and who do not draw omens, but put their trust totally in their Lord."

So 'Ukkashah bin Mihsan stood up and said, "Pray to Allah to make me among them."
He replied, "You are among them."
Another man stood up and said, "Pray to Allah to make me among them."
He replied, "'Ukkashah beat you to it."[1]

Of the false prophets who remained after the death of Aswad, the first to clash with the Muslims was Tulaiha bin Khuwailid. He was a chief of the tribe of Bani Asad, and had been opposing the Holy Prophet off and on for many years.

Tulaiha first showed his hostility to Islam three months after the Battle of Uhud. Believing that the Muslims had been badly hurt in that battle, he got his clan together with the intention of raiding Madinah and thus exploiting what he regarded as a fine opportunity; but the Prophet came to know of the concentration of the clan and sent a mounted column of 150 horsemen to deal with it. Before Tulaiha could get wind of this counter-move the Muslim horsemen were upon him. The infidels scattered without a fight, and the Muslims captured the flocks of the clan and drove them off to Madinah as spoils. This setback so discredited Tulaiha in the eyes of his tribe that he had to lie low for a while.

Then he took part in the Battle of the Ditch. Responding eagerly to the invitation of the Jews to take up arms against the Muslims, he got together a contingent from the Bani Asad and commanded it in the coalition that besieged Madinah. When Abu Sufyan withdrew from Madinah, the Bani Asad also returned to their settlements. Again Tulaiha got nowhere.

The next occasion on which he opposed the Muslims was their campaign against the Jews of Khaibar in 628 (7 Hijri). The Bani Asad, operating under Tulaiha, sided with the Jews. During the movement of the Muslim army towards Khaibar, Tulaiha fought a number of minor engagements with the Muslims but was worsted every time. Then he pulled out his forces and abandoned the Jews to their fate.

Two years later, during the 'Year of Delegations', the Bani Asad sent a delegation to Madinah which offered submission to the Prophet. The whole tribe accepted Islam, but like many other tribes of Arabia its conversion was a matter of political convenience rather than genuine belief. Outwardly Tulaiha also embraced Islam. Whether infidel or Muslim Tulaiha continued to enjoy considerable influence in his tribe as a chief and a soothsayer. He would foretell the future, dabble in clairvoyance and recite poetry.

During the illness of the Prophet, in fact a few days before the Prophet died, Tulaiha made a bid for independence. He declared himself a prophet! He called upon his people to follow him, and many did. When word arrived of the Holy Prophet's death, he intensified his efforts to establish himself as the new prophet, and as the contagion of the apostasy spread over Arabia, the entire tribe of Bani Asad flocked to his standard, accepting him as chief and prophet. To mark the severance of his ties with Madinah, Tulaiha expelled the Muslim tax collector of his area-a valiant young man by the name of Barar bin Al Azwar, of whom the account of the Campaign in Syria will have much to say.

Having proclaimed himself prophet, Tulaiha felt that he had to do something about religion in order to prove that he really was an apostle of Allah. He could think of no better way of creating a spectacular effect than by altering the form of prayer. He abolished prostration, which is an integral part of the Muslim prayer ritual. *"Allah does not want us to invert our faces"*, he declared, *"or bend our backs in an ugly posture. Pray standing!"* 2 And the Bani Asad prayed without prostration after their impostor.

With the spread of the apostasy the ranks of his followers swelled. He received offers of support from the major tribes of North-Central Arabia, the staunchest of which were the Ghatfan, followed by the Tayy, with both of which the Bani Asad had an old and abiding alliance. There was support also from the Hawazin and the Bani Sulaim, but this was lukewarm. Although these two great tribes also apostatised and fought the Muslims, they did not join Tulaiha and did not fight under his standard.

1. Bukhari, Muslim and Ahmad. Sahih Al-Jami Al-Saghir No. 3999 and Kitab Al-Tawhid of Sheikh Muhammad bin Abdul Wahhab, Chapter 3.
2. Ibn-ul-Aseer: Vol. 2, p. 131.

The most powerful single supporter of Tulaiha was Uyaina bin Hisn, the one-eyed chief of the Bani Fazara-a powerful clan of the Ghatfan. This was the man who had commanded the Ghatfan contingent at the Battle of the Ditch and whom the Holy Prophet had nicknamed the Willing Fool. Now he lived up to that name by following Tulaiha. He did not, however, believe whole-heartedly in the impostor, for he is known to have said, *"I would rather follow a prophet from an allied tribe than one from the Quraish. Anyway, Muhammad is dead and Tulaiha is alive."* 1 His support proved invaluable, for he brought the entire tribe of Ghatfan under the sway of Tulaiha.

Tulaiha gathered the Bani Asad at Samira. The Ghatfan lived in the neighbourhood of the Bani Asad and would join him soon. The Tayy also accepted him as chief-of-chiefs and prophet, but remained in their own region north and northeast of Khaibar, except for a small contingent, which joined him at Samira. Here Tulaiha began his preparations to fight the power of Islam.

When he heard of the gathering of the clans at Abraq and Zhu Qissa, he sent a contingent from his tribe to reinforce them under his brother, Hibal. The Muslim operations against Zhu Qissa and Abraq have already been described. While these operations were in progress, Tulaiha moved with his army to Buzakha, where he was joined some time later by the remnants of the apostates driven from Abraq.

At Buzakha, Tulaiha's preparations progressed rapidly. He sent couriers to many clans, inviting them to join him, and many clans responded to the call. Uyaina brought 700 warriors from the Bani Fazara. The largest groups were from the Bani Asad and the

Ghatfan. There also was a contingent from the Tayy, but the main part of the Tayy did not come to Buzakha.

Tulaiha was ready for battle when Khalid set out from, Zhu Qissa.

Before launching Khalid against Tulaiha, Abu Bakr sought ways and means of reducing the latter's strength, so that the battle could be fought with the maximum prospects of victory. Nothing could be done about the tribes of Bani Asad and Ghatfan which stood solidly behind Tulaiha; but the Tayy were another matter. They were not nearly so staunch in their support of the impostor, and their chief, Adi bin Hatim, was a devout Muslim. (A man, who was to live to the incredible age of 120 years, Adi was so tall that when he sat on his horse, his feet would touch the ground! [2]) When Adi had tried to prevent the apostasy of the Tayy, they had renounced him, with the result that he had left the tribe, along with a group of his faithful supporters, and joined the Caliph. Abu Bakr now decided to make an attempt at drawing the Tayy away from Tulaiha. And if they could not be persuaded to abandon the impostor, they should be fought and crushed quickly in their present location before they could join him at Buzakha. Thus Tulaiha would be denied the support of the Tayy.

Abu Bakr sent the Tayy chief to work on his tribe. With him marched Khalid, whose corps numbered about 4,000 men. *"If the efforts of Adi are not successful"*, Abu Bakr instructed Khalid, *"fight the Tayy in their present location."* [3] After dealing with the Tayy, Khalid was to, march on Buzakha.

Setting off from Zhu Qissa, Khalid marched in a northerly direction, making for Buzakha. When still a few marches from Buzakha, he turned left and approached the area south of the Aja Mountains, where the tribe of Tayy was gathered. Here Adi went forward and addressed the tribe: he spoke of Allah and His Messenger, of the fire of hell, of the futility of resistance; but in spite of his great eloquence he made no headway. The tribal elders rejected him, whereupon Adi warned them: *"Then prepare to meet an army that comes to destroy you and take your women. Do as you please."*

The warning had the desired effect. The elders reflected for a while and then said, *"Keep this army away from us until we have extricated our brethren who are with Tulaiha. We have a pact with him. If we break it, he will either kill our brethren or hold them as hostages. We must get them away from Tulaiha before openly renouncing him."*

1. Tabari: Vol. 2, p. 487.
2. Ibn Qutaiba: p. 313.
3. Tabari: Vol. 2, p. 483.

Adi returned to the Muslim camp and explained the position to Khalid, but Khalid was in no mood to waste time on negotiations. He held strong views about the apostasy and was not inclined to be kind to those who turned to disbelief-after-belief. *"Three days, O Khalid!"* Adi pleaded. *"Just three days! And I shall get you 500 warriors from my tribe to fight beside you. That is better than sending them to the Fire."* [1] Khalid agreed to wait.

The elders of the Tayy sent off a detachment of horsemen to Tulaiha, ostensibly as a reinforcement for their contingent. And there they started working secretly to get the Tayy contingent away from Tulaiha before Khalid's arrival at Buzakha. In this they succeeded. If any members of the Tayy remained with Tulaiha, and it appears that a few did, they took no part in the Battle of Buzakha.

Khalid had agreed not to attack the Tayy. Meanwhile he decided to turn on another apostate tribe which lived close by-the Jadila. The Caliph had said nothing about the Jadila, but Khalid did not need an invitation to fight. When he announced his intention of attacking the Jadila, Adi again came forward with an offer to persuade the tribe to submit without bloodshed. Khalid was not the man to worry about bloodshed, but in view of the possibility of augmenting his own strength with more warriors, he agreed to Adi's suggestion. The eloquence of Adi bore fruit. The Jadila submitted, and 1,000 warriors joined Khalid. With the strength of his corps augmented with the 500 horsemen from the Tayy and the 1,000 from the Jadila, Khalid, now much stronger than when he had left Zhu Qissa, marched for Buzakha. On his way he was to pick up more warriors.

When a day's march from Buzakha, Khalid sent forward two scouts on a reconnaissance mission. Both these men were Ansars, one of them a renowned Companion by the name of Ukasha bin Mihsan. These scouts met two apostates engaged on a similar mission for the enemy, one of whom was Hibal, the brother of Tulaiha. Hibal was killed, but the other escaped to carry the sad news to the impostor.

Enraged at the news of his brother's death, Tulaiha came forward in person with another brother, Salma. The two pairs met. There were two duels. Tulaiha and Ukasha were expert swordsmen and continued to fight long after Salma had killed the other Muslim. But at last Ukasha went down before Tulaiha. The bodies of the Muslims remained where they had fallen until the rest of the Muslims arrived to discover and bury them. The loss of these two Muslims was deeply mourned, for they were fine fighters and beloved comrades.

When Khalid got to the southern part of the plain of Buzakha, he went into camp a short distance from where the apostates were encamped. From these two camps the opposing forces would move out to battle. The battlefield consisted of the plain of Buzakha-a level, open plain with a few low, rocky hillocks on its western and northern edges. These hillocks were an extension of the south-eastern foothills of the Aja Range.

The stage for the Battle of Buzakha was set. The Muslims prepared for the morrow, as did the apostates. Khalid, the Sword of Allah, with about 6,000 men, faced Tulaiha the Impostor, the strength of whose army is not recorded but is believed to have been much more than that of the Muslims. It was now about the middle of September, 632 (late Jamadi-ul-Akhir, 11 Hijri).

On the morning after the arrival of Khalid, the two armies formed up for battle on the plain of Buzakha. Khalid commanded the Muslims in person and stood ahead of his corps. Tulaiha, however, appointed Uyaina to command his army, in the centre of which stood the 700 Bani Fazara (Uyaina's clan). The impostor himself sat in a tent a short distance behind his army, his head wrapped in a scarf and a cloak draped over his shoulders. He assumed a meditative posture and let it be known that he would receive guidance from Jibril, Allah's messenger angel, on the conduct of battle.

1. Tabari: Vol. 2, p. 483.
2. Nothing remains of Buzakha, but the plain which bears its name starts 25 miles south-west of the present Hail and runs in a south-westerly direction.

Soon after the two forces were arrayed for battle, Khalid launched an attack along the entire front. For some time the apostates resisted stubbornly, especially the Bani Fazara, but after a while the pressure of the Muslims began to tell and dents appeared in the apostate front line. Uyaina, alarmed at the severity of the Muslim attack, rode to Tulaiha's tent, hoping that divine guidance would come to their aid. *"Has Jibril come to you?"* he enquired. *"No"*, replied the impostor with a solemn expression. Uyaina returned to battle.

Some more time passed. Then Khalid was able to drive a wedge into the infidel centre, but it still held, and the fighting became more intense with every inch of ground hotly contested. Uyaina again rode to Tulaiha and asked, *"Has Jibril come to you?" "No, by Allah!"* replied the impostor. Again Uyaina returned to battle.

Scenting victory, the Muslims now attacked more fiercely and gained some more ground. It was all the apostates could do to prevent a complete rupture of their position. Seeing the situation turn hopeless Uyaina went for the third time to Tulaiha. There was a nervous impatience in his voice as he asked the familiar question: *"Has Jibril come to you?"* The impostor answered, *"Yes." "What did he say?"* asked Uyaina.

Calmly Tulaiha replied, *"He said 'You have a handmill just like his, and this is a day that you will not forget!'" "By Allah!"* Uyaina exploded as the scales fell from his eyes, *"This is a day that **you** shall certainly not forget."* He then dashed to his clan. *"O Bani Fazara!"* he shouted. *"This man is an impostor. Turn away from the fight!"*[1]

The Bani Fazara, the hard core of Tulaiha's centre, turned and rode away. With their departure the entire front gave way and the apostate opposition collapsed. Groups of infidels raced from the battlefield in all directions. The victorious Muslims cut those who resisted to pieces. Some hapless fugitives rushed to Tulaiha and asked, *"What are your commands?"* Tulaiha replied, *"Let those who can, do as I do and save themselves and their families."* [2]

With this parting instruction Tulaiha placed his wife on a fast camel, which he had kept ready saddled for just this eventuality. He himself sprang on to his horse, and man and wife disappeared in a cloud of dust.

The Battle of Buzakha was over. Khalid had been victorious. The second most powerful enemy of Islam had been defeated and his forces scattered.

Tulaiha fled to the border of Syria, where he took up residence among the Kalb. His imposturing days were over. But he had not been long with this tribe when he heard that the Bani Asad had re-entered Islam. Consequently he too became a Muslim and rejoined his tribe. He even visited Makkah for the pilgrimage during the time of Abu Bakr, but the Caliph, though informed of his visit, took no notice of him.

About two years later he visited Madinah and came to see Umar, who did not forgive easily. On seeing Tulaiha, Umar said to him, *"You killed two noble Muslims, including Ukasha bin Mihsan. By Allah, I shall never love you."*

Tulaiha had a subtle wit. He replied, *"Allah blessed them with paradise by my hand, while I did not benefit by theirs. I seek forgiveness from Allah."*

Umar, unrelenting, tried again. *"You lied when you said that Allah would do you no harm."*

"That", replied Tulaiha, *"arose from the mischief of disbelief which Allah has destroyed. I cannot now be blamed for it."*

Umar saw that he was not getting far with this exchange and made a last attempt. *"O trickster! What remains of your clairvoyance?"*

"Nothing but a gust or two from the bellows!" [3]

1. Tabari: Vol. 2, p. 485.
2. *Ibid*
3. Tabari: Vol. 2, p. 489; Balazuri: pp. 105-6.

A sense of humour was not one of Umar's strong points; and not being able to think of a suitable rejoinder, he turned away.

Tulaiha returned to his tribe and lived amongst them until the third invasion of Iraq. Then he volunteered for service in Iraq as a Muslim warrior and commander. He served with distinction, performing prodigies of valour and skill, and took part in the great battles of Qadissiyah and Nihawand, where he fell a martyr. Tulaiha thus more than earned his redemption.

As soon as the battle was over, Khalid sent out columns to pursue the fleeing apostates and subdue the neighbouring tribes. One column caught up with some apostates in the hilly region of Ruman, 30 miles south-south-east of Buzakha, but they submitted without a fight and became Muslims again. Khalid led a fast column in pursuit of Uyaina, who had fled to the south-east with his clan of Bani Fazara and some elements of the Bani Asad. Uyaina had only got as far as Ghamra, 60 miles away when Khalid overtook him. Uyaina then turned to fight again, for although he was now totally disillusioned about Tulaiha, he remained defiant and unrepentant. There was a sharp clash in which several apostates were killed and the rest fled. Uyaina was taken prisoner.

Uyaina's father had been a very prominent and highly respected chieftain of the Ghatfan, as a result of which Uyaina regarded himself as second to none in birth and rank. But this proud scion of a long line of chiefs, with whom the Holy Prophet himself had sought to negotiate peace at the Battle of the Ditch, was now put in irons and led as a humble captive to Madinah.

As he entered Madinah, the children, on discovering his identity and circumstances, crowded around him. They began to prod him with sharp sticks, chanting awhile *"O Enemy of Allah! You disbelieved-after-belief."* Uyaina protested piteously, *"By Allah, I never was a believer."* In other words, since he had never become a Muslim (as he now falsely claimed), he could not be accused of apostasy.

He pleaded his case before Abu Bakr, who pardoned him, and so Uyaina became a Muslim again and lived in peace amidst his tribe for many long years.

In the time of Caliph Uthman, Uyaina, now grown old, visited Madinah and called on the Caliph. It was well after sunset. Uthman, as always the generous host, asked him to stay for supper and was taken aback when Uyaina declined the invitation on the plea that he was fasting. (The Muslim fast begins at the first light of dawn and ends at sunset.) Seeing the look of surprise on Uthman's face, Uyaina exclaimed hastily, *"I find it easier to fast by night than by day!"* [2]

After the action at Ghamra, Khalid set off for Naqra where certain clans of the Bani Sulaim had gathered to continue the struggle against Islam. In command of this group of Bani Sulaim was a rash chieftain whose name was Amr bin Abdul Uzza, but who was more commonly known as Abu Shajra. This man had learnt no lesson from the defeat of Tulaiha, and in order to encourage his men to remain firm in their defiance of Muslim authority, he composed and recited the following lines:

> *My spear shall play havoc*
> *With the regiments of Khalid.*
> *And I trust thereafter*
> *It shall also crush Umar* [3]

1. Ghamra lies 15 miles north north-east of Samira, and a hill overlooking the present village is also named Ghamra. This place has been called Ghamr by Ibn Sad who places it at two stages from Feid (p. 590). It is actually 30 miles from Feid as the crow flies, and would be a little farther by caravan route.
2. Ibn Qutaiba: p. 304.
3. Tabari: Vol. 2, p. 494.

As soon as he arrived at Naqra, Khalid launched his column into a violent attack on the Bani Sulaim. Actually, he had pleasant memories of the Bani Sulaim. They had served under him during the conquest of Makkah and the Battle of Hunain and the advance to Taif. Except for their flight when ambushed in the Hunain defile (when most troops would have done the same), they had served him well. But now they had apostatised and deserved no mercy.

Fighting against their ex-commander, the Bani Sulaim resisted fiercely for some time and were able to kill several Muslims, but they too found the powerful blows of Khalid too hard to take and broke up. A large number of them were slaughtered before the rest found safety in flight. Their commander, Abu Shajra the soldier-poet, was taken prisoner and sent to Madinah, where he too pleaded his case with Abu Bakr and was pardoned. He also re-entered Islam.

In later years Abu Shajra fell upon bad times, he was impoverished. Hoping to get some help from Madinah, he rode thither, tied his camel outside the town and went in. Soon he came upon Umar who stood surrounded by the poor to whom he was distributing alms. Entering the throng, Abu Shajra, called, *"I too am in need."* Umar turned and looked at him but failed to recognise him. His appearance had changed much since the days of his apostasy. *"Who are you?"* Umar enquired.

"I am Abu Shajra."

Suddenly old memories flashed across the mind of Umar and he recalled the entire story of the wretched man. *"O Enemy of Allah!"* Umar roared. *"Was it not you who recited:*

> *My spear shall play havoc*
> *With the regiments of Khalid.*
> *And I trust thereafter*
> *It shall also crush Umar. . . !"*

Umar did not wait for a reply. He raised his whip, without which he never left his house, and struck at the man. Abu Shajra raised his arm to protect his head even as he pleaded, *"My submission to Islam has cancelled all that."* [1] Then the second blow fell!

Abu Shajra realised that no amount of pleading would stay the whip of Umar, who was clearly in a mood to strike first and ask questions later. He turned and ran as fast as his legs would carry him, with Umar in hot pursuit, brandishing his whip. But he outran Umar, got to his camel, leapt onto its back and sped away.

Abu Shajra never showed his face in Madinah again!

While the Battle of Buzakha was being fought, certain tribes had stood aside and watched. These were the tribe of Bani Amir and certain clans of the Hawazin and Bani Sulaim. Though inclined towards Tulaiha, they had wisely refrained from battle and preferred to sit on the fence until the outcome of battle was known.

The outcome was soon known. Peace and quiet had hardly returned to Buzakha when these tribes came to Khalid and submitted. *"We re-enter what we came out of"*, they declared. *"We believe in Allah and His Messenger. We shall submit to his orders with our lives and property."* [2]

Soon other sections of repentant Arabs began to pour into Buzakha. *"We submit!"* was the universal cry. But Khalid remembered the instructions of the Caliph-to kill all those who had killed Muslims. He refused to accept their submission (which meant that they could be attacked, killed, enslaved) until they had handed over every murderer in the tribe. To this the tribes agreed.

1. Balazuri: p. 107
2. Tabari: Vol. 2 pg. 486.

All the murderers were lined up. Khalid's justice was swift. He had each murderer killed in exactly the same manner as he had employed to kill his Muslim victim. Some were beheaded, some were burnt alive, and some stoned to death. Some were thrown from the tops of cliffs, while others were shot to death with arrows. A few were cast into wells.[1] An eye for an eye!

Having completed this task, Khalid wrote to Abu Bakr and gave him a complete account of all that had passed. The Caliph wrote him a complimentary letter in reply, congratulating him on his success, approving his actions and praying for his continued success.

After the action against the Bani Sulaim at Naqra, Khalid stayed at Buzakha for three weeks, receiving the submission of the tribes and punishing the murderers. Then he turned his steps towards Zafar, where a lady needed his attention. He looked forward eagerly to the rendezvous; and she awaited him with breathless anticipation!

Salma, alias Um Zhiml, was a first cousin of Uyaina. Her father too was a big chief, Malik bin Hudaifa, of the Ghatfan. Not only was her father a noted chief, but her mother, Um Qirfa, also was a great lady, held in esteem and veneration by the tribe. In the time of the Holy Prophet, the mother had fought against the Muslims and had been captured in battle and killed, but memories of the chieftainess had remained alive among the Ghatfan. Salma had been taken captive and led to Madinah, where the Prophet presented her as a slave to his wife, Aisha. But Salma was not happy, so Aisha set her free, and she returned to her tribe.

After the death of her parents, Salma rose in stature until she began to command the same respect and affection in her tribe as her mother had enjoyed. She became-and this was unusual among the Arabs-a chief in her own right. Her mother had owned a magnificent camel which was now inherited by Salma, and since the daughter looked just like the mother, whenever she rode the camel she reminded her people of the departed *grande dame*.

Salma became one of the leaders of the apostasy and an implacable enemy of Islam. After the Battle of Buzakha and the action at Ghamra, some of those who had lost to Khalid, along with many die-hards from the Hawazin and the Bani Sulaim, hastened to Zafar, at the western edge of the Sulma Range, and joined the army of Salma. She upbraided them mercilessly for their defeat and their abandonment of Uyaina, and such was the awe of this lady that they took it without a murmur. With her strong hand she whipped this motley collection into shape as a closely-knit, well-organised army, and within a few days she had become a threat to the authority of Islam. She knew that Khalid, now free of the problem of Buzakha, would come to deal with her, and she eagerly awaited a clash with the Sword of Allah.

Khalid marched his corps from Buzakha to Zafar where the army of Islam again came face-to-face with the army of disbelief. Again Khalid took the initiative and attacked.

But it proved a hard battle. While Khalid was able to drive back the wings, he could make no progress against the centre of the apostates. The centre stood firm. Here rode Salma in an armoured litter atop her mother's famous camel, and from this command post she personally conducted the battle. Around her camel were gathered the bravest of her warriors, determined to sacrifice their lives in defence of the noble animal and its venerated rider.

Khalid realised that in the person of Salma lay the moral strength of the enemy force, and that as long as she survived in her litter the battle would continue and turn into a bloodbath. She had to be eliminated. Consequently, leading a picked group of warriors, he made a determined thrust towards the camel, and after some vicious sword-fighting was able to get to the animal. With a few slashes the camel was brought down and with it fell the prized litter. Salma was killed immediately. Around her sprawled the bodies of 100 of her followers who had fought to the last in defence of their chief.

1. Tabari: Vol. 2, p. 490.
2. While the general location of Zafar can be established, its exact location is not certain.

Tabari gives Zafar as the scene of the battle and also mentions Ark as the town of the chieftainess Salma. Ark is now a village named Rakk, 35 miles from Hail, nestling at the foot of the northern spurs of the Salma Range. Twelve miles south of Rakk there is a hill called Zafar, on the western slope of the range, and I regard this as the Zafar where the battle was fought.

With the death of Salma, all resistance collapsed and the apostates scattered in all directions. Salma had given Khalid the hardest fight since Tulaiha.

The Salma Range-a range of black, rugged hills standing some 40 miles south-east of the town of Hail-is believed to have been named after Salma, Um Zhiml…a fitting tribute to a grand lady who had the courage to stand and fight against the greatest soldier of the day, and who went down fighting.

The Battle of Zafar was fought in late October 632 (late Rajab, 11 Hijri). For a few days Khalid rested his men. Then he gave orders for the march to Butah, to fight Malik bin Nuwaira.

The first phase of the Campaign of the Apostasy ended with the death of Salma. The major tribes of North-Central Arabia which had rebelled against Islam as followers of Tulaiha had now been defeated and subdued, and their leaders were either killed or captured or driven away. No more rebel chieftains raised their heads again in this region.

But one man remained, more of a bandit leader than a tribal chief, who was still causing anxiety to the Muslims. This man's name was Ayas bin Abd Yalil, but he was more commonly known as Al Faja'a. He was an adventurer.

At about the time when Khalid was consolidating his gains at Buzakha, Al Faja'a came to Abu Bakr. *"I am a Muslim."* he said. *"Equip me with weapons and I shall fight the infidels."* [1]

Abu Bakr was only too glad to hear the offer and equipped him with weapons. The man rode away from Madinah, formed a gang of bandits and began to waylay unwary travellers, many of whom were killed. The gangsters operated in the region lying to the east of Makkah and Madinah, and Muslim and infidel alike suffered at the hands of Al Faja'a.

When Abu Bakr heard of the depredations of Al Faja'a, he decided to make an example of him for the deceit he had practised on the Caliph. He sent a column to get the man alive, and a few days later the brigand was brought to Madinah in irons.

Abu Bakr ordered a large pile of faggots arranged in front of the mosque. When ready, the pile was set on fire. As the wood crackled and the flames rose skywards, Al Faja'a, still in irons, was thrown into the fire!

When Abu Bakr was dying, two years later, he expressed certain regrets. There were, he said, three things that he had done and wished he had not done, and three things that he had not done and wished he had. One of these related to Al Faja'a *"I wish"*, said Abu Bakr, *"I had had Al Faja'a killed outright and not burnt alive."* [2]

1. Tabari: Vol. 2, p. 492
2. *Ibid*: Vol. 2, p. 619; Balazuri: p. 112; Masudi: *Muruj*, Vol. 2, p. 308.

Chapter 14

(False Lords and Ladies)

"The Hour will not be established until there arise thirty impostors, liars, each one imagining that he is a messenger of Allah, a prophet, but I am the Seal of the Prophets: there is no prophet after me."
[Prophet Muhammad (SAWS)]1

Malik bin Nuwaira was a chief of the Bani Yarbu', a large section of the powerful tribe of Bani Tamim which inhabited the north-eastern region of Arabia, above Bahrain. Being close to Persia, some elements of the Bani Tamim had embraced Zoroastrianism, but by and large the tribe was pagan until Islam came to Arabia. The centre of Malik's clan was Butah.

Malik was a chief of noble birth. Famous for his generosity and hospitality, he would keep a light burning outside his house all night so that any traveller passing that way would know where to find shelter and food. He would get up during the night to check the light. A strikingly handsome man, he had a thick head of hair and his face, a contemporary has said, was "as fine as the moon." 3 He was skilful in the use of weapons and noted for his courage and chivalry, and he was an accomplished poet. Thus Malik possessed all the qualities which the Arabs looked for in the perfect male. He had everything!

Laila was the daughter of Al Minhal and was later also known as Umm Tamim. A dazzling beauty, she was one of the loveliest girls in Arabia, the fame of whose stunning good looks had spread far and wide. She was known especially for her gorgeous eyes and her lovely legs. She too had everything! 4

When she came of age she was pursued by every swain in the region but rejected the suit of one and all. Then she met Malik, with whom she was destined to enter the pages of history. Malik married Laila. Thus Malik, in addition to all his other enviable qualifications, also had as wife one of the loveliest women of the time.

Malik bin Nuwaira certainly had everything. Everything, that is, but faith.

During the Year of Delegations, when the tribe of Bani Tamim embraced, Islam, Malik also moved with the popular trend and became a Muslim. In view of his distinguished position in the tribe and his unquestionable talents, the Holy Prophet appointed him as an officer over the clan of Bani Handhalah. His main responsibility was the collection of taxes and their despatch to Madinah.

Malik performed these duties honestly and efficiently for some time. Then the Holy Prophet died. When news of his death reached Butah, Malik had just collected a good deal of tax, prior to its despatch to Madinah. Forgetting his oath of allegiance, he at once opened the coffers and returned the money to the taxpayers. *"O Bani Handhalah!"* he announced, *"your wealth is now our own."* [5] Malik had apostatised.

Sajjah was the daughter of Al Harith. Born in a family of chiefs, she had qualities of leadership, personality and intellect with which few women have been endowed. She was clairvoyant, would predict future events, and was so versatile a poetess that practically everything that she said was in verse. When people spoke to her, she rhymed back at them.

Later known as Um Sadira, she also belonged, on her father's side, to the Bani Yarbu' and thus was a kinswoman of Malik bin Nuwaira. On her mother's side, however, she belonged to the Taghlib, one of the tribes in the large group known as Rabi'a which inhabited Iraq. Sajjah lived mostly among the Taghlib who followed the Christian faith, and because of her mother's influence, Sajjah also had become a Christian, but Christianity did not have a very strong hold upon her, nor upon many members of the Taghlib, as we shall see.

When apostasy began to spread, Sajjah heard that Tulaiha and Musailima had proclaimed their prophethood. Her fertile imagination was intrigued by the possibilities that these false claims opened up. Why should only men be prophets? Why could a woman not enter the sacred precincts of prophethood? An adventuress at heart, she finally gave in to the temptation. *"I am a prophetess!"* she declared, and elucidated the point with a few appropriate verses.

1. Bukhari, Muslim, Abu Dawud, Tirmidhi, Ahmad, Hakim. Sahih Al-Jami' Al-Saghir No. 7417-8.
2. Butah is now nothing more than a tiny Bedouin settlement 14 miles south-south-west of the present Ras. It shows signs of having been a bigger place at one time.
3. Balazuri: p. 108.
4. Isfahani: Vol. 14, p. 65. "It used to be said that never had legs more beautiful than hers been seen."
5. Balazuri: p. 107.

Strangely enough, most of her mother's clan accepted her as prophetess and pledged to obey her. They had been Christians! She mustered many armed followers and came down into Arabia, where her father's tribe also flocked to her standard. No doubt many that followed her, elders and clansmen, were led by the temptation of plunder and the desire to settle old scores with some of the tribes in north-eastern Arabia which had old feuds with them.

Elated by her success in gathering followers, she arrived at Al Hazn with a fair-sized force and exchanged envoys with her kinsman, Malik bin Nuwaira. [1] She proposed a pact: they would operate jointly against the tribes that were their mutual feudal enemies and would thereafter war against the Muslim power at Madinah. In order to assure Malik that she had no aggressive designs upon the lands of the Bani Yarbu', she declared, *"I am only a woman of the Bani Yarbu'. The land is yours."* [2]

Malik accepted Sajjah's proposal and entered into a pact with her. However, he cooled her martial ardour somewhat and dissuaded her from warring against the Muslims. This happened in June 632.

The combined forces of Malik and Sajjah now turned upon the hapless tribes that had offended the Bani Tamim and the Taghlib. There was nothing religious in this operation, the underlying motives were revenge and the lust for loot. Any tribe that resisted was fought, subdued and plundered. Malik was joined to the impostress by the pact and his followers fought alongside hers in these raids. It appears, however, that he did not personally take part in these depredations.

Then Sajjah came to Nibbaj and began plundering the neighbourhood. [3] And here she suffered a serious setback. The local clans, driven by their common fear of the terrible lady, united in opposition to her and this resistance resulted in a battle. It was not by any means a decisive battle, but she got the worst of it; a few of her important officers were captured by her opponents, who refused to release them unless she pledged to depart from their area. To this she agreed.

The elders of the tribes which made up her following now gathered around their impostress. *"Where now?"* they asked.

"To Yamamah," she replied.

"But the people of Yamamah are mighty", they pointed out, *"and their Chief, Musailima, is a very powerful man."*

"To Yamamah", repeated Sajjah and then broke into verse:

> *Onward to Yamamah!*
> *With the flight of soaring pigeons;*
> *Where the fighting is the fiercest;*
> *And no blame shall fall upon you.*
> *Onward to Yamamah!* [4]

1. The location of Hazn is not certain, but according to local information in Hail, it is the same as the area of Hazm which lies between Samira and Butah. This seems to fit in with Yaqut's statement (Vol. 1. p. 661) that it was near Butah.
2. Tabari: Vol. 2, p. 496.
3. Nabbaj is the present Nabqiya (also called Nabjiya by the inhabitants) 25 miles north-east of Buraida. Now it is a village; then it was a sizeable town.
4. Tabari: Vol. 2, p. 498.

Musailima the Liar was the most formidable of the enemies of Islam who rose to threaten the existence of the new state. He was the son of Habib, of the Bani Hanifa, which was one of the largest tribes of Arabia and inhabited the region of Yamamah.

Musailima first mounted the stage of history in late 9th Hijri, the Year of Delegations' when he accompanied a delegation of the Bani Hanifa to Madinah. The delegation included two other prominent men who were to exercise a profound influence on Musailima and his tribe-one in aiding Musailima's rise to power and the other in saving the tribe from destruction. These men were, respectively, Nahar Ar-Rajjal bin Unfuwa [1] and Muja'a bin Marara.

The delegation arrived at Madinah. The camels were tied in a traveller's camp, and Musailima remained there to look after them while the other delegates went in. They had talks with the Prophet, submitted to him and embraced Islam. As was his custom, the Prophet presented gifts to the delegates, and when they had received their gifts one of them dropped a hint: *"We left one of our comrades in the camp to look after our mounts."* The Prophet gave them gifts for him also, and added, **"He is not the least among you that he should stay behind to guard the property of his comrades."** [2] These words were to be used by Musailima later to his own advantage.

On their return, this delegation passed on the message of Islam and established the new faith among the Bani Hanifa. The whole tribe was converted. They built a mosque at Yamamah and started regular prayers.

Thus some months passed. Then Musailima resiled from his faith and proclaimed his own prophethood. He gathered the people and, referring to Muhammad, addressed them: *"I have been given a share with him in this matter. Did he not say to our delegates that I was not the least among them? This could only mean that he knew that I had a share with him in this matter."* [3] (The matter was the prophethood.)

He then dazzled the crowd with his marvellous tricks. He was a superb conjuror and could do what no one had done before. He could put an egg in a bottle; he could cut off the feathers of a bird and then stick them on so the bird would fly again; and he used this skill to persuade the people that he really was divinely gifted. He took to addressing gatherings as an apostle of Allah, and would compose verses and offer them, as Quranic revelations. Most of his verses extolled the superiority of his tribe, the Bani Hanifa, over the Quraish. Some, however, were utterly ridiculous, like the following,

Allah has blessed my wisdom.
It is as strong as the gust that blows
From between the belly and the intestines! [4]

And the people marvelled at his wisdom and flocked to him. Strangely enough they did not doubt or dispute the divine mission of Muhammad. They accepted Muhammad as the apostle of Allah. But they also accepted Musailima as co-prophet-which is all that Musailima claimed.

Gradually the influence and authority of Musailima increased. Then one day, in late 10 Hijri, he wrote to Prophet Muhammad:

"From Musailima, Messenger of Allah, to Muhammad, Messenger of Allah. Salutations to you. I have been given a share with you in this matter. Half the earth belongs to us and half to the Quraish. But the Quraish are a people who transgress."

1. Some early historians have given this man's name as Rahhal.
2. Ibn Hisham: Vol. 2, pp. 576-7.
3. *Ibid*
4. *Ibid*

In reply the Holy Prophet wrote to Musailima:

"In the name of Allah, the Beneficent, the Merciful. From Muhammad, Messenger of Allah, to Musailima the Liar. Salutations to whosoever follows the Guidance. Lo! The earth belongs to Allah. He gives it to whomsoever He chooses from among His servants. And the Hereafter is for the virtuous." [1]

The impostor was henceforth known as Musailima the Liar!

Now Nahar Ar-Rajjal, whom we have mentioned earlier as a member of the Bani Hanifa delegation, came into action. This man had stayed behind at Madinah when the rest of the delegation returned home; and had attached himself to the Holy Prophet, from whom he acquired a great deal of knowledge about Islam. He learnt the Quran and rose in stature as a close and respected Companion of the Prophet. In a few months he had built up an enviable reputation as a devout and virtuous Muslim, and so he became known over most of Arabia.

When reports of the spread of Musailima's mischief became more alarming, the Holy Prophet began to consider ways and means of countering the influence of the Liar. Yamamah was too far away for a military operation, so he decided to send a man to work against Musailima amongst the people. And who could be better suited to this task than Rajjal? He was a chief of the Bani Hanifa, he had learnt the Quran; he had acquired wisdom and grace at the feet of the Prophet. And so Rajjal was sent by the Prophet to undo the mischief that Musailima had wrought at Yamamah.

As soon as he arrived at Yamamah, the rascal declared that Musailima was indeed a prophet. *"I have heard Muhammad say so"*, he lied [2] and who could doubt the words of this respected Companion! The arrival of the renegade proved a windfall for Musailima, and the Bani Hanifa came in even larger numbers to swear allegiance to 'Musailima, Messenger of Allah!'

Musailima and Rajjal now formed an evil and accursed partnership. Rajjal became the right hand man of Musailima, and the impostor made no important decision without consulting him.

With the death of the Holy Prophet, Musailima's hold over the Bani Hanifa became total. People flocked to him, and Musailima began making his own rules in matters of moral and religious conduct. He made alcohol lawful. He also ordered that once a man had fathered a son he would live in celibacy unless the son died, in which case women were permitted to him until he got another son.

His people began to believe that Musailima had miraculous powers, and Rajjal helped foster this image. Once Rajjal suggested that he stroke the head of every newborn babe, as Prophet Muhammad used to do as a form of blessing. Orders were issued accordingly. Thereafter every newborn babe in Yamamah was brought to Musailima to have its head stroked. Historians narrate that when these infants had grown to full manhood or womanhood, they did not have a single hair on their heads! But it this was not, of course, known till after Musailima's death. Many are the instances of Musailima emulating the acts of Muhammad with opposite and disastrous results.

Though all the Bani Hanifa followed him, not all believed in his divine mission, certainly not the intelligent ones. Some accepted him for political convenience or for reasons of personal advancement while many were motivated by feelings of tribal loyalty. One day Musailima appointed a new man as Muazzin, to call the men to prayer. This man, Jubair bin Umair, was a doubter. Instead of the words "I bear witness that Muhammad is the Messenger of Allah", in which the name of Musailima had to be substituted for that of Muhammad, this new *muazzin* [3] called, at the top of his voice: *"I bear witness that Musailima thinks he is the messenger of Allah."* [4]

Once a man-a clear headed fellow-who had never seen Musailima before came to visit the impostor. When he got to the door of Musailima's house, he asked the guard *"Where is Musailima?" "Silence!"* replied the guard. *"He is the messenger of Allah." "I shall not accept him as such until I have seen him"*, asserted the visitor, whose name was Talha.

1. Ibn Hisham: Vol. 2, pp. 600-1
2. Tabari: Vol. 2, p. 505.
3. One who calls the Adhan - the call to prayer
4. Balazuri: p. 100.

The visitor met the impostor. *"Are you Musailima?"* he enquired.

"Yes."

"Who came to you?"

"Rahman." (i.e. the Beneficent One).

"In light or in darkness?"

"In darkness."

"I bear witness", declared Talha, *"that you are an impostor and Muhammad is genuine. But an impostor from our tribe is preferable to a true one from the Quraish."* [1] This man later fought and died beside Musailima.

In appearance Musailima was terrible. A short-statured man, though immensely strong, he had a yellow complexion, small, close-set eyes and a flat nose. He was extremely ugly. But as often happens with very ugly and evil men, he had an irresistible fascination for women. They could not say "No!" He was such a talented and unscrupulous Casanova that no woman left alone with him could resist his advances or escape his devilish charm.

But Sajjah the impostress did not know this facet of Musailima's many-sided character as she set out for Yamamah. She would soon learn!

Sajjah marched with her army towards Yamamah. Musailima came to know of this move and was perturbed, for he did not know whether her intentions were hostile or friendly. He could certainly defeat her army in battle, but Ikrimah with his corps was camped some distance to the west, and Musailima had been waiting for several weeks for the Muslims to advance. If Ikrimah were to move at the time when he was engaged with the army of Sajjah, he would be in a most vulnerable position. It would mean simultaneous war with two enemies, Sajjah and the Muslims. Musailima decided to win over Sajjah and neutralise her. He knew how to deal with her: he would handle her as he would handle any woman, the art of which he knew so well.

He sent a message to Sajjah not to bring her warriors, as there was no work for them at Yamamah. She could come alone for talks. Consequently Sajjah left her army in camp and rode forward with 40 of her warriors to meet Musailima the Liar. She arrived at Yamamah to find the gate of the fort closed, and she received Musailima's instructions to leave her men outside and enter alone. Sajjah agreed, and leaving her 40 followers to bide their time in a camp, entered the fort.

Musailima had had a large tent pitched for her in the courtyard of his house. Since the weather was chilly, he had the tent properly heated so that she would be comfortable. And he had a certain incense burning in the tent that would affect Sajjah's senses in the way that he desired. It would make her shed her inhibitions!

She entered the tent. Some time later Musailima also entered. They were alone. The impostor began to talk, weaving a spell over the woman. He talked of Allah and of politics, of the trouble that he was having with the Quraish who were as numerous as 'the scales of a fish'.

After this preamble he said, *"Tell me of our revelations."*

"A woman should not begin", she replied. *"You tell me first what has been revealed to you."*
She gazed at him with awe as he intoned, as if reciting a Quranic verse:

Do you not see your Lord?
How he deals with pregnant women? He extracts a living being
From between the belly and the intestines.

1. Tabari: Vol. 2, p. 508.

"And what more?" she asked excitedly.

"He has revealed to me", continued Musailima, *"that He created woman a receptacle and, created man as her mate, to enter her and leave her at his pleasure. And then a little lamb is brought forth!"*

Sajjah was fascinated. *"You are indeed a prophet!"* she gushed.

Musailima moved closer. *"Do you feel like marrying me?"* he asked. *"Then with my tribe and yours I shall eat up the Arabs."*

"Yes", she answered. [1] Musailima had conquered again.

She stayed with him for three days, then he sent her back to her army. On arrival at her camp, she assembled the elders of the tribe. *"I have found the truth"*, she declared. *"I have accepted him as prophet and married him."*

The elders were not a little surprised. *"Has he given you a wedding gift?"* they asked. Sajjah confessed that she had received no wedding gift.

These elders knew a little more about Musailima than she did, and feared that their girl had been taken for a ride. *"Then go back to him"*, they insisted, *"and do not return without a wedding gift."*

Again Sajjah rode with her 40 companions to Yamamah. Musailima saw her coming and closed the gate of the fort. *"What is the matter?"* he asked angrily from within.

"Give me a wedding gift" she pleaded from outside.

Musailima thought for a moment, and then replied, *"I give you a wedding gift for all your people. Announce to your followers that I, Musailima bin Habib, Messenger of Allah, remit two of the prayers that Muhammad had imposed-the prayer of the early morning and the prayer of the night."* 2

With this wedding gift Sajjah returned to her army.

A few days later, wishing to establish more durable ties with her people than those of the tent in his courtyard, Musailima sent an envoy to Sajjah. He offered her political and economic partnership: she could have half the grain of Yamamah. Sajjah refused. But Musailima sent his envoy again to insist that she accept at least a quarter of the grain. She accepted this and left for Iraq. This happened around late October 632 (late Rajab, 11 Hijri), shortly before Ikrimah's clash with Musailima.

Musailima had finished with her. And she had finished with politics and prophethood. She took up residence amongst her mother's tribe and lived in obscurity for the rest of her life. Later she embraced Islam and was believed to be a pious and virtuous Muslim. During the caliphate of Muawiyah she moved to Kufa, where she died at a ripe old age.

1. Tabari Vol. 2, p. 499.
2. *Ibid.*

Chapter 15

(The End of Malik ibn Nuwaira)

"We were like the drinking-mates of Jadhimah
For a time, till it was said we would never separate.
We spent the best days of our lives, but before us
Death had destroyed the nations of Kisra and Tubba'.
When we parted it was as though Malik and I,
Despite long association, were never together for even a night."
[Mutammim bin Nuwayrah, mourning the death of his brother Malik.]1

When, after finishing with Salma and her followers, Khalid gave orders for the march to Butah against Malik bin Nuwaira, he had no suspicion that some of his own men would oppose his plan. Preparations for the move were carried out as ordered, but when the time to march came, a large group of his soldiers refused to move.

These were the Ansars. Their elders came to Khalid and said that they would not march to Butah. *"What you plan now"*, they asserted, *"was not included in the instructions of the Caliph. His instructions were to fight at Buzakha and free this region of apostasy. Thereafter we were to await his instructions."*

Khalid was surprised at this statement. He had no intention of letting this group, even if it was a highly honoured group of Companions, deter him from conducting operations as he saw fit. *"That may be the Caliph's instructions to you,"* he replied, *"but his instructions to me were to operate against the infidels. In any case I am the commander of this force. I am better informed of the situation than you are. If I see an opportunity for which I have received no instructions, I shall certainly not let it slip by. Should we be faced with a challenge for which there were no instructions from the Caliph, would we not accept it? Malik bin Nuwaira is there, and I shall go to fight him. Let the Emigrants and those who are willing follow me. The others I shall not compel."* [2]

Khalid marched off without the Ansars.

Hardly an hour had passed when the Ansars realised the seriousness of their error in refusing to march with the rest of the corps. *"If they meet with success, we shall be left out of it"*, said one. Others added, *"And if they come to grief, nobody will ever talk to us again."* Their minds were soon made up. They sent a fast rider after Khalid to say, *"Wait! We are coming."* Khalid waited until they had joined him and then resumed the march to Butah.

During the first week of November 632 (mid-Shaban, 11 Hijri) Khalid arrived at Butah, all set for battle. But Butah had no opposition to offer. There was not a single warrior in sight.

When Sajjah the impostress left Arabia for Iraq, Malik bin Nuwaira began to have second thoughts about the part that he had played in the conspiracy against Islam. He received reports of how the Sword of Allah had destroyed the army of Tulaiha, and also heard of the swift and severe punishment Khalid had meted out to the murderers of Muslims. Malik was afraid. With the departure of Sajjah he had lost a strong ally, and he felt abandoned, betrayed.

He began to realise the seriousness of his action in making a pact with the impostress. His guilt of apostasy was clear and could not be disputed. Then came reports that Khalid had defeated Salma and was now marching in the direction of Butah. Malik was a brave man, but he did not feel up to fighting Khalid.

1. Ibn Kathir, Al-Bidayah wan-Nihayah, Dar Abi Hayyan, Cairo, 1st ed. 1416/1996, Vol. 6 P.394.
2. Tabari: Vol. 2, p. 501. From this exchange it would appear that Khalid's decision to march to Butah was his own and not part of the over-all plan of the Caliph, but again according to Tabari (Vol. 2, pp. 480, 483) Abu Bakr's instructions to Khalid definitely included Malik bin Nuwaira at Butah as the next objective after Tulaiha had been dealt with. Perhaps Khalid's men did not know that the Caliph had given this task to their commander.

Feeling helpless and forsaken, Malik decided to save what he could from the wreckage. He would atone for his crimes by repentance and submission, which was also a political necessity, for there was nothing else that he could do. Malik gathered the clan of Bani Yarbu' and addressed them as follows:

"O Bani Yarbu'! We disobeyed our rulers when they called upon us to remain steadfast in faith. And we prevented others from obeying them. We have come to no good."

"I have studied the situation. I see the situation turning in their favour while we have no control over it. Beware of fighting them! Disperse to your homes and make peace with them." 1

Under these orders his warriors dispersed. Malik then went quietly to his house, not far from Butah, to be consoled by the charming Laila.

In one more gesture to show his change of heart, Malik collected all the tax that was due to Madinah and sent it to Khalid, who was on the march to Butah when the envoys bringing the tax met him. Khalid took the tax, but did not accept this as sufficient atonement, for the tax was in any case due as an obligation.

"What made you enter into a pact with Sajjah?" Khalid asked the envoys. *"Nothing more"*, they replied, *"than a desire for tribal revenge against our feudal enemies."* 2

Khalid did not question the envoys further, but retained his suspicions. This could be a trick to lull him into a false sense of security and draw him unsuspecting into an ambush. Ever since the ambush at Hunain, Khalid had never relaxed his vigilance. He continued the advance as a military operation against an armed opponent.

Khalid found Butah undefended and unmanned. There was no army to fight-not even an occasional group of soldiers. He occupied Butah and sent out mounted detachments to scour the countryside and deal with the apostate clans of the tribe of Bani Tamim. To the commanders of these detachments, he repeated the instructions of the Caliph-on approaching any clan, they would call the *Adhan*, if the clan responded with the *Adhan*, it would be left alone; if it did not, it would be attacked.

The following day a detachment commanded by Dhiraar bin Al Azwar got to the house of Malik bin Nuwaira, where Dhiraar seized Malik and Laila and a few men of the Bani Yarbu'. The other detachments had no trouble, for all the clans submitted without opposition.

Malik and Laila were ushered into the presence of Khalid, Malik appearing as a rebel and apostate chief on trial for crimes against the State and Islam. He looked defiant, true to the nature of a proud, noble?born chieftain who faced the trials of life with dignity. He could not be humble.

Khalid began to talk. He spoke of the crimes that Malik had committed and the damage that he had done to the cause of Islam. Then Khalid asked him some questions. In his reply, Malik referred to the Holy Prophet as "your master". Khalid was angered by the unrepentant and supercilious attitude of the accused. He said, *"Do you not regard him as your master?"* 3

Khalid felt convinced that Malik was guilty, that he remained an unbeliever. He gave the order for his execution. Dhiraar took Malik away and personally carried out the sentence. And it was the end of Malik bin Nuwaira.

Laila became a young widow, but not for long. That same night Khalid married her! She had hardly made up her mind to mourn her departed husband when she became a bride again, this time of the Sword of Allah!

When Khalid announced his intention of marrying Laila some Muslims did not take kindly to the announcement. Some even began to suggest that perhaps Malik was not really an unbeliever but had returned to the Faith, that perhaps Khalid had ordered his execution in order to be able to have Laila for himself. One man in particular, Abu Qatadah, a Companion of high standing, remonstrated with Khalid, but Khalid put him in his place with a few well-chosen words. Feeling slighted and angry at what he regarded as Khalid's high-handedness, Abu Qatadah next day mounted his horse and set off at a gallop for Madinah. On arrival at the capital, he went straight to Abu Bakr and told him that Malik bin Nuwaira, was a Muslim and that Khalid had killed him in order to be able to marry the beautiful Laila. This Abu Qatadah was the same man who, shortly after the conquest of Makkah, had ridden to the Holy Prophet and complained that Khalid had ruthlessly killed the Bani Jazima despite their surrender. His disapproval of Khalid was not new.

1. Tabari: Vol. 2, Pp. 501-2.
2. *Ibid*
3. *Ibid*: Vol. 2, p. 504.

Abu Bakr, however, was not pleased to see Abu Qatadah, especially as he had left the army without his commander's permission. *"Return at once to your post!"* ordered the Caliph; and Abu Qatadah rode back to Butah.[1]

But even before he had gone his words were all over Madinah. They were heard by Umar who leapt to his feet and rushed to Abu Bakr. *"You have appointed a man to command"*, he said, *"who kills Muslims and burns men alive."* [2] Abu Bakr was not impressed. He had clear evidence of Malik's distributing the tax money on getting news of the Prophet's death and of his pact with Sajjah. There was no doubt about Malik's apostasy. As for burning men alive, the Caliph had himself ordered that those apostates who had burnt Muslims alive would be treated in like manner. [3] Khalid had burnt no others.

Umar continued: *"There is tyranny in the sword of Khalid. He should be brought home in fetters. Dismiss the man!"*

Abu Bakr knew that there was little love lost between these two great men. *"O Umar"*, he replied firmly, *"keep your tongue off Khalid. I shall not sheathe the sword that Allah has drawn against the infidels."* By now Khalid was being commonly referred to as the Sword of Allah.

Umar persisted: *"But this enemy of Allah has killed a Muslim and taken his wife!"* [4] Abu Bakr agreed to go into the matter. He sent for Khalid.

By now Khalid had come to know of the resentment that his actions had aroused. He shrugged it off with the words: *"When Allah decides a matter, it is done."* 5 Anyway, a little criticism did not worry Khalid. Then came the summons of the Caliph to present himself at Madinah. Khalid guessed that this was connected with the allegations against him, and was now more than a little worried.

On arrival at Madinah, Khalid went straight to the mosque. In those early days the mosque was not merely a place of worship. It was also a meeting place, an assembly hall, a school, a place of rest, and the centre of civic activity. Khalid was wearing an arrow in his turban as an adornment, and this made him look a bit of a dandy, for most Muslims preferred simplicity in their dress and avoided all forms of ostentation.

Umar was in the mosque and saw Khalid. Livid with anger, he walked up to Khalid, tore the arrow from Khalid's turban and broke it in two. *"You killed a Muslim and snatched his wife"*, Umar shouted. *"You ought to be stoned to death."* 6 Khalid knew that Umar had much influence with Abu Bakr, and fearing that the Caliph might have similar opinions, he turned away in silence.

He next went to see Abu Bakr, who demanded an explanation. Khalid told him the whole story. After due consideration, the Caliph decided that Khalid was not guilty. He did, however, upbraid his general for marrying Laila and thus leaving himself open to criticism, and since there was some possibility of a mistake, as certain people believed that Malik was a Muslim, Abu Bakr ordered the payment of blood-money to the heirs of Malik.

1. Tabari: Vol. 2, pp. 501-2.
2. Balazuri: p. 107.
3. Tabari: Vol. 2, p. 482.
4. *Ibid*: Vol. 2, pp. 503-4; Balazuri: p. 107.
5. Tabari: Vol. 2, p. 502.
6. Tabari: Vol. 2, p. 504.

Khalid came out of the house of the Caliph. His step was light and his manner carefree as he walked to the mosque where Umar sat conversing with some friends. This time Khalid was surer of his position and could afford to repay the compliment. He called to Umar, *"Come to me, O left-handed one!"* 1 Umar guessed that the Caliph had acquitted Khalid. He stood up and without a word marched off to his house.

This matter of Malik and Laila has been the subject of much dispute in Muslim history. Some, quoting sources like Abu Qatadah, have said that the household of Malik had called the *Adhan* and that Malik had returned to the faith before he was taken captive. Others have said that Khalid never ordered the killing of Malik, that the weather was chilly and Khalid had said, *"Warm your prisoners"*, that in certain dialects the same word is used to denote 'warming', and 'killing', thus Dhiraar misunderstood Khalid's order and went and killed Malik.

These versions of the story are, in all probability, not true. They have been offered by factions-one to explain away Umar's hostility towards Khalid and the other to clear Khalid of the possible guilt of murdering a Muslim.

There is no doubt about the apostasy and sedition of Malik bin Nuwaira, his distribution of the tax money, his pact with Sajjah, and the participation of his warriors, on his orders, in the depredations of Sajjah. All historians have, without exception, reported these incidents as facts. There is also no doubt, in the mind of this writer that Khalid ordered the killing of Malik and did so with the honest and sincere conviction that Malik was an apostate and a traitor. But suspicion continued to lurk in the minds of some Arabs, certainly in the mind of Umar, that this was a crime de passion. Umar was further encouraged in this belief by the brother of Malik, who came to see him and told him what a wonderful man Malik was and how tragic it was that he had fallen a victim to Khalid's lust!

The long and short of the whole affair was that Malik was killed and the beautiful Laila with the gorgeous eyes and the lovely legs became the wife of Khalid bin Al Waleed. He would one day pay a very high price for the pleasure!

1. Tabari: Vol. 2, p. 504.

Chapter 16

(The Battle of Yamamah)

"Musaylimah! Recant - do not contend,
For in prophethood you have not been given a share.
You lied about Allah regarding His revelation
And your desires are the whims of a stupid fool.
Your people have indulged you instead of preventing you,
But if Khalid comes to them you will be abandoned.
Then you will have no stairway to the heavens
And no path to travel in the earth."
[Thumamah bin Uthal, a Companion from the tribe of Musaylimah][1]

When Abu Bakr organised the Muslim forces into 11 corps at Zhu Qissa, he appointed Ikrimah, son of Abu Jahl, as the commander of one of them. Ikrimah's orders were to advance and make contact with the forces of Musailima the Liar at Yamamah, but not to get involved in battle with the impostor. Abu Bakr knew better than most of his generals the power and ability of Musailima, and did not wish to risk fighting him with insufficient forces. Since Khalid was his finest general, the Caliph had made up his mind to use him to deal with Musailima after he had finished with the other enemies of Islam.

Abu Bakr's intention in giving Ikrimah this mission was to tie Musailima down at Yamamah. With Ikrimah on the horizon, the Liar would remain in expectation of a Muslim attack and thus not be able to leave his base. With Musailima so committed, Khalid would be free to deal with the apostate tribes of North-Central Arabia without interference from Yamamah. In selecting Ikrimah for this task Abu Bakr had picked a valiant man. Moreover, Ikrimah was anxious to prove his devotion to Islam and atone for his violent hostility to the Holy Prophet before his entry into the new faith.

Ikrimah advanced with his corps and established a camp somewhere in the region of Yamamah. The location of his camp is not known. From this base he kept the forces of the Bani Hanifa under observation while awaiting instructions from the Caliph, and the presence of Ikrimah had the desired effect of keeping Musailima in Yamamah. However, whether or not he had any intention of ever leaving Yamamah we do not know.

When Ikrimah received reports of the defeat of Tulaiha by Khalid, he began to get impatient for battle. The waiting irked his fiery temperament. Ikrimah was a fearless man and a forceful general, but he lacked Khalid's cool judgement and patience-qualities which distinguish the bold commander from the rash one.

The next development that Ikrimah heard of was that Shurahbil bin Hasanah was marching to join him. Shurahbil too had been given a corps by the Caliph with orders to follow Ikrimah, and await further instructions. In a few days Shurahbil would be with him.

Then came news of how Khalid had routed the forces of Salma the queenly leader of men. Ikrimah could wait no longer. Why let Khalid win all the glory? Why wait for Shurahbil? Why not have a crack at Musailima himself? If he could defeat Musailima single-handed, he would win glory and renown such as would eclipse the achievements of all the others. And what a delightful surprise it would be for the Caliph! Ikrimah set his corps in motion. This happened at the end of October 632 (end of Rajab, 11 Hijri).

A few days later he was back in his camp, having received a sound thrashing from Musailima. Chastened and repentant, he wrote to Abu Bakr and gave him a complete account of his actions, including the inglorious outcome. Shurahbal also heard the bad news and stopped some distance short of Ikrimah's camp.

Abu Bakr was both pained and angered by the rashness of Ikrimah and his disobedience of the orders given to him. He made no attempt to conceal his anger in the letter that he wrote to Ikrimah. *"O son of the mother of Ikrimah!"* he began. (This was a polite way of expressing doubt regarding the identity of the man's father!) *"Do not let me see your face. Your return under these circumstances would only weaken the resolve of the people. Proceed with your force to Oman to assist Hudaifa. Once Hudaifa has completed his task, march to Mahra to help Arfaja and thereafter go to the Yemen to help Muhajir. I shall not speak to you until you have proved yourself in further trials."* [2] The three men to be assisted were among the 11 corps commanders.

1. *Mukhtasar Sirat al-Rasul sall-Allahu 'alayhi wa sallam*, of Sheikh Muhammad bin Abdul Wahhab.
2. Tabari: Vol. 2, pp. 504, 509.

Smarting under the shame of his ignominious repulse at the hands of Musailima and the harsh words of the Caliph, Ikrimah took his corps and, bypassing Yamamah, marched to Oman.

Shurahbil remained in the region of Yamamah. To ensure that he did not fall into the error of Ikrimah, Abu Bakr wrote to him: *"Stay where you are and await further instructions."* [1]

Having ordered the payment of blood money to the heirs of Malik bin Nuwaira, the Caliph sent for Khalid and gave him the mission of destroying the forces of Musailima the Liar at Yamamah. In addition to his own large corps, Khalid would have under command the corps of Shurahbil. Another body of Ansars and Emigrants was being scraped together by Abu Bakr at Madinah, and this too would be sent to Butah shortly to join the forces of Khalid. Thus Khalid would command the main army of Islam.

Khalid rode to Butah where his old corps awaited him. Meanwhile the Caliph wrote to Shurahbil: *"You will come under Khalid's command as he joins you. When the problem of Yamamah has been solved, you will proceed with your men to join Amr bin Al Aas and operate against the Quza'a."* 2 This was the apostate tribe, which Usama had punished but not subdued, near the Syrian frontier.

Khalid waited at Butah until the arrival of the Ansars and Emigrants from Madinah, then marched for Yamamah. He was glad to think that the fresh troops of Shurahbil would also be available to him. He did take them under his command, but they were not all that fresh. A few days before Khalid's arrival Shurahbil had given in to the same temptation as Ikrimah, seeking glory, he had advanced and clashed with Musailima. Feeling sorry about the whole affair, Shurahbil expressed his regrets to Khalid, who rebuked him severely.

Khalid was still some distance from Yamamah when his scouts brought word that Musailima was encamped in the plain of Aqraba, on the north bank of the Wadi Hanifa through which, the road led to Yamamah. Not wishing to approach his enemy through the valley, Khalid left the road a few miles west of Aqraba, moved from the south and appeared on the high ground which rose a mile south of the *wadi* opposite the town of Jubaila. 3 From this high ground Khalid could see the entire plain of Aqraba, on the forward border of which stretched the camp of the Bani Hanifa. Khalid established his camp on the high ground. The strength of his army amounted to 13,000 men.

Khalid had not gone many days from Butah when Musailima's agents informed him of the march of the Muslims and of the fact that this was the main army of Islam. The route from Butah to Yamamah came through the Wadi Hanifa, and on the north bank of this wadi, behind Jubaila, lay the plain of Aqraba which marked the outer limit of the fertile region that stretched from Aqraba to Yamamah and further south-east. It was a region of farms and orchards and cultivated fields. Yamamah itself, to be more accurate, was a province rather than a place, with its capital at Hijr, which was also generally, called Yamamah. The Hijr of old stood where Riyadh stands today. 4

Musailima had no intention of letting the Muslims play havoc with the towns and villages of his people. Consequently he took his army forward to Jubaila, 25 miles north?west of Yamamah, and established his camp near Jubaila, where the plain of Aqraba began. From this location Musailima could not only defend the fertile plains of Yamamah but also threaten Khalid's route of advance, so that should Khalid blunder through the Wadi Hanifa, the Bani Hanifa would fall upon his left flank. And Khalid could not avoid battle here and proceed to Yamamah, because Musailima would then pounce upon his back. (The principle here was the same as applied by the Holy Prophet at Uhud.)

Musailima was ready for battle on the plain of Aqraba with an army of 40,000 warriors, all eager for combat. The two successful actions fought by them against Ikrimah and Shurahbil, both of whom had recoiled from the blows of Musailima, had increased their confidence in themselves and created an aura of invincibility around the Liar. His men were now prepared to sacrifice their very lives in defence of their leader and his cause. And Musailima had no doubt that he would inflict the same punishment upon Khalid as he had inflicted upon his two predecessors.

1. Tabari: Vol. 2, p. 522.
2. *Ibid*: Vol. 2, p. 509.
3. Jubaila is now a small village. According to local tradition, it was then a large town.
4. The village of Yamamah which exists about 50 miles south-east of Riyadh, near Al Kharj, is not the Yamamah of history; not the Yamamah of this battle.

A few days before the arrival of Khalid, Musailima lost one of his ablest commanders-the chief, Muja'a bin Marara, who has been mentioned as one of the important members of the Bani Hanifa delegation to the Holy Prophet. This man had set off with 40 riders to raid a neighbouring clan with which he had an old feud. On its way back from the raid, the group stopped for the night at a pass called Saniyat-ul-Yamamah, a day's march from Aqraba. Muja'a's men slept soundly, but it was their last sleep, for early in the morning the entire group was captured by one of the mounted detachments which preceded the army of Khalid. The apostates were taken before the Sword of Allah.

Khalid questioned them about their faith. In whom did they believe? In Muhammad or in Musailima? Without exception they remained unrepentant. Some sought to meet Khalid half way by suggesting: *"Let there be a prophet from among you and a prophet from among us!"* [1] Khalid was not going to waste his time on such trash, he had them all beheaded with the exception of the leader, Muja'a who was kept in chains as a prisoner. He was a prominent chief and might come in useful as a hostage. With this captive chief in tow, the Muslim army arrived near Aqraba and pitched camp as has already been described. Both armies were now ready for battle.

The actual valley of Wadi Hanifa marked the battle front. On the northern side the bank rose to about 100 feet, rising gently at places, steeply at others, and precipitously at yet others. On the southern side it rose more gently and continued to rise up to a height of 200 feet, a mile away from the valley where Khalid had pitched his camp. On the north bank also lay the town of Jubaila and on the western edge of the town a gully ran down to the wadi. The Muslim front ran along the southern bank for a length of about 3 miles, on the northern bank stood the apostates. The town and the gulley marked the centre of Musailima's army. Behind the apostates stretched the plain of Aqraba; and on this plain, about 2 miles from the *wadi*, stood a vast walled garden known as Abaz. As a result of this battle it was to become known as "The Garden of Death."

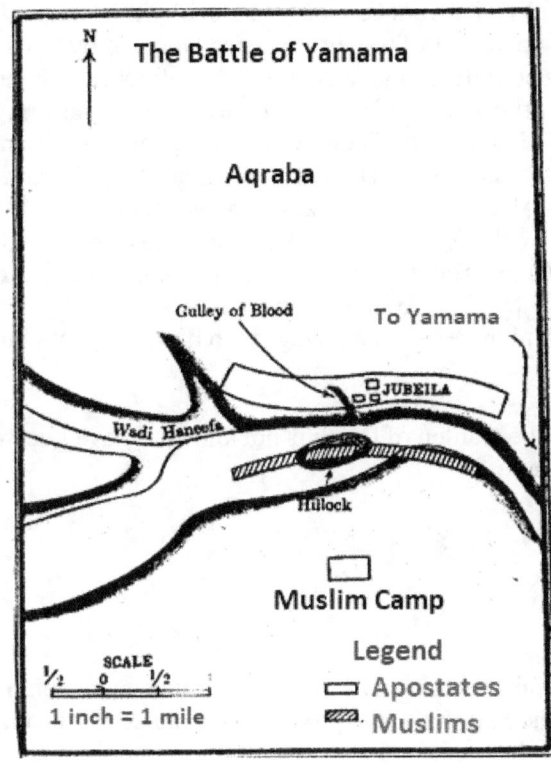

On the following morning the two armies deployed for battle. Musailima organised his 40,000 men into a centre, a left wing and a right wing. The left was under the command of Rajjal, the renegade, the right under Muhakim bin Tufail, and the centre directly under the Liar. In order to strengthen the determination of the men, the son of Musailima, also named Shurahbil, rode in front of all the regiments exhorting them to fight with courage. *"O Bani Hanifa"* he called. *"Fight today for your honour. If you are defeated your women will be enslaved and ravished by the enemy. Fight to defend your women!"* [3]

Musailima decided to await the attack of Khalid. He would fight on the defensive initially, and go on to the offensive when he had blunted the attack of his adversary and thrown him off balance.

The Muslims had spent the night in prayer. This was the largest and most fanatical enemy force they had ever faced and its commander was the most vicious and cunning of men. After the prayer of dawn Khalid drew up his 13,000 men for battle on the south bank, and he too organised his army into a centre and two wings. The left was commanded by Abu Hudeifa, the right by Zaid (elder brother of Umar), while the centre was directly under Khalid. For this battle Khalid formed his men not in tribal groups, as had been the custom heretofore, but in regiments and wings as required for battle, with tribal contingents intermingled.

Khalid planned, as was usual with him, to attack at the very outset, throw his opponent on the defensive and keep him that way. Thus Musailima would be robbed of his freedom of manoeuvre and could do no more than react helplessly to the thrusts of the attacker. But Khalid had no illusions about the trial that faced the Muslims. This was going to be a bitter and bloody battle as had never been fought before by the forces of Islam. The rebels had a numerical superiority of three-to-one and were led by a wily and brave general. But Khalid was confident of victory. He had confidence in himself and in the skill and courage of his officers and men. As he rode in front of his army, he looked with pride and satisfaction at his stalwarts. There were famous men in this army, and some who would rise to fame in later years. There was Zaid, brother of Umar, and Abdullah, son of Umar. There was Abu Dujanah, who at Uhud had shielded the Holy Prophet from the arrows of the enemy with his body. There was the Caliph's son, Abdur-Rahman. There was Muawiyah, son of Abu Sufyan, who would become the first caliph of the Umayyid Dynasty. There was Um Ammarah, the lady who had fought beside the Prophet at Uhud, with her son. And there was the Savage with his deadly javelin.

1. Tabari: Vol. 2, p. 510.
2. The exact location of the Garden of Death is not known. I have guessed its location from the course of the battle.
3. Tabari: Vol. 2, p. 509.

The officers of the Muslim army paced in front of the regiments, reciting verses of the Quran. They reminded the Faithful of the promise of paradise for the martyrs and of the threat of hell for the faint-hearted.

Early on a cold morning in the third week of December 632 (beginning of Shawal, 11 Hijri), began the Battle of Yamamah.

Khalid ordered a general attack, and the entire Muslim front surged forward with cries of *Allah-o-Akbar*. Khalid led the charge of the centre while Abu Hudaifa and Zaid led the charge of the wings. The two armies clashed and the air was rent by shouts and screams as strong men slashed and thrust at each other. Khalid cut down every man who came before him. The Muslim champions performed prodigies of valour and Khalid felt that his warriors would soon break through the army of disbelief.

But the army of disbelief stood as firm as a rock. Many fell before the onslaught of the Faithful, but there was no break in the infidel front. The apostates fought fanatically, preferring death to giving up an inch of ground; and the Muslims realised with some surprise that they were making no headway. After some time spent in hard slogging, a slight lack of order became apparent in the Muslim ranks as a result of their forward movement and their attempts to pierce the front of the infidels. But this caused no concern. So long as they remained on the offensive and the enemy on the defensive, a certain amount of disorder did not matter.

Then Musailima, realising that if he remained on the defensive much longer the chances of a Muslim break-through would increase, ordered a general counter-attack all along the front. The apostates moved forward like a vast tidal wave, and the Muslims now found to their horror that they were being pressed back. The fighting became more savage as they struggled desperately to stem he advance of the apostates, who paid heavily in blood for every yard of ground that they gained, but strengthened by their belief in the Liar's promise that paradise awaited those who fell, they pressed on relentlessly. Some lack of cohesion was now felt in the Muslim regiments due to the mixture of tribal contingents which were not yet accustomed to fighting side by side.

Gradually the numerical superiority of the apostates began to tell. Fighting in massed, compact bodies against the thinner Muslim ranks, they increased their pressure. The Muslims proceeded to fall back steadily. Then the pace of withdrawal became faster. The apostate assaults became bolder. And the Muslim withdrawal turned into a confused retreat. Some regiments turned and fled, others soon followed their example, causing a general exodus from the battlefield. The officers were unable to stop the retreat and were swept back with the tide of their men. The Muslim army passed through its camp and went on some distance beyond it before it stopped.

As the Muslims left the plain of Aqraba, the apostates followed in hot pursuit. This was not a planned manoeuvre, but an instinctive reaction, like the reaction of the Muslims to the Quraish flight in the first part of the Battle of Uhud. And like those Muslims, the apostates stopped at their opponents' camp and began to plunder it. Again as at Uhud, his opponents' occupation with looting gave Khalid time to prepare and launch a riposte. But more of that later.

In the Muslim camp stood the tent of Khalid and in this tent sat his latest wife, Laila, and the captive chief, Muja'a, still in irons. A few infidels, flushed with success and excited by thoughts of the orgy of plunder that awaited them, entered the tent of Khalid. They saw and recognised Muja'a. They saw Laila and wanted to kill her, but were restrained by the chief. *"I am her protector"*, he warned them. *"Go for the men!"* [1] In their haste to lay their hands on the booty the infidels did not stop to release their chief.

1. Tabari: Vol. 2, p. 511.

For some time the devastation of the camp proceeded at a horrible pace as the infidels snatched what they could carry and smashed what they could not. They cut the tents to shreds. Then, as quickly as it had started, the looting stopped. The apostates hastened back to the plain of Aqraba, for in the south they could see the Muslim army, formed in perfect order with solid ranks, advancing again.

Amazingly, as they stopped to regain their breath and think about what had happened, there was no fear in the hearts of the Muslims. There was only anger at their own disorganisation and the consequent retreat. Just how had this happened? How could it have happened? They had certainly inflicted greater losses on the enemy than they themselves had suffered.

Their courage remained steadfast, but they also felt baffled. Their frustrated anger found an outlet in mutual tribal recrimination-tribe against tribe, clan against clan, city against desert. They blamed each other for the debacle. *"We know more about war than you"*, said the city dwellers. *"No"*, replied the desert Arabs, *"we know more."* A clamour went up: *"Let us separate into our tribal groups. Then we shall see who vindicates his honour."* [1]

Khalid could see what had gone wrong. The apostate front had not given way under the terrible onslaught of the Muslims, as all fronts had done before this. What is more, the apostates had counter-attacked while the Muslims were somewhat disorganised. The

Muslims had lost their balance and under the pressure of the counter-attack were unable to regain it. There had been no lack of bravery.

Khalid saw that forming regiments out of mixed tribal contingents had been a mistake, for the clan feeling was still very strong among the Arabs. It added another pillar of strength to the Islamic zeal and the individual courage and skill which distinguished the Muslim army. In face of the three-to-one superiority of the enemy and the blind, fanatical determination of Musailima's followers, the absence of tribal loyalty had resulted in a weakening of cohesion in the Muslim regiments.

Khalid corrected this mistake and regrouped the army. He deployed it in the same battle formation with the same commanders, but the soldiers were now formed into clan and tribal units. Thus every man would fight not only for Islam but also for the honour of his clan. There would be healthy rivalry among the clans.

Once the reorganisation was complete, Khalid and his senior commanders went about the regiments. They spoke to the men and strengthened their resolve to punish Musailima for the disgrace that they had suffered. The men swore that if necessary they would fight with their teeth.

Khalid also picked a handful of warriors and formed them into a personal bodyguard. It was his intention to set an example for his men by throwing himself into the thick of the fighting. This bodyguard would prove useful. *"Stay close behind me"*, he told these men.

Thus reorganised and reformed into orderly ranks, the Muslims once again advanced to the plain of Aqraba. They returned to battle not like lions, but like hungry lions!

Meanwhile Musailima the Liar had redeployed his army in the same battle formation as before. He awaited the second strike of the Sword of Allah, confident that he would once again send the Muslims reeling from the battlefield.

On the orders of Khalid, the Muslim army again swept forward with cries of *Allah-o-Akbar* and the war cry of this battle: *"Ya Muhammad!"*[2] The smaller army again engaged the superior massed forces of the apostates. The wings clashed with the wings and the centre with the centre. The commander of the Muslim right, Zaid, confronted Rajjal the renegade who commanded the infidel left. Wishing to save the renegade from the fire of hell, Zaid called, *"O Rajjal! You left the true faith. Return to it. That would be more noble and virtuous."* [3] The renegade refused, and in the fierce duel that followed Zaid despatched Rajjal to the Fire.

1. Tabari: Vol. 2, p. 513.
2. This is a misunderstanding. The actual slogan, as recorded by Ibn Kathir in Al-Bidayah wan-Nihayah, Vol. 6 P. 397, was '*Ya Muhammadah!* (O for Muhammad!)', rather than '*Ya Muhammad!* (O Muhammad!)'; this is like the cry '*Ya Islamah!* (O for Islam!)'. The *Ya* here is for exclamation, not for prayer, as in the Prophet's statement (SAWS), '*Ya tuba lil-Sham!* (O joy for Syria!)', and this is further confirmed by the suffix *"ah"*. The Companions understood Islam far too well to pray to the Prophet (SAWS)!
3. Tabari: Vol. 2, p. 511.

The Muslims launched violent assaults all along the front, and the apostates were hard put to hold their ground; yet hold it they did. Their front would not break. Apostates fell in hundreds, and Muslim casualties also began to mount. With the apostates superior in numbers and the Muslims superior in skill and courage, the two sides were evenly matched. Parts of the two fronts, locked in mortal combat, heaved back and forth. The dust from thousands of stamping feet rose and hung like a cloud over the heads of the belligerents. Broken swords and spears littered the wadi and the plain as mangled and torn bodies fell in heaps on the blood-sodden earth. The most dreadful carnage took place in the gulley in which human blood ran in a rivulet down to the wadi. As a result, this gulley became known as the Gulley of Blood-*Shueib-ud-Dam*-and it is still known by that name. But the battle hung in the balance and gave no promise of a decision.

Khalid now realised that with their fanatical faith in their false prophet the apostates would not give in. It was evident that only the death of Musailima could break the spirit of the infidels, it would be a moral setback, which would lead quickly to physical defeat. But Musailima was not duelling in front like Khalid. He would have to be drawn out of the safety of the apostate ranks in which he stood surrounded by his faithful followers.

As the first violent spasm of combat spent itself, the warriors stopped to regain their breath. There was a lull. Then Khalid stepped out towards the enemy centre and threw a challenge to single combat: *"I am the son of Al Waleed! Will anyone duel?"* Several champions came out of the apostate ranks to accept the challenge of Khalid and advanced towards him one by one. Khalid took perhaps a minute to dispose of each opponent. After each duel he would recite his own extemporised verses:

> *I am the son of many chiefs.*
> *My sword is sharp and terrible.*
> *It is the mightiest of things*
> *When the pot of war boils fiercely.* [1]

Slowly and steadily Khalid advanced towards Musailima, killing champion after champion. Then there were none left brave enough to come forth against him. But by now he was close enough to Musailima to talk to him without shouting. The Liar, however, was surrounded by his guards, and Khalid could not get at him.

Khalid proposed talks. Musailima agreed. He stepped forward cautiously and halted just outside duelling distance of Khalid. *"If we agree to come to terms, what terms will you accept?"* [2] enquired Khalid.

Musailima cocked his head to one side as if listening to some invisible person who stood beside him and would talk to him. It was in this manner that he 'received revelations'! Seeing him thus reminded Khalid of the words of the Holy Prophet, who had said that Musailima was never alone, that he always had Satan beside him, that he never disobeyed Satan, and that when worked up he foamed at the mouth. Satan forbade Musailima to agree to terms, and the Liar turned his face to Khalid and shook his head.

Khalid had already determined to kill Musailima. The talks were only bait to draw him close enough. He would have to work fast before Musailima withdrew to the safety of his guards. Khalid asked another question. Again Musailima turned his head to one side, intently listening to 'the voice.' At that instant Khalid sprang at him.

Khalid was fast. But Musailima was faster. In a flash he had turned on his heels and was gone!

1. Tabari: Vol. 2, p. 513.
2. *Ibid*: Vol. 2, p. 514.

Musailima was safe once again in the arms of his guards. But in that moment of flight something meaningful happened to the spirit of the two armies, depressing one and exalting the other. The flight of their 'prophet' and commander from Khalid was a disgraceful sight in the eyes of the apostates, the Muslims rejoiced. To exploit the psychological opportunity which now presented itself, Khalid ordered an immediate renewal of the offensive.

With shouts of *Allah-o-Akbar* the Muslims again went into the attack. They fought with fresh vigour and dash, and at last victory beckoned. The apostates began to fall back as the Muslims struck with sword and dagger. The retrograde movement of the apostates gathered speed. The spirits of the Muslims rose as they redoubled their efforts. Then the infidel front broke into pieces.

Musailima could do nothing. His top commander, Rajjal, was dead. It was now the commander of his right wing, Muhakim, who came to the rescue of the apostates. *"Bani Hanifa!"* he shouted. *"The garden! The garden! Enter the garden and I shall protect your rear."*

But the disintegration of the apostates had gone too far to be halted. The bulk of the army broke and fled, scattering in all directions. Only about a fourth of Musailima's army remained in fighting shape, and this part hastened to the walled garden while Muhakim covered its retreat with a small rear-guard. This rear-guard was soon cut to pieces by the Muslims, and Muhakim fell to the arrow of the Caliph's son, Abdur-Rahman.

The Muslims now pursued the fleeing apostates across the plain of Aqraba, striking down the stragglers left and right. Soon they arrived at the walled garden where a little over 7,000 apostates, Musailima among them, had taken shelter. The infidels had closed the gate, and as they looked at the high wall that surrounded the vast garden, they felt safe and secure. Little did they know!

The major portion of the Muslim army assembled in the vicinity of the Garden of Death. It was now afternoon, and the Muslims were anxious to get into the garden and finish the job that they had started early that morning, before darkness intervened. But no way could be found into the garden. The wall stretched on all sides as an impenetrable barrier, with the gate securely bolted from within. There was no siege equipment, nor time to spend on a siege.

While Khalid searched his brain for ideas, an old warrior by the name of Baraa bin Malik, who stood in the group that confronted the gate, said to his comrades *"Throw me over the wall into the garden."* [1] His comrades refused, for Baraa was a distinguished and much-respected Companion, and they hesitated to do something which would certainly result in his death. But Baraa insisted. At last his comrades agreed to his request and lifted him on their shoulders near the gate. He got his hands onto the edge of the wall, swung himself up and jumped into the garden. In a minute or so he had killed two or three infidels who stood between him and the gate, and before others could intercept him, he had loosened the heavy bolt. The gate was flung open and a flood of Muslims roared through it like water thundering through a breach in a dam. The last and most gory place of the Battle of Yamamah had begun.

Initially the infidels were able to contain the advance of the Muslims, who were confined by the gate to a narrow front and lacked elbow-room. But steadily the Muslims cut their way through the apostates, who began to fall in heaps under the attacker's blows. The apostates stepped back as the Muslims poured into the garden in ever-increasing numbers.

The fighting became more vicious. Since there was no room for manoeuvre, both sides engaged in a straight slogging match. Gradually the ranks of the apostates thinned as they fell in combat. But Musailima was still fighting: he had no intention of giving up. As the front moved closer to him, he drew his sword and joined in the combat, surprising the Muslims by his strength and dexterity. The wily general was also a brave and skilful fighter. He began to foam at the mouth, for desperation had turned the ugly impostor into an awesome demon.

The last phase of the battle now entered its climax. The Muslim army pressed the apostates everywhere and it was only the endeavours of Musailima which prevented a general collapse. The Muslims cut, slashed and stabbed with wild fury. Maimed and mutilated bodies covered the ground. Those who fell suffered a painful death under the trampling feet of those who would not give in. The carnage was frightful and the dust on the ground turned into red mud.

1. Tabari: Vol. 2, p. 514.

Many apostates ran in despair to Musailima. *"Where is the victory that you promised?"* they asked. *"Fight on, O Bani Hanifa!"* was the impostor's set answer. *"Fight on till the end!"* [1]

Musailima knew that he would get no quarter from Khalid, that he was doomed, and evil genius that he was, he decided to take his tribe down with him. The blood of several Muslims dripped from his sword, and his guards, as fanatical as ever, fought around him. Then he came under the hawk-like gaze of the Savage. The Savage was one of the 'war criminals' whose names had been announced by the Holy Prophet on the eve of the conquest of Makkah. Fearing the worst, he had fled Makkah and gone to Taif, where he lived among the Thaqeef for some time. In 9 Hijri, when the Thaqeef submitted to the Prophet, he too embraced Islam and went personally to swear allegiance to the Prophet.

The Prophet had not seen him for many years, and was not certain if he was the man. **"Are you the Savage?"** he asked.

"Yes, O Messenger of Allah!"

"Tell me how you killed Hamza." [2]

The Savage recounted the whole story from beginning to end. It never occurred to him that there was an ethical angle too to this episode, that he had killed one of the noblest and most gallant of the Faithful. He narrated the story as a proud veteran would regale his audience with tales of his daring exploits. And the killing of a matchless warrior like Hamza was undoubtedly a military achievement. The Savage excelled himself as a story-teller.

But there was no applause. On the face of the Prophet was a look of deep sorrow as he said *"Never let me see you again."* [3] Something inside him warned the Savage that to remain in Madinah, where the memory of Hamza was deeply cherished, might be unhealthy for him. He left at once.

For the next two years he lived in various settlements around Taif, seeking obscurity and avoiding travellers. He was troubled by his conscience and feared for his life. It was a wretched existence. Then came the apostasy. The Savage remained loyal to his new faith and elected to fight for Islam against the unbelievers. Now he was serving under the banner of the Sword of Allah.

The Savage tightened his grip on his javelin when he saw Musailima-the javelin that had sent so many men to their death. The Liar was fighting ferociously. In beating off the assaults of Muslims who strove to get to him, he would fight now in front of his guards, now amongst them. At times he was covered by his guards, but he was never lost to the unblinking gaze of the black killer. The Savage had chosen his next victim-one whose death might ease the gnawing pain in his heart.

From his position some distance behind the Muslim front, the Savage stealthily moved forward to get within javelin range of his target. The throng of swearing, sweating, blood-covered warriors around Musailima seemed to disappear from his sight. In the terrible mind of the Savage only his victim remained.

1. Tabari: Vol. 2, p. 514.
2. Ibn Hisham: Vol. 2, p. 72.
3. *Ibid*.

The Savage saw Um Ammarah, the grand lady of Uhud (though at this moment there was nothing ladylike about her appearance or actions), struggle to get to Musailima. She was duelling with an infidel who barred her way. Suddenly the infidel struck at her and cut off her hand. Her son, who stood next to her, felled the infidel with one mortal blow and helped his mother away. She was heart-broken at being unable to get to Musailima.

The Savage moved closer. In his mind appeared a vision of the noble martyr of Uhud, Hamza, whose killing had been the cause of all his troubles. He could picture the fine, strong, handsome features of Hamza. With an effort he drove the memory of that painful episode from his mind and looked again at Musailima. He was shocked at the contrast. The ugly, yellow, flat-nosed face of the impostor, distorted with rage and hate, with foam discolouring his mouth, was a frightening sight. All the evil in this demoniac man seemed to have come out on his face.

With a practised eye the Savage measured the distance. The range was just right. As he poised for the throw and aimed his javelin, he noticed Abu Dujanah (the human shield of the Prophet at Uhud) slashing away with his sword to get to Musailima. Abu Dujanah was a superb swordsman and would soon reach his objective. With a grunt the Savage hurled his weapon.

The javelin struck Musailima in the belly. The false prophet fell, his face twisted with pain, his hands clawing at the shaft. The next moment Abu Dujanah was upon him. With one neat stroke of his sword he severed the evil head of the Liar. As Abu Dujanah straightened up to announce the good news, a flashing infidel sword struck him down. One apostate, looking at the Liar, shouted, *"A black slave has killed him."* The cry was taken up by Muslim and infidel and rang across the garden: *"Musailima is dead!"* [1]

The Savage later served in the Syrian Campaign under Khalid. When Syria had been conquered and established as a province of the Muslim State, the Savage settled down at Emessa and lived to a ripe old age. But he spent most of his days in a drunken stupor. He was even awarded 80 stripes by Umar for drinking (he was the first Muslim to be punished for this offence in Syria), [2] but refused to stay away from the bottle. Umar gave up, with the philosophical remark, *"Perhaps the curse of Allah rests on the Savage for the blood of Hamza."* [3]

In Emessa, in later years, the Savage became a famous figure and a tourist attraction. Visitors would go to his house, hoping to find him sober, and ask him about Hamza and Musailima. If sober, he would recount in detail first the killing of Hamza and then the killing of Musailima. Coming to the end of his story, he would raise his javelin with fierce pride and say, *"With this javelin, in my days of unbelief I killed the best of men, and in my days of belief I killed the worst!"* [4]

The news of the death of Musailima the Liar brought about a rapid collapse of the apostates. Some turned in suicidal desperation to greater violence, but they could only prolong their agony, not save their lives. Most of the apostates ceased to struggle, and in total despair waited for a Muslim sword to end their suffering. With one last superhuman effort the Muslims charged into the confused, helpless mass of apostates, and with their swords fulfilled the promise of the wrath of Allah against the unbelievers. Now it was no longer a battle, it was plain slaughter.

By the time the sun set, peace and quiet had returned to the Garden of Death. The Muslims were too tired to raise their swords. And there was no one left to kill.

1. Ibn Hisham: Vol. 2, p. 73.
2. Ibn Qutaiba: p. 330.
3. Ibn Hisham: Vol. 2, p. 73.
4. *Ibid.*

For the night the Muslims dropped where they stood, and escaped from the nightmare of the battle into the sleep of the victorious.

Next morning Khalid walked about the battlefield. Everywhere he saw the wreckage of battle. Broken, twisted bodies, lying in grotesque shapes, littered the wadi and the plain of Aqraba and the Garden of Death. In places he picked his way over blood-soaked earth.

All the important leaders of the apostasy in Yamamah had been killed-all save the captive Muja'a who now, still in irons, dragged his feet beside the victor. Khalid had taken him along so that he could identify some of the dead leaders and also feel the full impact of the defeat of the Bani Hanifa.

The state of the Muslims too was appalling. The battle had taken a heavy toll, and right now they were in no condition even to defend themselves, let alone fight a battle. Exhausted and worn-out, they lay where they had dropped the night before, resting their weary limbs. But Khalid had reason to be satisfied with the outcome of the battle: Musailima was dead and his army had been torn to pieces. A glow of pleasure warmed the heart of Khalid. But Muja'a soon dispelled it.

"Yes, you have won a victory", he conceded. *"But you should know that you have fought only a small portion of the Bani Hanifa-all that Musailima could hastily gather. The major portion of the army is still in the fort at Yamamah."*

Khalid stared at him incredulously, *"May Allah curse you! What are you saying?"*

"Yes, that is so", Muja'a went on. *"I suggest that you accept a peaceful surrender. If you will state your terms, I shall go into the fort and try to persuade the army to lay down its arms."*

It did not take Khalid long to realise the impossibility of fighting, with his exhausted men, an even larger army than the one he had just tackled. *"Yes"*, he replied, accepting Muja'a's proposal. *"Let there be peace."*

The terms of surrender were worked out by the two leaders. The Muslims would take all the gold, the swords, the armour and the horses in Yamamah, but only half its population would be enslaved. Muja'a was released from his fetters and, on giving his word to return, allowed to proceed to the fortified city. After some time he returned, shaking his head sadly. *"They do not agree. They are all set to fight. In fact they turned against me. You can attack now if you wish."*

Khalid decided to take a look at the city himself. Leaving the bulk of his weary army to bury the martyrs and gather the spoils, he took a mounted detachment and rode to Yamamah, accompanied by Muja'a. As he got near the northern wall of the fortified city he stopped in amazement, for the battlements were crowded with warriors whose armour and weapons glinted ominously in the sun. How on earth would he deal with this fresh army in an impregnable fort? His men were in no state to fight, they wanted nothing but rest.

The voice of Muja'a broke the silence. *"They might be prepared to surrender the fort if you do not enslave any of them. You could have all the gold, the swords, the armour, the horses."*

"Have they agreed to this?" asked Khalid.

"I have discussed the matter, but they gave no decision."

Khalid was prepared to go so far and no further. He looked at Muja'a sternly. *"I will give you three days"*, he said. *"If the gates are not opened on these revised terms, I shall attack. And then there shall be no terms of any kind."*

Muja'a again went into the fort. This time he returned smiling. *"They have agreed."* [1] he announced.

1. Tabari: Vol. 2, pp. 515-7; Balazuri: pp. 99-100.

The pact was drawn up accordingly. It was signed on behalf of the Muslims by Khalid and on behalf of the Bani Hanifa by Muja'a bin Marara. [1]

When the pact had been signed, Muja'a returned to the fort, and soon after the gates of the fort were thrown open. Khalid, accompanied by his mounted warriors and Muja'a, rode into the city, expecting to see hordes of armed warriors, but wherever he looked, he saw nothing but women and old men and children. He turned in amazement to Muja'a. *"Where are the warriors I saw?"*

Muja'a pointed at the women. *"Those are the warriors you saw"*, he explained. *"When I came into the fort I dressed these women in armour, gave them weapons, and made them parade on the battlements. There are no warriors!"*

Furious at being tricked, Khalid swore at Muja'a, *"You deceived me, O Muja'a!"*

Muja'a merely shrugged his shoulders. *"They are my people. I could do nothing else."*

But for the pact, Khalid would have torn Muja'a apart with his bare hands. However, the pact had been signed and its terms had to be respected. The Bani Hanifa, those of them who were in the city, were safe. Soon they had come out of their city roamed freely in the neighbourhood.

A day or two later a message arrived from the Caliph, who was not yet aware of the end of the Battle of Yamamah, instructing Khalid to kill all the apostates of the Bani Hanifa. Khalid wrote back explaining that the Caliph's order could not be implemented because of the pact that he had signed. Abu Bakr agreed to the observance of the terms of the pact.

But the pact only applied to those who had been in the fort. The rest of the vast tribe of Bani Hanifa-tens of thousands of people living in the region around Yamamah-were not covered by the pact. The most important element of the Bani Hanifa now was the remnants of the army of Musailima which had fled from the plain of Aqraba. These warriors, amounting to more than 20,000 men, were moving at random in clans and groups. After Musailima's death they posed no great danger to Islam, but they could nevertheless cause considerable mischief. They had to be crushed. Under the harsh laws of war, they had no claim to immunity from attack until they had fully submitted.

Khalid was determined to wipe out all resistance among the Bani Hanifa so that undisturbed peace might prevail in the region. He allowed his army a couple of day's rest: then he divided it into several columns which he despatched to subdue the region around Yamamah and to kill or capture all who resisted. These columns fanned out in the countryside.

The fugitives were sought out wherever they had taken shelter. Thousands remained unrepentant and defiant, these were attacked and wiped out, and their women and children taken captive. But other thousands submitted and were spared. Eventually all the survivors re-entered Islam.

Khalid set up his headquarters near Yamamah, where he was to stay about two months before receiving his next military task from the Caliph.

With the successful conclusion of the Battle of Yamamah, most of Arabia was freed of the mischief of the apostasy. Some of it still remained on the fringes of the peninsula, but this posed no serious threat. Some battles were still to be fought, but they were minor affairs compared with the great clashes which have been described in this and the preceding chapters.

1. There is some difference of opinion among early historians about the exact terms of the pact, but the details are not important.

The Battle of Yamamah was the fiercest and bloodiest battle so far fought in the history of Islam. Never before had the Muslims been faced with such a trial of strength, and they rose gloriously to the occasion under the leadership of the Sword of Allah. By crushing the vastly superior forces of the Bani Hanifa led by the redoubtable Musailima, the Muslims proved themselves to be men of steel. Half a century later old men would describe this battle in vivid detail to their grandchildren and end the account with the proud boast: *"I was at Yamamah!"*

The casualties were staggering. Of the apostates 21,000 were killed-7,000 in the plain of Aqraba, 7,000 in the Garden of Death, and 7,000 in the mopping up operations of the columns sent out by Khalid.

The Muslims suffered lightly in comparison with the apostates, but compared with their own past battle losses, their casualties were heavy indeed. Twelve hundred Muslims fell as martyrs-most of them in or near the *wadi*. [1] Half this loss was suffered by the Ansars and the Emigrants-the closest and most revered Companions of the Prophet. It is also said that the martyrs included 300 of those who knew the whole Quran by heart. Some of the finest of Muslims fell in this battle-Abu Dujanah, Abu Hudaifa (the commander of the left wing), Zaid (brother of Umar and commander of the right wing). While Zaid fell, Umar's son, Abdullah, survived.

When Abdullah returned to Madinah he went to pay his respects to his father, but there was no welcome in the eyes of Umar as he looked at his son. *"Why were you not killed beside Zaid? Zaid is dead and you live! Let me not see your face again!"*

"Father", pleaded this brave young man, *"my uncle asked for martyrdom and Allah honoured him with it. I also sought martyrdom but did not attain it."* [2]

In the Battle of Yamamah, Abu Bakr's campaign against the apostates reached its high-water mark. This was the climax. Abu Bakr's strategy of using Khalid as his right arm to fight the main apostate chiefs in turn, going from nearer to farther objectives, had met with admirable success. Henceforth things would be easier.

One episode remains to be narrated before we finish with the Battle of Yamamah. On the day that the city of Yamamah opened its gates, Khalid sat outside his tent in the evening. Beside him sat Muja'a. They were alone.

Suddenly Khalid turned to Muja'a. "*I want to marry your daughter!*"

Muja'a stared in amazement at Khalid. He could not possibly have heard right!

Khalid, his tone more insistent, repeated, *"I want to marry your daughter!"*

Muja'a now realised that Khalid was not mad, that he knew what he wanted. Yet in view of the occasion, the whole idea seemed utterly ridiculous. *"Steady, O Khalid!"* he replied. *"Do you want the Caliph to break your back and mine also?"*

"I want to marry your daughter", repeated Khalid. And that very evening he married the beautiful daughter of Muja'a bin Marara.

A few days later Khalid received an angry letter from Abu Bakr. *"O son of the mother of Khalid!"* wrote the Caliph. *"You have time to marry women while in your courtyard the blood of 1,200 Muslims is not yet dry!"* When he had read the letter Khalid muttered, *"This must be the work of that left-handed one!"* [3]

However he continued to enjoy his new bride. It seems that he had discarded the glamorous widow of Malik bin Nuwaira. We do not know what happened to that lady, for history makes no further mention of the beautiful Laila with the gorgeous eyes and the lovely legs.

1. The visitor to Jubaila today is shown a graveyard on the southern bank of the wadi where the Muslim martyrs lie buried, and on the northern bank he is shown a low mound between the village and the gully, where the apostate dead were buried.
2. Tabari: Vol. 2, pp. 512-3.
3. Tabari: Vol. 2, p. 519.

Chapter 17

(The Collapse of the Apostasy)

"There was not born to Adam among his descendants anyone better than Abu Bakr, after the Prophets and Messengers. Abu Bakr took a stance in the Days of Apostasy which was like the stance of one of the Prophets."
[Abu Husain][1]

What remained of the apostasy in the less vital areas of Arabia was rooted out by the Muslims in a series of well planned campaigns within five months.

Amr bin Al Aas had been sent with his corps to the Syrian border to subdue the apostates in that region. The most important tribes that needed chastisement were the Quza'a and the Wadi'a, the latter being a section of the large tribe of Kalb. While Khalid was fighting in Central Arabia, Amr struck at the apostates in the north, but achieved only limited success. He was not able to beat the tribes into submission.

When the Battle of Yamamah was over, Shurahbil bin Hasanah proceeded with his corps, on the orders of the Caliph, to reinforce Amr, and the two commanders operated in unison to bring about the subjugation of the northern tribes. Most of the apostates were concentrated in the region of Tabuk and Daumat-ul-Jandal, and it was here that Amr and Shurahbil struck their hardest blows. In a few weeks the apostasy was destroyed and the tribes re-entered Islam. Peace returned to Northern Arabia.

The main tribe inhabiting Oman was the Azd. The chief of this tribe was Laqeet bin Malik, known more commonly as Dhul Taj, the Crowned One. These Arabs, like those whose apostasy is described later in this chapter, had embraced Islam in the time of the Prophet and agreed to abide by the terms imposed by the Muslim State.

On hearing the news of the Holy Prophet's death, the bulk of the Azd, led by Dhul Taj, revolted and renounced Islam. It is not certain that this man was an impostor. Going by a brief comment of Tabari that he "claimed what prophets claim", [2] we could assume that he probably did make some claim to prophethood. Be that as it may, while Abu Bakr was busy dealing with the immediate threat to Madinah, Dhul Taj declared himself King of Oman and established himself as its undisputed ruler with his headquarters at Daba.

After Khalid had left Zhu Qissa to seek Tulaiha, the Caliph despatched Hudaifa bin Mihsan (one of the corps commanders) to tackle the apostasy in Oman. Hudaifa entered the province of Oman, but not having strong enough forces to fight Dhul Taj, he decided to await reinforcements. He wrote to the Caliph accordingly, who, as has already been noted, instructed Ikrimah to march from Yamamah to the aid of Hudhaifa. On his arrival, the two generals combined their forces and set out to fight Dhul Taj at Daba.

The Battle of Daba was fought towards the end of November 632 (early Ramadan, 11 Hijri). At first the battle went badly for the Muslims; but at a critical moment a force of local Muslims, who had clung to their faith in spite of Dhul Taj, appeared on, the battlefield in support of their co-religionists. With this fresh addition to their strength the Muslims were able to defeat the infidel army. Dhul Taj was killed in battle.

Being appointed governor of Oman, Hudaifa next set about the re-establishment of law and order. Ikrimah, having no local administrative responsibility, used his corps to subdue the neighbourhood of Daba; and in a number of small actions, succeeded in breaking the resistance of those of the Azd who had continued to defy the authority of Islam. Thereafter the Azd once again became peaceful, law-abiding Muslims and gave no further trouble to Madinah.

1. *Tarikh Al-Khulafaa* of As-Suyuti
2. Tabari: Vol. 2, p. 529.

From Oman, following the orders of Abu Bakr, Ikrimah marched to Mahra. Here too the germs of apostasy had infected the local population, though not in such virulent a form as in some other provinces. Mahra actually was the objective of Arfaja bin Harsama (one of the corps commanders) and Ikrimah's instructions were to assist Arfaja; but as the latter had not yet arrived, Ikrimah decided that instead of waiting for him he would tackle the local apostasy on his own.

The army of local rebels that had gathered at Jairut consisted of two unequal factions. Ikrimah arrived at Jairut and confronted the infidels in early January, 633 (mid-Shawal, 11 Hijri). When ready to engage the enemy, he called upon the apostates to return to the fold of Islam. Of the two apostates factions, the larger rejected the call, but the smaller one accepted it and came over to join the Muslims, whereupon Ikrimah attacked and defeated the rebels. Their commander was killed, and a large quantity of booty came into Ikrimah's hands.

Having re-established Islam in Mahra, Ikrimah moved his corps to Abyan, where he rested his men and awaited further developments.

In Bahrain an independent action against the rebels was fought by the corps of Ala bin Al Hadhrami. It was after the Battle of Yamamah that Abu Bakr had sent this general to crush the apostasy in Bahrain, telling him that he would get no help from other Muslim forces, that he would be entirely on his own.

Ula arrived in Bahrain to find the apostate forces gathered at Hajr and entrenched in a strong position. (This was the only instance of entrenchment being used in these campaigns.) Ula mounted several attacks and the battle continued for some days but without success, as he found it difficult to cross the trench line. Whenever he managed to get some forces across they were repulsed. Ula began to wonder just how he was going to crack this virtually impregnable position.

Then early one night Ula heard wild, joyous shouts coming from the rebel position. At a loss to understand this phenomenon, he sent spies to investigate. These spies returned soon after to inform him that there was wild revelry in the enemy camp and that everybody was drunk. Ula at once ordered a night attack. As the Muslims went into the assault they found no sentries and caught the enemy completely by surprise. They plunged into the rebels, and hundreds of them were killed before they realised that their celebration had been disturbed! Hundreds, more were slain before the rest could come to their senses and escape.

The following day Ula pursued the fugitives to the coast, where they made one more stand but were decisively defeated. Most of them surrendered and re-entered Islam.

This operation was completed at about the end of January 633 (second week of Dhul Qad, 11 Hijri).

The Yemen had been the first province to rebel against the authority of Islam when the tribe of Ans rose in arms under the leadership of its chief and false prophet-Aswad, the Black One. The affair of Aswad has already been described. He was killed by Fairoz the Persian, while the Holy Prophet still lived, and thereafter Fairoz had acted as governor at San'a.

When word arrived that the Holy Prophet had died, the people of the Yemen again revolted, this time under the leadership of a man named Qais bin Abd Yaghus. The avowed aim of the apostates was to drive the Muslims out of the Yemen, and they decided to achieve this objective by assassinating Fairoz and other important Muslim leaders, thus rendering the Muslim community leaderless. Its subsequent expulsion would then, present no difficulty.

To implement this perfidious plan Qais invited Fairoz and other Muslim officers to his house for talks. Some Muslims fell into the trap and were speedily despatched by the assassins; but at the eleventh hour Fairoz got wind of the plot and of the organisation behind it. Having no military force at his disposal for immediate use, Fairoz sought safety in flight. He left San'a. Qais came to know of his departure and pursued him, but Fairoz was able to evade his pursuers and reach the hills where he found a safe refuge. This happened in June or July 632 (Rabi-ul-Awwal or Akhir, 11 Hijri).

For the next six months Fairoz remained in his mountainous stronghold, where over the months he was joined by thousands of Muslims who were prepared to shed their blood to oust Qais and restore Muslim rule in the Yemen. Fairoz organised these Muslims into an army. When he felt strong enough to face Qais in the field, he marched to San'a with this army. Qais awaited him here, and in mid-January 633 (late Shawal, 11 Hijri) they joined battle just outside the town. The Muslims were victorious, and Qais fled to Abyan, where Ikrimah was to rest later, after subduing Mahra.

At Abyan, Qais was joined by other apostate chiefs, but they fell out amongst themselves. Seeing no hope of further successful opposition to Madinah, they all surrendered to the Muslims and were subsequently pardoned by the Caliph. Some of these apostate chiefs, after re-entering Islam, fought bravely in Iraq and Syria during the years that followed.

The last of the great revolts of the apostasy was that of the powerful tribe of Kinda, which inhabited the region of Najran, Hadhramaut and Eastern Yemen. The progress of events in this revolt followed much the same pattern as elsewhere.

On the death of the Prophet, the Kinda became restive, though they did not break into revolt immediately. The governor of Hadhramaut was Ziyad bin Lubaid who lived at Zafar, the capital of Hadhramaut. An honest, Allah-fearing Muslim, he was extremely strict in the collection of taxes, which caused some heart-burning among the Kinda. All their attempts at evading full payment of taxes were thwarted by Ziyad.

In January 633 (Shawal, 11 Hijri), the discontent of the Kinda came to a head. One of their minor chieftains had handed in a rather fine camel as part of the tax. He later changed his mind and asked to have it back but Ziyad rejected the request. This chieftain then sent some of his men to steal the camel.

In return Ziyad sent a few soldiers to catch the camel-lifters. Shortly afterwards the camel and the culprits were brought in and locked up. Next morning a riotous assembly of the Kinda demanded the return of their imprisoned comrades. Ziyad refused to release the thieves, announcing that they would be tried under Muslim law. At this the situation exploded.

Large sections of the Kinda revolted and apostatised. They not only refused to pay taxes or abide by the laws of Islam but also took up arms to oppose the authority of Madinah with violence. Several other dissident elements joined them in this purpose, and together they established military camps and prepared for war.

One of these rebel camps was at Riyaz, not far from Zafar. To this Ziyad sent a column on a night raid which turned out to be eminently successful. Some apostates were killed, several captured, and the rest driven away. As the captives were being taken to War, they passed the greatest of the Kinda chiefs, Ash'as bin Qais, who had not yet turned apostate. *"O Ash'as"*, the captives called to him, *"We are of your mother's clan."* The tribal loyalty of Ash'as proved itself stronger than his faith or his respect for central authority. Accompanied by many of his warriors, he intercepted the Muslim column, liberated the captives, and sent the Muslims home empty-handed.

This marked the beginning of the revolt of Ash'as. The Kinda flocked to his standard in large numbers and prepared for battle, but the strength of the two forces, apostate and Muslim, was so well balanced that neither side felt able to start serious hostilities. Ziyad waited for reinforcements before attacking Ash'as.

1. Tabari: Vol. 2, p. 545.

Reinforcements were on the way. Muhajir bin Abi Umayyah, the last of the corps commanders to be despatched by Abu Bakr, had just subdued some rebels in Najran and was to go on to the Yemen. Abu Bakr directed him to proceed instead to Hadhramaut to join Ziyad and deal with the apostasy of the Kinda. Similar instructions reached Ikrimah, who was now at Abyan.

The forces of Muhajir and Ziyad combined at Zafar, under the overall command of the former, and set out to fight Ash'as.

Ash'as bin Qais was one of the most remarkable men of his time. Coming from a princely family of the Kinda, he was a man of many parts. An able general, a clever chief, a bold warrior and an accomplished poet, he had a fertile imagination and a smooth tongue. A man of charm and wit, he was the most colourful of the many colourful personalities thrown up by the apostasy. But he had one big flaw; he was treacherous! Historians have noted that his was the only family that produced four breakers of pacts in an unbroken line, Ash'as, his father, his son, and his grandson.

Ash'as lived close to the borderline between virtue and evil, between faith and unbelief, but never quite crossed that line. Practising a kind of moral and spiritual brinkmanship, he was clever enough to get away with it. And now, in late January 633 (the second week of Dhul Qad, 11 Hijri), he faced the Muslim army in battle.

The battle did not last long. Ash'as was defeated, though the defeat was not decisive. He speedily withdrew his army from the battlefield and retreated to the fort of Nujair, where he was joined by other dissident clans. Here Ash'as prepared for a siege.

Just after this battle the corps of Ikrimah also arrived. The three Muslim corps, under the over-all command of Muhajir, advanced on Nujair and laid siege to the fortified city. There were three routes leading into the city. The generals deployed their forces on all three of these routes, completely surrounding and isolating the city. Reinforcements and provisions coming to Ash'as were either captured or driven back.

The siege continued for several days. A number of sallies were made by the beleaguered garrison, but all were repulsed with losses. Yet the Kinda remained firm in their determination to fight on.

Some time in mid-February 633 (early Dhul Hajj, 11 Hijri) Ash'as realised that the situation was hopeless. There was no possibility of success. It was only a matter of time before the fort fell to the Muslims, and then there would be a blood-bath. The next action of Ash'as was characteristic of the man: he decided to sell his tribe to save himself!

He sent a message to Ikrimah proposing talks. Ash'as knew Ikrimah well, for in their days of unbelief they had been good friends. As a result of the proposal talks were arranged with Ikrimah and Muhajir on one side and Ash'as on the other. Accompanied by a few men, Ash'as came out of the fort secretly to the rendezvous.

"I shall open the gates of the fort to you if you will spare the lives of 10 men and their families", Ash'as offered. To this the Muslims agreed. *"Write down the names of the 10 men,"* said Muhajir, *"and we shall seal the document."*

Ash'as went aside with his men and began to write down the names. It was his intention first to write the name of nine favoured ones and then add his own as the tenth, but he did not notice that one of his men was looking over his shoulder and reading the names as he wrote. This man, named Jahdam, was not one of the favoured nine. As Asha's wrote the ninth name, Jahdam drew his dagger. *"Write my name,"* he hissed, *"or I kill you."* Hoping to save himself later by his wits, Ash'as wrote down Jahdam as the tenth name. The list was complete. Muhajir sealed the document.

1. Tabari: Vol. 2, p. 547.

Ash'as and his men returned to the fort. At the agreed time he opened one of the gates, and the Muslims poured into the fort and fell upon the unsuspecting garrison. There was a terrible slaughter, and it continued until everyone in the fort had laid down his arms. Ash'as and a group of men and families that remained near him were scared.

The fort of Nujair had now fallen. When Muhajir checked the list prepared by Ash'as, he noticed that the name of Ash'as was not in it. He was delighted. *"O Enemy of Allah!"* he said to Ash'as. *"Now I have a chance to punish you."* [1] He would have killed Ash'as, but Ikrimah intervened and insisted that Ash'as be sent to Madinah, where Abu Bakr could decide his fate. Consequently Ash'as was put in chains.

Within the fort the Muslims had taken many captives who were to be sent to Madinah as slaves, and these included a large number of attractive young women. They were led out of the fort and passed Ash'as, of whose perfidy they had by now come to know. As they slowly filed past him, the captive women looked at him reproachfully and wailed, *"You traitor! You traitor!"* [2] To add to his discomfiture, Ash'as was sent with this same group of captives to Madinah. It could not have been a very pleasant journey!

Ash'as was no stranger to Madinah. He had visited the place during the Year of Delegations, when the Kinda submitted to the Holy Prophet and embraced Islam. During that visit he had also married Um Farwa, sister of Abu Bakr, but when leaving Madinah he had left her behind with Abu Bakr, with the promise of picking her up on his next visit. This next visit was now taking place under very different and uncongenial circumstances.

The Caliph charged Ash'as with all his crimes against Islam and the State. He expressed his low opinion of the way Ash'as had betrayed his own tribe. Was there any reason why the accused should not be beheaded at once?

Attention has been drawn to the smooth tongue of Ash'as. This time he excelled himself. Not only did he win a pardon, he also persuaded Abu Bakr to return his wife to him! He remained in Madinah, though, unwilling to return to his own tribe. In later years he fought with distinction in Syria, Iraq and Persia, and in the time of Uthman he was made governor of Azerbaijan.

But his treacherousness never left him. Many were the people, including Abu Bakr, who wished that he had not been pardoned after his apostasy. In fact when Abu Bakr was dying, and spoke to his friends of his regrets about things he had not done and wished he had, he said, *"I wish I had had Ash'as beheaded."* 3

Students of Muslim history might recollect that Imam Hassan's wife, who poisoned him 4 at the instigation of Caliph Muawiyah, for which service he paid her 100,000 dirhams, was the daughter of Ash'as. 5

With the defeat of the Kinda at Nujair the last of the great apostate movements collapsed. Arabia was safe for Islam. The unholy fire that had raged across the land was now dead. Arabia would see revolt and civil war many times in its stormy history, but it would never again see apostasy.

The Campaign of the Apostasy was fought and completed during the eleventh year of the Hijri. The year 12 Hijri dawned, on March 18, 633, with Arabia united under the central authority of the Caliph at Madinah. 6

This campaign was Abu Bakr's greatest political and military triumph. Although the Caliph would launch bold ventures for the conquest of Iraq and Syria, it was by his able and successful conduct of the Campaign of the Apostasy that he rendered his greatest service to Islam. And this would not have been possible without the arm of the Sword of Allah.

1. Tabari: Vol. 2, p. 548.
2. *ibid*.
3. Tabari: Vol. 2, p. 619, Masudi: *Muruj*, Vol. 2, p. 308; Balazuri: p. 112
4. Qadi Abu Bakr ibn al-'Arabi says in *Al-'Awasim min Al-Qawasim* (Al-Maktabah al-'Ilmiyyah, Beirut, pp. 213-4), *"If it is said that Muawiyah intrigued against Al-Hassan in order to poison him. We replied that this is impossible for two reasons. One of them is that he did not fear any power of Al-Hassan once the latter had surrendered authority. The second is that this is an unseen matter which only Allah knows. How can you state it without proof and accuse any of His creatures in a distant time when we do not have any sound transmission about it? Moreover, this occurred in the presence of the people of sects who were in a state of sedition and partisanship. Each of them ascribed what he should not ascribe to his companion. Only the pure is accepted from it. Only the resolute just man is listened to regarding it."*

Muhibb al-Din al-Khatib comments on the above as follows, *"In the Minhaj as-Sunnah (2:225) Ibn Taymiyya spoke about the Shi'a claim that Mu'awiyah had poisoned Al-Hassan, 'That was not established by any clear proof in the Shariah nor by any plausible statement nor by a clear transmission. This is part of what it is not possible to know.*

Hence this saying is a statement without knowledge ... In our time, we saw people among the Turks and others who said that he was poisoned and died of poisoning. People disagree about that and even the place where he died and the fortress in which he died. You will find each of them relating something different from what the other people related.' After Ibn Taymiyyah mentioned that Al-Hassan died in Madinah while Muawiyyah was in Syria, he mentioned the possibilities of the report, assuming it to be sound. One of them is that Al-Hassan was divorced and did not remain with a wife ..." (see also *Al-Muntaqa*, [al-Dhahabi's abridgement of the Minhaj], p. 266)" See also *Defence against Disaster*, Madinah Press, 1416/1995, pp. 203-4.

A report mentioned by As-Suyuti in *Tarikh Al-Khulafaa* says that Yazid bin Muawwiyah bribed Al-Hassan's wife to poison him by offering to marry her in return. This illustrates the conflicting and contradictory nature of these reports, which are almost certainly Shi'ite fabrications. Ibn Hajr Al-Asqalani merely says in his biography of Al-Hassan in *Al-Isabah*, "It is said that he was poisoned to death."

5. Ibn Qutaiba: p. 212; Masudi: *Muruj*, Vol. 3, p. 5. This is Masudi's figure. Some historians have given the sum as 150,000 dirhams.
6. For an explanation of the chronology of the Campaign of the Apostasy, which is subject to some possible sources of error, see Note 3 in Appendix B.

Part III: The Invasion of Iraq

Chapter 18

(The Clash with Persia)

"Say to the desert Arabs who lagged behind: 'You shall be summoned against a people given to vehement war: you shall fight them, or they shall submit.
Then if you show obedience, Allah will grant you a goodly reward, but if you turn back as you did before, He will punish you with a painful Punishment."
[Quran 48:16]

The fort of Nujair, the last stronghold of apostasy, had fallen to the Muslims in about the middle of February 633. Soon after, Abu Bakr wrote to Khalid, who was still at Yamamah: *"Proceed to Iraq. Start operations in the region of Uballa. Fight the Persians and the people who inhabit their land. Your objective is Hira."* [1]

It was a big order. Abu Bakr was taking on the mightiest empire of the time, before which the world had trembled for more than a thousand years.

The Persian Empire was unique in many ways. It was the first truly great empire of history, stretching, in the time of the early Achaemenians, from Northern Greece in the west to the Punjab in the east. It was unique also in the length of time over which it flourished-from the Sixth Century BC to the Seventh Century AD, except for a gap caused by the Greek conquest. [2] No other empire in history had lasted so long in all its greatness as a force of culture and civilisation and as a military power. It had known reverses, but after each reverse it had risen again in its characteristic glory and brilliance.

The last golden age of Persia had occurred in the Sixth Century AD when Anushirwan the Just restored the empire to its earlier level of greatness. Anushirwan reigned for 48 years and was a contemporary of Justinian. He wrested Syria from the Romans, the Yemen from the Abyssinians, and much of Central Asia from the Turks and other wild tribes of the steppes. This magnificent emperor died in 579, nine years after the birth of Prophet Muhammad.

As often happens when a great ruler passes away, Anushirwan was followed by a number of lesser mortals and the glory and prosperity of the empire began to fade. Civil war and intrigue sapped the strength of the state. The decline approached its climax in the time of Shiruya (Ciroes) a great-grandson of Anushirwan, who first imprisoned and then killed his father, Chosroes Parwez. Not content with this heinous crime, he turned to worse cruelties. So that none may dispute his right to the throne or pose a challenge to his authority, he had all the male members of his family killed with the exception of his son, Ardshir. The estimate of those of the house of Anushirwan who lost their lives to the maniacal fury of Shiruya, adult and child, varies from 15 to 18. And Shiruya reigned for only seven months before he too was dead.

With his death the confusion became worse. And there is confusion also in the accounts of the early historians about the order in which various emperors followed Shiruya and the duration of their respective reigns. All that is certain and unanimously accepted is the position of Yazdjurd bin Shahryar bin Perwez, who somehow escaped the assassins of Shiruya and became the last Persian Emperor of the line of Sasan. This ill-starred young man was to see the final disintegration of the great empire of the Chosroes.

1. Tabari: Vol. 2, pp. 553-4.
2. The Parthians, who overthrew the Seleucid power, though not Persians, were nevertheless Iranians. Thus the Greek interlude lasted less than two centuries until its end at the hands of the Parthians in the middle of the Second Century BC, The Persian Sasanids came to power in 220 AD.

Between Shiruya and Yazdjurd there were about eight rulers in a period of four or five years, and these included two women-Buran and Azarmidukht, both daughters of Chosroes Parwez. The first of these, Buran, proved a wise and virtuous monarch but lacked the strong hand that was needed to arrest the decline in imperial affairs. She was crowned during the lifetime of the Holy Prophet, who, when he heard of her coronation, made his famous remark: *"A nation will never prosper that entrusts its affairs to a woman!"* [1]

We will not go into a description of all the countries which comprised the geographical domain of the Persian Empire, but will confine ourselves to Iraq. Iraq then was not a sovereign State; it was substantially less than that. It was not merely a province; it was considerably more than that. Iraq was a land-one of the lands of the Persian Empire; and in its western and southern parts it was an Arab land.

The Arabs had been known in Iraq since the days of Bukht Nassar, [2] but did not then enjoy any power in the land. It was not until the early part of the Christian era, when a fresh migration of Arab tribes came to Iraq from the Yemen, that they began to command authority and influence. One of the great chiefs of these migrating Arabs, a man by the name of Malik bin Fahm, proclaimed himself king and began to rule over the western part of Iraq. Two generations after him the throne passed to Amr bin Adi, of the tribe of Lakhm, who started the Lakhmid Dynasty which was also at times called the House of Munzir. The kings of this dynasty ruled for many generations as vassals of the Persian Emperor.

The last of the House of Munzir was Numan bin Mundhir, who committed an act of disloyalty against Chosroes Parwez for which he was sentenced to death. The sentence was carried out in style-he was trampled to death by an elephant! This led to a revolt by the Arabs of Iraq, which was soon crushed by the Emperor, and with this abortive revolt ended the House of Munzir.

Chosroes then appointed a new king, Iyas bin Qubaisa of the tribe of Tayy, to rule over Iraq. For some years the new king enjoyed a reasonable degree of autonomy. Then most of his authority was taken away and Persian generals and administrators took over the entire government of the land. Iyas remained a titular king.

A land of culture, wealth and abundance, Iraq was the most prized possession of the Persian Empire. To the Arabs from the barren wastes of Arabia it was a green jewel, a land flowing with milk and honey. Its two mighty rivers, the Euphrates and the Tigris, were the greatest known rivers of the time-west of the Indus and north of the Nile. But these rivers did not then flow as they flow now, nor were the cities of Iraq then its cities of today. Kufa and Basra did not exist (they were founded in 17 Hijri). Baghdad was a small though much-frequented market town on the west bank of the Tigris. The then glorious cities of Ctesiphon and Hira are now turned to dust. Ctesiphon was the capital-a mighty metropolis and the seat of glory of the Persian Empire. Reportedly built by Ardsheer bin Babak (also, known as Ardsheer Babakan and Artaxerxes, the founder of the Sasanid Dynasty) it sprawled on both sides of the Tigris and was known to the Muslims as Madain, literally the Cities, for it consisted of several cities in one. 3 Hira was the capital of the Arab Lakhmid Dynasty. Situated on the west bank of the Euphrates, it was a glittering, throbbing city with many citadels. 4 And there was Uballa, the main port of the Persian Empire which was visited by ships from India and China and other maritime countries of the East. Uballa was also the capital of the military district of Dast Meisan. 5

1. Masudi: *Tanbeeh*, p. 90; Ibn Qutaiba: p. 666.
2. Nebuchadnezzar, Seventh-Sixth Century BC.
3. According to some sources, Ctesiphon existed before Ardsheer and was used by the Parthians as a winter residence.
4. The site of Hira is 12 miles south-east of Nejef and half a mile south of the present Abu Sukheir. Nothing remains of the ancient city except some traces of the White Palace which stood at the northern end of Hira. According to Gibbon (Vol. 5, p. 299), Hira was founded in 190 AD.
5. Uballa stood where the part of modern Basra known as Ashar stands today.

The Euphrates and the Tigris have been known to change their course more than once since the time of Babylon. The maps in this book indicate the course which these rivers followed in the early days of Islam. The main difference from today is in the course of the Tigris. In pre-Islamic times it had flowed in what is its present channel, which is known as Dijlat-ul-Aura (the One-Eyed Tigris), but then it had abandoned this channel and adopted a new course from Kut downwards, along the Dujaila (the Little Tigris) and the Akhzar, to enter a region of lakes and marshes comprising an area about 100 miles square, just north-west of Uballa. The old bed of the river had then become a dry, sandy bed. The marshes extended much farther north than they do today; and the Tigris picked its way through these marshes to rejoin the bed of the One-Eyed Tigris in the region of Mazar (the present Azeir), whence it flowed south and south-east into the Persian Gulf. [1] But the Tigris changed its course again in the Sixteenth Century and returned to its old bed. This, however, is not the largest branch of the Tigris, for the Gharraf, taking off from Kut and joining the Euphrates at Nasiriya, is larger. The Dujaila, which in the early days of Islam was the main channel, is now a modest river-the third largest branch of the Tigris, after the Gharraf and the One-Eyed Tigris.

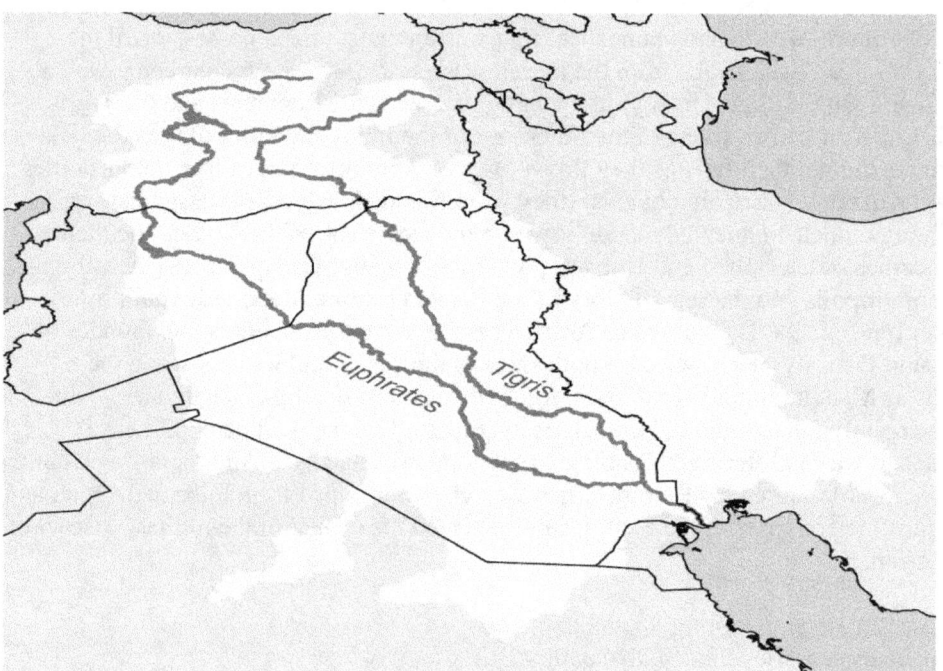

The Euphrates followed a clear course down to the present Hindiya, whence it split into two main channels as it does today-both sizable rivers: the Hilla branch and the main Euphrates. The main branch (the western one) again split up, flowing generally in one large and several subsidiary channels, which over the centuries have changed course several times, though not as drastically as the Tigris. The two main branches reunited at Samawa, whence the Euphrates flowed towards the region of lakes and marshes. While some of the water of the river lost itself in the marshes, one clear channel marked on today's maps as the Euphrates retained its distinct identity as it flowed eastwards to join the Tigris at Qurna. The marshes were drained by a large river known as Maqil, which emptied into the Tigris a little north of Basra; and from this junction all these waters flowed into the Persian Gulf as one great river, known today as the Shatt-ul-Arab..

Many changes have taken place in the bends and twists of these rivers. I have not shown these details on the maps as there is no way of knowing how they appeared then. Hence, only the main branches of these rivers are shown on our maps, and without all the twists and turns which must undoubtedly have existed.

This then was how Iraq stood politically and geographically, when the Caliph launched Khalid on it. It was a land occupied by Persians and Arabs, and ruled by the Persian court. The Empire had begun to decline politically, but it would be wrong to imagine that it had declined militarily. The military effectiveness of an empire may remain at a high level for decades after its political disintegration has set in. And so it was with the Persians in the year 633.

The Persian army, including its Arab auxiliaries, was the most formidable and most efficient military machine of the time. Led by experienced and dedicated veterans, it was a proud, sophisticated and well-tried force which gloried in its past achievements and its present might. The Persian soldier was the best-equipped warrior of his day. He wore a coat of mail or a breast-plate; on his head rested a helmet of either chain mail or beaten metal; his forearms were covered by metal sleeves, and his legs, were protected by greaves (like leg-guards covering the front part of the leg). He carried a spear, lance or javelin, a sword, and either an axe or an iron mace (the latter was a favourite and much-feared Persian weapon). He also carried one or two bows with 30 arrows and two spare bowstrings hanging from his helmet. [2] Thus, powerfully equipped and armed was the Persian soldier. But, and this was inevitable, he lacked mobility. In the general, set-piece battle he acknowledged no equals; and in this he was right, until Khalid's lightly armed and fast-moving riders came along.

It all started with Muthanna bin Harithah. A tiger of a man who later died of wounds suffered in battle with the Persians, Muthanna was a chief of the tribe of Bani Bakr, which inhabited the north-eastern part of the Arabian peninsula and southern Iraq. It is not certain that Muthanna had become a Muslim during the time of the Prophet. He probably had, because a delegation from the Bani Bakr had travelled to Madinah during the Year of Delegations and had accepted Islam at the hands of the Prophet. But there is no actual record of Muthanna's conversion.

1. Ibn Rusta: pp. 94-5. At Mazar (Azeir) today only a small river flows into the Tigris from the west-certainly too small to form the bed of the old Tigris. The old bed has probably silted up and ceased to be discernible.
2. Dinawari: p. 73.

Shortly after the Battle of Yamamah, Muthanna turned his attention towards Iraq. Seeking adventure and spoils, and encouraged by the disarray which was apparent in the political affairs of the Persian Empire, Muthanna took a band of his followers and began to raid into Iraq. At first he stuck to the periphery of the desert so that he could withdraw quickly into the safety of the sandy wastes, but gradually his incursions became bolder. He varied his objectives, striking now in the east, now in the west. Most of his raids, however, were in the region of Uballa, and he returned from these raids with spoils to dazzle the hungry Arab of the desert. The Persian garrisons were helpless against Muthanna's ghostlike riders, who vanished as rapidly as they struck.

Encouraged by his successes, Muthanna approached Abu Bakr. This was early in February 633 (late Dhul Qad, 11 Hijri). He painted a glowing picture-the vulnerable state of Iraq, the riches that waited to be plundered, the prolonged political crisis which bedevilled the Persian court, the inability of the Persian garrisons to fight mobile, fast-moving engagements. *"Appoint me as commander of my people"*, said Muthanna, *"and I shall raid the Persians. Thus I shall also protect our region from them."* 1

The Caliph agreed and gave him a letter of authority appointing him commander over all the Muslims of the Bani Bakr. With this letter of authority Muthanna returned to North-Eastern Arabia. Here he converted more tribesmen to Islam, gathered a small army of 2,000 men and resumed his raids with even greater enthusiasm and violence.

Muthanna was gone from Madinah, but his words continued to ring in the ears of the Caliph. He had planted a seed in the mind of Abu Bakr which germinated in a few days into a decision to take Iraq. He would not fight the entire Persian Empire, for that would be too big an objective in present circumstances. He would just take the Iraq of the Arabs, which meant the region west of the Tigris. Thus he would enlarge the boundaries of Islam and spread the new faith. At home there was peace, for with the defeat of the Kinda at Fort Nujair, Islam had been re-established in the land of Arabia.

Islam is a religion of peace, but not the peace of the timid and the submissive. It believes in peace, but the peace of the just and strong. **"Fight in the way of Allah"**, says the Quran, **"against those who fight you, but do not transgress." [Quran 2:190]**... **"And fight them until mischief is no more and religion is all for Allah."[Quran 8:39]**. And so it would be war with the fire-worshipping Persians.

Abu Bakr had made up his mind to invade Iraq; but he would have to proceed with great care, for the Arab feared the Persian-with a deep, unreasoning fear which ran in the tribal consciousness as a racial complex and was the result of centuries of Persian power and glory. In return the Persian regarded the Arab with contempt. It was important not to suffer a defeat, for that would confirm and strengthen this instinctive fear. To make certain of victory, Abu Bakr decided on two measures: (a) the invading army would consist entirely of volunteers; (b) Khalid would be the commander of the army.

With this in view, he sent orders to Khalid to invade Iraq and fight the Persians. He further instructed Khalid to call to arms those who had fought the apostates and remained steadfast in their faith after the death of the Messenger of Allah, and to exclude from the expedition those who had apostatised. Finally, he added (referring to the soldiers): *"Whoever wishes to return to his home may do so."* 2

When Khalid announced to his troops that the Caliph had given them permission to return home if they wished to do so, he was shocked by the result: thousands of his army left the army and returned Madinah and other places whence they had come. Whereas at the Battle of Yamamah he had commanded an army of 13,000 men, he was now left with only 2,000 men. Khalid wrote in haste to the Caliph, informing him of this alarming state of affairs and asking for reinforcements. When the letter reached Abu Bakr, he was sitting among his friends and advisers. He read the letter aloud so that all present might hear what it said. Then he sent for a young stalwart by the name of Qaqa bin Amr.

1. Tabari: Vol. 2, p. 552.
2. *Ibid*: Vol. 2, p. 553.

The young man arrived in the presence of the Caliph, armed and equipped for travel. The Caliph ordered him to proceed forthwith to Yamamah as a reinforcement to the army of Khalid. The Companions stared in amazement at the Caliph. *"Are you reinforcing one whose army has left him, with one man?"* they asked. 1

Abu Bakr looked for a moment at Qaqa. Then he said, *"No army can be defeated if its ranks possess the likes of this man."* 2 And Qaqa bin Amr rode away to reinforce the army of Khalid!

But this was not the only action that Abu Bakr took to build up Khalid's forces. He also wrote to Muthanna and Mazhur bin Adi (an important chief in North-Eastern Arabia), instructing them to muster their warriors and consider themselves and their men under the command of Khalid for the invasion of Iraq.

Having issued these instructions, Abu Bakr sat back and relaxed. He had given Khalid his mission to invade Iraq and fight the Persians; he had laid down a starting-point for the campaign, the region of Uballa; he had given Khalid his objective - Hira; and he had placed under Khalid's command whatever force he could muster. There was nothing else that he could do. It was up to Khalid to accomplish his mission. And Khalid, now in the forty-eighth year of his life, set about the conquest of Iraq. 3

1. Tabari: Vol. 2, pp. 553-4.
2. *Ibid.*
3. There are two main versions of campaign of Iraq: the first of Ibn Ishaq and Waqidi, the second of Saif bin Umar. Tabari favours the latter version, and this is the one here used in the account of Khalid's invasion of Iraq. In this also there are two versions of Abu Bakr's plan for the invasion. For an explanation, see Note 4 in Appendix B.

Chapter 19

(The battle of Chains)

"We did trample Hormuz with fury restrained..."
[Al-Qa'qa' bin Amr, commander in Khalid's army]1

On receiving the orders of the Caliph, Khalid at once undertook preparations to raise a new army. His riders galloped far and wide in the region of Yamamah and in Central and Northern Arabia, calling brave men to arms for the invasion of Iraq. And brave men assembled in thousands, many of them his old comrades of the Campaign of the Apostasy who, having visited their homes, decided to return to his standard for fresh adventure and glory. Khalid's name was now a magnet that drew warriors to him. Fighting under Khalid meant not only victory in the way of Allah, but also spoils and slaves ... in fact the best of both worlds! Within a few weeks an army of 10,000 men was ready to march with Khalid. 2

There were four important Muslim chiefs with large followings in North-Eastern Arabia, Muthanna bin Harithah, Mazhur bin Adi, Harmala and Sulma. The first two of these have already been mentioned in the preceding chapter. The Caliph had written to them to muster warriors and operate under the command of Khalid. Now Khalid wrote to all four of them, informing them of his appointment as commander of the Muslim army and of the mission which he had received from the Caliph. He ordered them to report to him, along with their men, in the region of Uballa. It is believed that Muthanna, who was at, Khaffan at the time (a place 20 miles south of Hira) 3 was displeased with the arrangement. He had hoped that the Caliph would give him a large independent command in Iraq, as he certainly deserved; but he came as ordered, and placed himself and his men at the disposal of Khalid. He was to prove the best of subordinate commanders.

Each of these four chiefs brought 2,000 men. Thus Khalid entered Iraq with 18,000 warriors 4 -the largest Muslim army yet assembled for battle.

In about the third week of March 633 (beginning of Muharram, 12 Hijri), Khalid set out from Yamamah. But before doing so he wrote to Hormuz, the Persian governor of the frontier district of Dast Meisan:
Submit to Islam and be safe. Or agree to the payment of the Jizya, and you and your people will be under our protection, else you will have only yourself to blame for the consequences, for I bring a people who desire death as ardently as you desire life. 5

Hormuz read the letter with a mixture of anger and contempt, and informed the Persian Emperor Ardsheer of Khalid's threat. He made up his mind to teach these crude Arabs a lesson that they should never forget.

Khalid began his advance from Yamamah with his army divided into three groups. He did this in order not to tire his men or waste time by having too many troops in the same marching column. Each group set off a day apart. Thus each group was a day's march from the next, far enough for ease of movement, and yet close enough to be swiftly concentrated for battle if required. Khalid himself moved with the third group on the third day-D plus 2. The whole army would concentrate again near Hufair 5 ; and before leaving Yamamah he promised his men a great battle with Hormuz.

1. Ibn Kathir, Al-Bidayah wan-Nihayah, Dar Abi Hayyan, Cairo, 1st ed. 1416/1996, Vol. 6 P. 425.
2. Tabari: Vol. 2, p. 554.
3. Musil (p. 284) places Khafran 20 kilometres south-east of Qadissiyah. It was at or near the present Qawam which is six miles west of Shinafiya.
4. Tabari: Vol. 2, p. 554.
5. *Ibid*: Vol. 2, p. 554.
6. *Ibid*: Vol. 2, p. 555.

Hormuz was the military governor of Dast Meisan. An experienced veteran and a trusted servant of the Empire. Hormuz was given this district to govern and protect because of its vital importance, which was both political and economic. It was a frontier district and lately had had a good deal of trouble with the Arab raiders of Muthanna. It was also a wealthy district in natural produce and commerce. Its chief city, Uballa, was the main port of the Persian Empire and thus vital to its commercial prosperity. Uballa was also a junction of many land routes-from Bahrain, from Arabia, from Western and Central Iraq, from Persia proper, which gave it a decisive strategical importance. It was a gateway, which it was the job of Hormuz to govern as an administrator and defend as a general.

The Persian society of the time had an imperial and aristocratic character. As is inevitable in such societies, it had an elaborate system of ranks to indicate a man's social and official position at the court. The outward symbol of rank was the cap; as a man rose in rank, his cap became more costly. The highest rank below the Emperor carried a cap worth 100,000 dirhams, which was studded with diamonds and pearls and other precious stones. Hormuz was a 100,000 dirham-man! 1

A true imperialist, he was of a proud and arrogant nature and held the local Arabs in contempt, which he did nothing to conceal. He was harsh and highhanded in his treatment of the Arabs, who in return hated and feared him. In fact his heavy hand became the cause of a saying amongst the Arabs: More hateful than Hormuz. 2

Soon after receiving the letter of Khalid, which he knew came from Yamamah, Hormuz informed the Emperor of the imminent invasion of Iraq by Khalid and prepared to fight this insolent upstart! He gathered his army and set out from Uballa, preceded by a cavalry screen.

The direct route from Yamamah to Uballa lay through Kazima (in modern Kuwait) and thither went Hormuz, expecting Khalid to take this route. On arrival at Kazima, he deployed his army facing south-west, with a centre and two wings, and ordered that men should be linked together with chains. So deployed, he awaited the arrival of Khalid. But of Khalid there was no sign. And the following morning his scouts brought word that Khalid was not moving towards Kazima; he was making for Hufair. 3

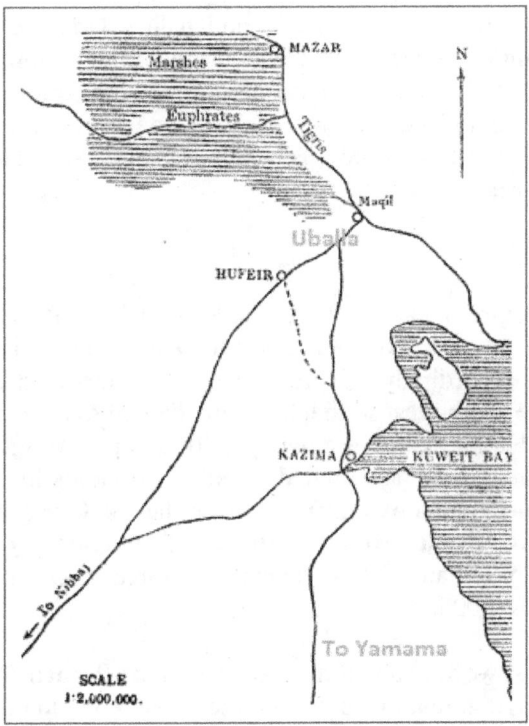

Khalid had, already before he left Yamamah, arrived at a broad conception of how he would deal with the army of Hormuz. He had been given the mission of fighting the

Persians, and a defeat of the Persian army was essential if the invasion of Iraq was to proceed as intended by the Caliph. With the Persian army intact at Uballa, Khalid could not get far. The direction given to him by the Caliph, i.e. Uballa, was by itself certain to bring the Persians to battle, for no Persian general could let Uballa fall.

Khalid knew the fine quality and the numerical strength of the Persian army and the courage, skill and armament of the Persian soldier. Heavily armed and equipped, he was the ideal man for the set-piece frontal clash. The only weakness of the Persian soldier and army lay in their lack of mobility; the Persian was not able to move fast, and any prolonged movement would tire him. On the other hand, Khalid's troops were mobile, mounted on camels with horses at the ready for cavalry attacks; and they were not only brave and skilful fighters, but also adept at fast movement across any terrain, especially the desert. Moreover, thousands of them were veterans of the Campaign of the Apostasy.

Khalid decided to use his own mobility to exploit its lack in the Persian army. He would force the Persians to carry out march and counter-march till he had worn them out. Then he would strike when the Persians were exhausted. Geography would help him. There were two routes to Uballa, via Kazima and Hufair, whose existence would facilitate his manoeuvre.

1. Tabari: Vol. 2, p. 556.
2. *Ibid*: Vol. 2, p. 555.
3. Kazima was on the northern coast of the Kuwait Bay, 11, 5 miles from the present Basra-Kuwait road. It was a fairly large city, over a mile in diameter, of which nothing remains but some castle-like ruins on a tongue of land jutting into the sea. These ruins may, however, be of a later period than Khalid's. No trace remains of Hufair nor is there any local tradition regarding its location. According to Ibn Rusta (p. 180) it was 18 miles from Basra on the road to Madinah. Since the old Arab mile was a little longer than the current mile, I place it at present-day Rumaila, which is 21 miles from old Basra. (Some later writers have confused this Hufair with Hafar-ul-Batin, which is in Arabia, 120 miles south-west of Kazima.)

Having written to Hormuz from Yamamah, Khalid knew that the Persian would expect him to advance on the direct route from Yamamah to Uballa, via Kazima, and would make his defensive plans accordingly. Khalid decided not to move on that route, but to approach Uballa from the south-west so that he would be free to manoeuvre on two axes-the Kazima axis and the Hufair axis-thus creating a difficult problem for the less mobile Persians. With this design in mind he marched to, Nibbaj, dividing his army into three groups as already explained, and took under command the 2,000 warriors of Muthanna, who, along with their intrepid chief, were awaiting Khalid at Nibbaj. 1 From Nibbaj he marched in the direction of Hufair, picking up the other three chiefs on the way, and approached Hufair with 18,000 men.

Khalid was not in the least worried about the presence of the Persian Army at Kazima. Hormuz at Kazima posed no threat to Khalid, for the Persians could not venture into the desert to disrupt his communications, apart from the fact that a mobile force like Khalid's operating in the desert did not present particularly vulnerable lines of communication. Khalid made no attempt to rush through Hufair and make for Uballa, because with Hormuz's large army on his flank his forward movement beyond Hufair might spell serious trouble. Hormuz could fall upon his rear and cut his line of retreat. No Arab would ever accept interference, or even a threat of interference, with his route back into the friendly, safe desert where he alone was master. Hence Khalid waited in front of Hufair, while light detachments of his cavalry kept Hormuz under observation. He knew, that his presence near Hufair would cause near-panic in the mind of Hormuz.

This is just what happened. The moment Hormuz got word of Khalid's movement towards Hufair, he realised the grave danger in which his army was placed. The Arab was not so simple after all! As an experienced strategist, he knew that his base was threatened. He immediately ordered a move to Hufair, 50 miles away, and his army, weighed down with its heavy equipment, trudged along the track. The two days' march was tiring, but the tough and disciplined Persian soldier accepted his trials without complaint. On arrival at Hufair, however, Hormuz found no trace of Khalid. Expecting the Muslims to arrive soon, he deployed for battle as he had done at Kazima, chains and all; but hardly had his men taken up their positions when his scouts came rushing to inform him that Khalid was moving towards Kazima!

And Khalid was indeed moving towards Kazima. He had waited near, Hufair until he heard of the hurried approach of Hormuz. Then he had withdrawn a short distance and begun a counter-march through the desert towards Kazima, not going too far into the desert so as not to become invisible to Persian scouts. Khalid was in no hurry. His men were well mounted, and he took his time. He had no desire to get to Kazima first and occupy it, for then he would have to position himself for battle and his opponent would be free to manoeuvre. Khalid preferred to let the Persians position themselves while he himself remained free to approach and attack as he liked, with the desert behind him.

The Persians again packed their bags and set off for Kazima, for Hormuz could not leave the Kazima route to the Muslims. Hormuz could have fought a defensive battle closer to Uballa; but having experienced the terrible havoc wrought by Muthanna in his district, he was in no mood to let Khalid approach close enough to let his raiders loose in the fertile region of Uballa. He was determined to fight and destroy Khalid at a safe distance from the district which it was his duty to protect, and he rejoiced at the prospect of a set-piece battle against the desert Arabs. Moreover, armies act as magnets: they attract each other. Sometimes an area which is not otherwise strategically important becomes so through the presence of a hostile army. Now Hormuz was drawn to Kazima not only by the strategical importance of the place but also by the army of Khalid.

This time the forced march did not go down so well with the Persians and there was grumbling, especially amongst the Arab auxiliaries serving under Hormuz, who cursed the Persian for all the trouble that he was causing them. The Persians arrived at Kazima in a state of exhaustion. Hormuz, the professional regular soldier, wasted no time and at once deployed the army for battle in the normal formation of a centre and wings. The generals commanding his wings were Qubaz and Anushjan. The men again linked themselves with chains.

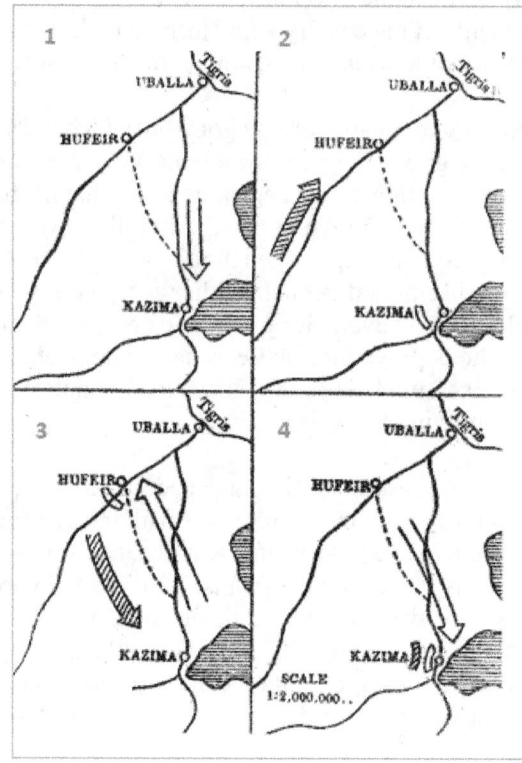

1. The old Nibbaj is the present Nabqiyya, 25 miles north-east of Buraida.

Chains were often used by the Persian army to link their men in battle. They were normally of four lengths, to link three, five, seven or ten men, [1] and were supposed to act as a source of strength to the army. It would not be correct to say, as some critics have suggested, that the chains were used by the officers for fear that their men would run away. The chains were used as a manifestation of suicidal courage, confirming the soldiers' willingness to die on the battlefield rather than seek safety in flight. They also lessened the danger of a breakthrough by enemy cavalry, as with the men linked together in chains it was not easy for cavalry groups to knock down a few men and create a gap for penetration. And since the Persian Army was organised and trained for the set-piece battle, this tactic enabled it to stand like a rock in the face of enemy assault. But the chains had one major drawback: in case of defeat the men were incapable of withdrawal, for then the chains acted as fetters. Men chained to fallen comrades, lost all power of movement and became helpless victims of their assailants.

It was the use of chains in this battle that gave it the name of the Battle of Chains.

The Arab auxiliaries, however, did not approve of these chains and never resorted to their use. When on this occasion the Persians chained themselves, the Arabs said, *"You have bound yourselves for the enemy. Beware of doing so!"* To this the Persians retorted, *"We can see that you wish to be free to run!"* [2]

Now Khalid came out of the desert and approached the Persians. He had made up his mind to fight a battle here and now before the Persian army recovered from its fatigue. But the Muslim army had no water, and this caused some alarm among the men, who informed Khalid of their misgivings. *"Dismount and unload the camels"*, ordered Khalid. *"By my faith, the water will go to whichever army is more steadfast and more deserving."* [3] Their confidence in their leader unshaken, the Muslims prepared for battle. They had not been at this for long when it began to rain, and it rained enough for the Muslims to drink their fill and replenish their water-skins.

Hormuz had deployed his army just forward of the western edge of Kazima, keeping the city covered by his dispositions. In front of the Persians stretched a sandy, scrub-covered plain for a depth of about 3 miles. Beyond the plain rose a complex of low, barren hills about 200 to 300 feet high. This range was part of the desert, running all the way to Hufair, and it was over this range that Khalid had marched to Kazima. Emerging from these hills, Khalid now moved his army into the sandy plain; and keeping his back to the hills and the desert, formed up for battle with the usual centre and wings. As commanders of the wings, he appointed Asim bin Amr (brother of Qaqa bin Amr) and Adi bin Hatim (the very tall chief of the Tayy, who has been mentioned earlier, in Part II). Some time in the first week of April 633 (third week of Muharram, 12 Hijri) began the Battle of Chains.

The battle started in grand style with a duel between the two army commanders. Hormuz was a mighty fighter, renowned in the Empire as a champion whom few would dare to meet in single combat. (In those chivalrous days no one could be a commanding general without at the same time being a brave and skilful fighter.) He urged his horse forward and halted in the open space between the two armies, though closer to his own front rank. Then he called, *"Man to man! Where is Khalid?"* [4] From the Muslim ranks Khalid rode out and stopped a few paces from Hormuz. The two armies watched in silence as these redoubtable champions prepared to fight it out.

Hormuz dismounted, motioning to Khalid to do the same. Khalid dismounted. This was brave of Hormuz, for a dismounted duel left little chance of escape; but on this occasion Hormuz was not being as chivalrous as one might imagine. Before coming out of the Persian ranks Hormuz had picked a few of his stalwarts and placed them in the front rank near the site which he had chosen for the duel. He instructed them as follows: he would engage Khalid in single combat; at the appropriate time he would call to the men; they would then dash out, surround the combatants and kill Khalid while Hormuz held him. The chosen warriors watched intently as the two generals dismounted. They felt certain that Khalid would not get away.

1. Tabari: Vol. 3, p. 206. According to Abu Yusuf (p. 33) the chain lengths were: five, seven, eight and 10 men.
2. *Ibid*: Vol. 2, p. 555.
3. *Ibid*.
4. Tabari: Vol. 2, p. 555.

The generals began to fight with sword and shield. Each struck several times at his adversary, but none of the blows made any impression. Each was surprised at the skill of the other. Hormuz now suggested that they drop their swords and wrestle. Khalid, unaware of the plot, dropped his sword as Hormuz dropped his. They began to wrestle.

Then, as they were locked in a powerful embrace, Hormuz shouted to his men, who rushed forward. Before Khalid realised what was happening he found himself and Hormuz surrounded by several fierce looking Persians.

Now Khalid knew. He was without his sword and shield, and Hormuz would not relax his iron grip. There seemed to be no way out of the predicament; but then, being a stronger man than Hormuz, Khalid began to whirl his adversary round and round, thus making it practically impossible for the Persians to strike at him.

A storm of sound arose over the battlefield as the two armies shouted-one with delight, the other with dismay. In this noise, their attention riveted on the wrestlers, the Persian killers did not hear the galloping hooves that approached them. They did not know what hit them. Two or three of them sprawled on the ground as headless trunks, before the others realised that the number of combatants in this melee had increased by just one more. The extra man was Qaqa bin Amr-the one-man reinforcement sent by Abu Bakr.

Qaqa had seen the Persian killers rush towards the two generals, and in a flash understood the perfidy of the enemy general and the peril which faced Khalid. There was no time to tell this to anyone; no time to explain or gather comrades to support him. He had spurred his horse into a mad gallop, and arriving in the nick of time, had set upon the Persians with his sword. Qaqa killed all of them! [1]

Khalid, freed of the menace of the Persian killers, turned his entire attention to Hormuz. After a minute or two Hormuz lay motionless on the ground, and Khalid rose from his chest with a dripping dagger in his hand.

Khalid now ordered a general attack, and the Muslims, incensed by the treacherous plot of the enemy commander, went into battle with a vengeance. The centre and the wings swept across the plain to assault the Persian army. The Persians had suffered a moral setback with the death of their commanding general; but they were more numerous than the Muslims and, their iron discipline held them together. They fought hard. For some time the battle hung in the balance with the fast-moving Muslims assailing the front and the steady, chain-linked Persian infantry repulsing all assaults. But soon the superior skill and courage of the Muslims and the fatigue of the Persians began to tell, and after several attempts the Muslims succeeded in breaking the Persian front in a number of places.

Sensing defeat, the Persian generals commanding the wings-Qubaz and Anushjan-ordered a withdrawal and began to pull their men back. This led to a general retreat, and as the Muslims struck still more fiercely, the retreat turned into a rout. Most of the Persians who were not chained managed to escape, but those who were chain-linked found their chains a death trap. Unable to move fast, they fell an easy prey to the victorious Muslims and were slain in thousands before darkness set in to put an end to the slaughter. Qubaz and Anushjan managed to escape and succeeded in extricating a large portion of the army from the battlefield.

The first battle with the power of Persia was over. It had ended in an overwhelming victory for the Muslims.

The following day was spent in attending to the wounded and collecting the spoils-weapons, armour, stores, costly garments, horses, captives-of which Khalid distributed four-fifths among his men. The share of each cavalryman came to a thousand dirhams, while the infantryman's share was a third of that. This ratio was a tradition of the Prophet.

The cavalryman was given three shares because he had to maintain his horse as well and was more valuable for the mobile, fast-moving operations which the Arabs loved.

One-fifth of the spoils was sent to the Caliph as the share of the state, and this included the 100,000 dirham cap of Hormuz. By right it belonged to Khalid, for in a duel all the belongings of the vanquished were taken by the victor; and for this reason Abu Bakr returned the cap to Khalid, who, preferring cash, sold it!

1. There is no record of the actual number of Persians who took part in this plot and were killed by Qaqa.

The Muslims appear also to have captured an elephant in the Battle of Chains, and this animal was sent to Madinah along with other spoils. The city of the Prophet had never before seen an elephant and there was tremendous excitement in the capital when the behemoth arrived. The people marvelled at this greatest of land animals; but Abu Bakr could not think of any use for the unfamiliar beast and returned it to Khalid. What happened to it thereafter we do not know.

While the families of the Persians and those of the Iraqi Arabs who had supported them were taken captive, the rest of the population of the district was left unmolested. This population consisted mainly of small farmers, peasants and shepherds, and they all agreed to pay the Jizya and come under Muslim protection.

For a few days Khalid remained busy with organisational matters. Then he set his army in motion towards the north. Ahead of the main body of the army he sent Muthanna and his 2,000 riders to reconnoitre the country and kill any stragglers behind by the retreating Persians.

Muthanna reached a small river just north of where Zubair stands today, on the bank of which stood a fort known as Hisnul-Mar'at, i.e., the Fort of the Lady, so called because a lady ruled over it. [1] Muthanna laid siege to the fort; but in order to avoid delay in his advance, he left his brother, Mu'anna, in charge of the siege operations with a few hundred men and himself proceeded north with the rest of his column.

Two or three days of siege operations were enough to convince the lady of the fort of the futility of resistance. The Persian army of Uballa had been defeated and she could expect no help from any quarter. Mu'anna offered to accept a peaceful surrender without bloodshed, without plunder, without enslavement. The lady agreed; the defenders surrendered. Mu'anna and the lady of the fort appear to have found much pleasure in their meeting with each other. First the lady became a Muslim, and then, without any further delay, Mu'anna married her!

Meanwhile Khalid was advancing northwards from Kazima with the main body of the army.

1. The river is still there and is known as the River of the Lady, but there is no trace of the fort.

Chapter 20

(The Battle of the River)

"...We crushed the two horns of Qarin at Thaniyy, with violence unleashed."
[Al-Qa'qa' bin Amr, commander in Khalid's army]1

When the Persian Emperor received the message of Hormuz regarding the Muslim advance from Yamamah, he organised a fresh army at Ctesiphon and placed it under the command of a top-ranking general by the name of Qarin bin Qaryana. Qarin too was a 100,000 dirham-man. The Emperor ordered him to proceed to Uballa with the new army to reinforce Hormuz, and with this mission Qarin set off from Ctesiphon.

Marching along the left bank of the Tigris, Qarin reached Mazar, crossed the Tigris and moved south along the right bank until he reached the Maqil River. He crossed this river and then another largish river a little south of the Maqil. He had hardly done so when he received reports of the disaster of Kazima. These reports were followed by the remnants of the Persian army which had survived the Battle of Kazima and now came streaming into Qarin's camp under the two generals, Qubaz and Anushjan. The survivors included thousands of Arab auxiliaries; and as is usual in such cases, the two partners-Persian and Arab-began to blame each other for the defeat. Their spirits were not as high as at Kazima; but they were brave, men and reacted more with anger than fear at the reverse they had suffered.

Qubaz and Anushjan were eager for battle again. They and Qarin found it difficult to believe that a regular imperial army could be defeated in battle by a force of uncultured and unsophisticated Arabs from the desert. They did not realise that the Battle of Kazima had been fought with not an uncivilised Arab force but a fine Muslim army, purified and strengthened by the new faith. However, Qarin was prudent enough not to advance beyond the south bank. Here he could fight with his back to the river and thus ensure the safety of his rear. By limiting the possibilities of manoeuvre, he would fight the frontal set-piece battle which the Persians loved and for which their training and discipline were ideally suited.

The remnants of the Persian army of Uballa were followed by the light cavalry detachments of Muthanna; and once contact was established with the Persians, the Muslim horsemen scoured the countryside for supplies while Muthanna kept the Persians occupied and carried out reconnaissances. The Persians made no attempt to sally out of their camp. Muthanna sent a messenger to Khalid to inform him that he had made contact with a powerful enemy force at *Sinyy*.2

The word *sinyy* was used by the Arabs to denote a river. Muthanna had contacted the Persians on the south bank of a river, and for this reason the battle which will now be described is called *the Battle of the River*.

On leaving Kazima, Khalid marched north until he reached some ruins in the vicinity of the present Zubair, about 10 miles south-west of Uballa. He had already decided not to turn towards Uballa, where there was no enemy left to fight, when Muthanna's messenger brought the news about the concentration of Qarin's army and the survivors of Kazima. Khalid was anxious to contact and destroy the new Persian army while the impact of

Kazima was still fresh in the Persian mind. Consequently, while he sent Maqal bin Muqarrin with a detachment to enter Uballa and gather spoils (which Maqal did), Khalid marched towards the River with the main body of the army. He caught up with Muthanna in the third week of April 633 (beginning of Safar, 12 Hijri).

Khalid then carried out a personal reconnaissance of the Persian position. Since the Persians had their backs to the river there was no possibility of outflanking them; and Khalid could think of no way of manoeuvring the Persians away from their position as he had done with Hormuz. Khalid accordingly decided to fight a general set-piece battle, in the imperial Persian style. This was unavoidable, because with Qarin poised for action as he was, Khalid could neither cross the river to enter deep into Iraq nor proceed westwards towards Hira.

1. Ibn Kathir, Al-Bidayah wan-Nihayah, Dar Abi Hayyan, Cairo, 1st ed. 1416/1996, Vol. 6 P. 425.
2. It is difficult to express this word in English. In Arabic it is written as (Yaqut: Vol. 1, p. 937) or as Tabari puts it (Vol. 2, p. 557)

The two armies formed up for battle. Qubaz and Anushjan commanded the wings of the Persian army while Qarin kept the centre under his direct control and stood in front of it. Detachments of Arab auxiliaries were deployed in various parts of the army. Qarin was a brave but wise general. He deployed with the river close behind him, and saw to it that a fleet of boats was kept ready at the near bank ... just in case! Khalid also deployed with a centre and wings, again appointing Asim bin Amr and Adi bin Hatim as the commanders of the wings.

The battle began with three duels. The first to step forward and call out a challenge was Qarin. As Khalid urged his horse forward, another Muslim, one by the name of Maqal bin Al Ashi, rode out of the Muslim front rank and made for Qarin. Maqal reached Qarin before Khalid, and since he was an accomplished swordsman and quite able to fight in the top class of champions, Khalid did not call him back. They fought, and Maqal killed his man. Qarin was the last of the 100,000 dirham men to face Khalid in battle.

As the Persian commander went down before the sword of Maqal, the other two Persian generals, Qubaz and Anushjan, came forward and gave the challenge for single combat. The challenge was accepted by the commanders of the Muslim wings, Asim and Adi. Asim killed Anushjan. Adi killed Qubaz. As these Persian generals fell, Khalid gave the order for a general attack, and the Muslims rushed forward to assault the massed Persian army.

In those days the personal performance of the commander was a particularly important factor in battle. His visible success in combat inspired his men, while his death or flight led to demoralisation and disorganisation. The Persian army here had now lost its three top generals; yet the men fought bravely and were able to hold the Muslim attacks for a while. But because of the absence of able generals, disorder and confusion soon became apparent in the Persian ranks. Eventually, under the violence of continued Muslim attacks, the Persian army lost all cohesion, turned about and made for the river bank.

This disorganised retreat led to disaster. The lightly armed Muslims moved faster than the heavily equipped Persians and caught up with their fleeing adversaries. On the river bank confusion became total as the Persians scrambled into the boats in a blind urge to get away from the horror that pursued them. Thousands of them were slain as other thousands rowed away to safety. Those who survived owed their lives to the caution of Qarin, who had wisely kept the boats ready by the river bank. But for these boats not a single Persian would have got away. The Muslims having no means of crossing the river, were unable to pursue the fugitives.

According to Tabari, 30,000 Persians were killed in this battle. [1] The spoils of the battle exceeded the booty taken at Kazima, and four-fifths of the spoils were again promptly distributed among the men while one-fifth was sent to Madinah.

Khalid now turned more seriously to the administration of the districts conquered by the Muslims and placed this administration on a more permanent footing. Submitting to Khalid, all the local inhabitants agreed to pay the Jizya and come under Muslim protection. They were left unmolested. Khalid organised a team of officials to collect taxes and placed Suwaid bin Muqarrin in command of this team with his headquarters at Hufair.

But while these administrative matters were engaging Khalid's attention, his agents had slipped across the Euphrates to pick up the trial of the vanquished army of Qarin. Yet other agents were moving along the Euphrates towards Hira to discover further movements and concentrations of the imperial army of the Chosroes. [2]

1. Tabari: Vol. 2, p. 558.
2. Tabari also calls this battle the Battle of Mazar, which I feel is incorrect. For an explanation see Note 5 in Appendix B.

Chapter 21

(The Hell of Walaja)

The people would say in those days,
"In the month of Safar,
Is killed every tyrant ruler,
At the junction of the river."[1]

The news of the debacle at the River inflamed Ctesiphon. A second Persian army had been cut to pieces by this new and unexpected force emerging out of the barren wastes of Arabia. Each of the two Persian army commanders had been an illustrious imperial figure, a 100,000 dirham-man. And not only these two, but two other first-rate generals had been slain by the enemy. It was unbelievable! Considering that this new enemy had never been known for any advanced military organisation, these two defeats seemed like nightmares-frightening but unreal.

Emperor Ardsheer decided to take no chances. He ordered the concentration of another two armies; and he gave this order on the very day on which the Battle of the River was fought. This may surprise the reader, for the battlefield was 300 miles from Ctesiphon by road. But the Persians had a remarkable system of military communication. Before battle they would station a line of men, picked for their powerful voices, at shouting distance one from another, all the way from the battlefield to the capital. Hundreds of men would be used to form this line. Each event on the battlefield would be shouted by A to B; by B to C; by C to D; and so on. 2 Thus every action on the battlefield would be known to the Emperor in a few hours.

Following the orders of the Emperor, Persian warriors began to concentrate at the imperial capital. They came from all towns and garrisons except those manning the western frontier with the Eastern Roman Empire. In a few days the first army was ready.

The Persian court expected the Muslims to proceed along the Euphrates to North-Western Iraq. The Persians understood the Arab mind well enough to know that no Arab force would move far from the desert so long as there were opposing forces within striking distance of its rear and its route to the desert. Expecting the Muslims army to move west, Ardsheer picked on Walaja as the place at which to stop Khalid and destroy his army.

The first of the new Persian armies raised at Ctesiphon was placed under the command of Andarzaghar, who until recently had been military governor of the frontier province of Khurasan and was held in high esteem by Persian and Arab alike. He was a Persian born among the Arabs of Iraq. He had grown up among the Arabs and, unlike most Persians of his class, was genuinely fond of them.

Andarzaghar was ordered to move his army to Walaja, where he would soon be joined by the second army. He set off from Ctesiphon, moved along the east bank of the Tigris, crossed the Tigris at Kaskar, 4 moved south-west to the Euphrates, near Walaja, crossed the Euphrates and established his camp at Walaja. Before setting out from the capital he had sent couriers to many Arab tribes which he knew; and on his way to, Walaja he picked up thousands of Arabs who were willing to fight under his standard. He had also met and taken command of the remnants of the army of Qarin. When he arrived at Walaja he was delighted with the strength of his army. Patiently he waited for Bahman who was to join him in a few days.

Bahman was the commander of the second army. One of the top personalities of the Persian military hierarchy, he too was a 100,000 dirham-man. He was ordered by the Emperor to take the second army, when ready, to Walaja where Andarzaghar would await him. Bahman would be in over-all command of both the armies, and with this enormous might would fight and destroy the Muslim army in one great battle.

1. Ibn Kathir, Al-Bidayah wan-Nihayah, Dar Abi Hayyan, Cairo, 1st ed. 1416/1996, Vol. 6 P. 421.
2. Tabari: Vol. 3, p. 43.
3. No trace remains of Walaja. According to Yaqut (Vol. 4, p. 939), it was east of the Kufa-Makkah road, and a well-watered region stretched between it and Hira. Musil (p. 293) places it near Ain Zahik, which, though still known by that name to the local inhabitants of the region, is marked on maps as Ain-ul-Muhari and is 5 miles south-south-west of Shinafiya. The area of Walaja, now completely barren, was then very fertile.
4. This was the place where Wasit was founded in 83 Hijri. In fact Kaskar became the eastern part of Wasit.

Bahman moved on a separate route to Andarzaghar's. From Ctesiphon he marched south, between the two rivers, making directly for Walaja. But he left Ctesiphon several days after the first army, and his movement was slower.

The Battle of the River had been a glorious victory. With few casualties to themselves, the Muslims had shattered a large Persian army and acquired a vast amount of booty. But the battle left Khalid in a more thoughtful mood; and only now did he begin to appreciate the immensity of the resources of the Persian Empire. He had fought two bloody battles with two separate Persian armies and driven them mercilessly from the battlefield, but he was still only on the fringes of the Empire. The Persians could field many armies like the ones he had fought at Kazima and the River.

It was a sobering thought. And Khalid was on his own. He was the first Muslim commander to set out to conquer alien lands. He was not only the military commander but also the political head, and as such had to govern, on behalf of the Caliph in Madinah, all the territories conquered for Islam. There was no superior to whom he could turn for guidance in matters of politics and administration. Moreover, his men were not as fresh as on the eve of Kazima. They had marched long and fast and fought hard, and were now feeling more than a little tired. Khalid rested his army for a few days.

By now Khalid had organised an efficient network of intelligence agents. The agents were local Arabs who were completely won over by the generous treatment of the local population by Khalid, which contrasted strikingly with the harshness and arrogance of the imperial Persians. Consequently they had thrown in their lot with the Muslims and kept Khalid apprised of the affairs of Persia and the movements of Persian forces. These agents now informed him of the march of Andarzaghar from Ctesiphon; of the large Arab contingents which joined him; of his picking up the survivors of Qarin's army; of his movement towards Walaja. They also brought word of the movement towards Walaja. They also brought word of the march of the second army under Bahman from Ctesiphon and its movement in a southerly direction. As more intelligence arrived, Khalid realised that the two Persian armies would shortly meet and then either bar his way south of the Euphrates or advance to fight him in the region of Uballa. The Persians would be in such overwhelming strength that there could be no possibility of his engaging in a successful battle. Khalid had to get to Hira, and Walaja was smack on his route.

Another point that worried Khalid was that too many Persians were escaping from one battle to fight another day. The survivors of Kazima had joined Qarin and fought at the River. The survivors of *the River* had joined Andarzaghar and were now moving towards Walaja. If he was to have a sporting chance of defeating all the armies that faced him, he would have to make sure that none got away from one battle to join the army preparing for the next.

These then were the two problems that faced Khalid. The first was strategical: two Persian armies were about to combine to oppose him. To this problem he found a masterly strategical solution, *i.e.* to advance rapidly and fight and eliminate one army (Andarzaghar's) before the other army (Bahman's) arrived on the scene. The second problem was tactical: how to prevent enemy warriors escaping from one battle to fight another. To this he found a tactical solution which only a genius could conceive and only a master could implement-but more of this later.

Khalid gave instructions to Suwaid bin Muqarrin to see to the administration of the conquered districts with his team of officials, and posted a few detachments to guard the lower Tigris against possible enemy crossings from the north and east and to give warning of any fresh enemy forces coming from those directions. With the rest of the army-about 15,000 men-he set off in the direction of Hira, moving at a fast pace along the south edge of the great marsh.

If Andarzaghar had been given the choice, he would undoubtedly have preferred to wait for the arrival of Bahman before fighting a decisive battle with the Muslims. But Andarzaghar was not given the choice. A few days before Bahman was expected, the Muslim army appeared over the eastern horizon and camped a short distance from Walaja. However, Andarzaghar was not worried. He had a large army of Persians and Arabs and felt confident of victory. He did not even bother to withdraw to the river bank, a mile away, so that he could use the river to guard his rear. He prepared for battle at Walaja.

For the whole of the next day the two armies remained in their respective camps, keeping each other under observation, while commanders and other officers carried out reconnaissances and made preparations for the morrow. The following morning the armies deployed for battle, each with a centre and wings. The Muslims armies were again commanded by Asim bin Amr and Adi bin Hatim.

The battlefield consisted of an even plain stretching between two low, flat ridges which were about 2 miles apart and 20 to 30 feet in height. The north-eastern part of the plain ran into a barren desert. A short distance beyond the north-eastern ridge flowed a branch of the Euphrates now known as the River Khasif. The Persians deployed in the centre of this plain, facing east-south-east, with the western ridge behind them and their left resting on the north-eastern ridge. Khalid formed up his army just forward of the north-eastern ridge, facing the Persians. The centre of the battlefield, i.e. the mid-point between the two armies, was about 2 miles south-east of the present Ain-ul-Muhari and 6 miles south of the present Shinafiya.

Andarzaghar was surprised at the strength of the Muslim army. Only about 10,000 he guessed. From what he had heard, Andarzaghar had expected Khalid's army to be much larger. And where was the dreaded Muslim cavalry? Most of these men were on foot! Perhaps the Persian survivors of Kazima and the River had exaggerated the enemy's strength, as defeated soldiers are wont to do. Or perhaps the cavalry was fighting dismounted. Andarzaghar did not know that the Muslims who faced him were also surprised at their numbers, for they did not seem to be as many as they had been the day before. But the matter did not worry them. The Sword of Allah knew best!

The situation put Andarzaghar in high spirits. He would make mincemeat of this small force and clear the land of Iraq of these insolent desert-dwellers. He would at first await the Muslim attack. He would hold the attack and wear down the Muslims; then he would launch a counter-attack and crush the enemy.

When Khalid's army advanced for a general attack, Andarzaghar was overjoyed. This was just what he wanted. The two armies met with a clash of steel, and the men lost all count of time as they struggled mightily in combat.

For some time the battle raged with unabated fury. The agile, skilful Muslims struck at the heavily armed Persians, but the Persians stood their ground, repulsing all attacks.

After an hour or so both sides began to feel tired-the Muslims more so because they were fewer in number and each of them faced several Persians in combat. The Persians had reserves which they employed to replace their men in the front line. However, the example of Khalid kept Muslim spirits undaunted. He was fighting in the front rank.

In particular, during this first phase of the battle, the Muslims gained further confidence from the thrilling spectacle of Khalid's duel with a Persian champion of gigantic proportions known as Hazar Mard, who was said to have been the equal of a thousand warriors. [1] This giant of a man stepped forward and extended a general challenge which was accepted by Khalid. After a few minutes of duelling, Khalid found an opening and felled the man with his sword. When the Persian's body lay quite still, Khalid sat down on his great chest and called out to his slave to bring him his food. Then, seated on this grisly bench, Khalid ate a hearty lunch! [2]

The first phase was over. The second phase of the battle began with the counter-attack of the Persians. The experienced eye of Andarzaghar could see clear signs of fatigue on the faces of the Muslims. He judged that this was the right moment for his counter-stroke; and in this he was right. At this command the Persians surged forward and struck at the Muslims. The Muslims were able to hold them for some time, but the Herculean efforts that they were called upon to make placed an almost unbearable strain on their nerves and limbs. Slowly they fell back, though in good order. The Persians launched furious charges, and the Muslims looked to Khalid for any sign of a change in plan or anything to relieve the tension. But from Khalid they got no such sign. He was fighting like a lion and urged his men to do likewise. And his men did likewise.

1. In Persian, *Hazar Mard* means a thousand men, and this was an appellation given to especially formidable warriors in recognition of their prowess and strength.
2. Tabari: Vol. 2, p. 560. Abu Yusuf: p. 142.

The Persians were paying heavily for their advance, but they exulted in the success that they were gaining. Andarzaghar was beside himself with joy. Victory was just round the corner. He had not reached the top rung of the Persian socio-military ladder, but now he saw visions of a 100,000 dirham-cap. The Muslims continued to fight with the suicidal desperation of wild animals at bay. They had reached the limits of human endurance; and some even began to wonder if Khalid had at last met his match. A little more of this and the front would shatter into a thousand pieces.

Then Khalid gave the signal. We do not know just what this signal was, but it was received by those for whom it was intended. The next moment, over the crest of the ridge which stretched behind the Persian army appeared two dark lines of mounted warriors-one from the Persian left-rear, the other from the right-rear. Cries of Allah-o-Akbar rent the air as the Muslim cavalry charged at a gallop; and the plain of Walaja trembled under the thundering hooves of the Arab horse.

The joy of the Persians turned to terror. While a moment before they had been shouting with glee, they now screamed in panic as the Muslim cavalry rammed into the rear of the Persian army. The main body of Muslims under Khalid, refreshed and strengthened by the sight they beheld, resumed the attack against the Persian front, at the same time extending its flanks to join hands with the cavalry and completely surround the Persians. The army of Andarzaghar was caught in a trap from which there could be no escape.

In an instant the disciplined Persians turned into a rabble. When groups of soldiers turned to the rear they were pierced by lances or felled by swords. When they turned to the front they were struck down by sword and dagger. Recoiling from the assaults that came from all directions, they gathered in an unwieldy mass, unable to use their weapons freely or avoid the blows of their assailants. Those who wanted to fight did not know whom to fight. Those who wanted to flee did not know where to go. In a mad urge to get away from the horror they trampled each other and fought each other. The battlefield of Walaja became a hell for the army of Andarzaghar.

The ring of steel became tighter as the furious charges of the Muslims continued. The very helplessness of the Persians excited the Muslims to greater violence, and they swore that they would not let the Persians and Iraqi Arabs escape this time.

In this the Muslims succeeded. A few thousand imperial warriors did get away; for no army can be so completely destroyed that not a single survivor remains, but the army as a whole ceased to exist. It was as if a vast chasm had opened under it and swallowed it up. While the armies of Hormuz and Qarin had suffered crushing defeats, the army of Andarzaghar was annihilated. The army of Andarzaghar was no more.

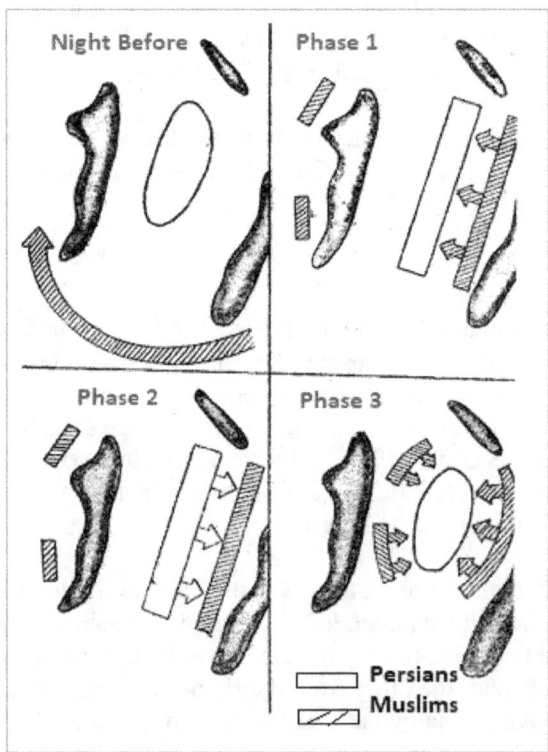

Andarzaghar himself, strangely enough, managed to escape. But the direction of his escape was towards the desert rather than the Euphrates, and having no desire but to put as much distance as possible between himself and the hell of Walaja, he went deep into the desert. In the desert the ill-fated man lost himself and died of thirst.

After the battle Khalid got his exhausted men together. He realised that this battle had imposed a terrible strain upon them. It had been the fiercest of the three fierce battles which they had fought in Iraq; and he wanted to make certain, that their spirits were not dampened by memories of the trial, for more trials awaited them.

He addressed the men. He started by praising Allah and calling His blessings upon the Holy Prophet. Then he continued:

"Do you not see the wealth of the land of the Persians? Do you not remember the poverty of the land of the Arabs? Do you not see how the crops in this land cover the earth? If the holy war were not enjoined by Allah, we should still come and conquer this rich land and exchange the hunger of our deserts for the abundant eating which is now ours." 1

1. Tabari: Vol. 2, p. 559.

And the warriors of Khalid agreed.

The day before the Battle of Walaja was fought, Khalid had sent for two of his officers, Busr bin Abi Rahm and Saeed bin Marra. 1 He made each of them the commander of a mobile striking force of about 2,000 cavalry and instructed them as follows:

a. They would take their horsemen out during the night and move wide round the south of the Persian camp.
b. On arrival on the far side of the ridge which stretched behind the Persian camp, they would conceal their men but keep them ready to move at short notice.
c. When battle was joined in the morning, they would keep their men mounted behind the crest of the ridge and position observers to watch for the signal of Khalid.
d. When Khalid gave the signal, the two striking forces would charge the Persian army in the rear, each group echeloned a bit to one flank.

Necessary orders were issued by Khalid to those who had to be in the know of the plan, so that the organisation and preparation of the striking forces could be carried out without a hitch; but the utmost secrecy was maintained and the Muslim rank and file knew nothing of the planned manoeuvre. In the morning, the cavalry comprising these striking forces was nowhere to be seen; and Khalid formed up the rest of his army, about 10,000 men in front of the Persians.

This was the plan of the Battle of Walaja, fought in early May 633 (third week of Safar, 12 Hijri). It was a frontal holding attack combined with a powerful envelopment. The operation went, down to the smallest detail, as planned by Khalid. Only a master could have done it.

This is not the first time in history that this brilliant manoeuvre was carried out. It had been done before. The most famous example of this type of manoeuvre was the Battle of Cannae in 216 BC, at which Hannibal did much the same to the Romans. After Hannibal's battle this type of manoeuvre became known as a *Cannae*.

But Khalid had never heard of Hannibal. With Khalid this was an original conception. 2

1. Tabari: Vol. 2, p. 559.
2. There is a difference between Walaja and Cannae in that Hannibal's cavalry moved out on both flanks, drove off the Roman cavalry, and then, at the appropriate time, fell upon the rear of the Romans, while Khalid's cavalry moved (as we reconstruct the battle) round one flank. But this is a matter of pre-battle movement. The pattern of battle was the same.

Chapter 22

(The River of Blood)

*"Allah watered the dead with the relentless Euphrates,
And others in the middle areas of Najaf."*
[Al-Qa'qa' bin Amr, commander in Khalid's army]1

The third great battle with the Persians had been won, and Khalid was nearer his ultimate objective-Hira. But he still had far to go and had no illusions about the journey. It was unlikely that the proud Persians would withdraw from his path. Much blood must yet be shed.

In spite of his masterly manoeuvre and his best efforts, a few thousand enemy warriors did manage to escape from the Battle of Walaja. They were mainly Christian Arabs from the tribe of Bani Bakr (Muthanna's tribe-those elements which had not accepted the new faith and had clung to Christianity). Much of this tribe lived in Iraq, as Persian subjects. They had responded to the call of Andarzaghar and with him they had fought and suffered at Walaja.

These Arab survivors of Walaja, fleeing from the battlefield, crossed the River Khaseef and moved between it and the Euphrates (the two rivers were about 3 miles apart, the former being a branch of the Euphrates). Their flight ended at Ullais, about 10 miles from Walaja. Here they felt reasonably safe, as the place was on the right bank of the Euphrates, and on the other side of Ullais ran the Khaseef, which actually took off from the Euphrates just above Ullais. Ullais could only be approached frontally, i.e. from the south-east. 2

For a few days Khalid rested his exhausted troops and himself remained busy with the distribution of the spoils and preparations for the onward march. Knowing of the existence of Bahman's army, he could appreciate that another bloody battle would have to be fought before he got to Hira. Since the centre of gravity of the campaign in Iraq had now shifted from the Tigris to the Euphrates, he recalled the Muslim detachments which he had left on the lower Tigris.

Khalid knew from his agents about the presence of hostile Arabs at Ullais; but since they were only the survivors of Walaja he did not consider them a military problem. In any case, he did not wish to over-strain his men by rushing them into another battle before they had recovered from their great trial of strength with Andarzaghar. But when about 10 days later he was informed of the arrival of more Arab forces at Ullais, it became evident that he would have to deal with a complete and almost new army. The hostile concentration was large enough to promise a major battle. As soon as his detachments from the lower Tigris had joined him, Khalid set off from Walaja with an army whose strength, as at the time of its entry into Iraq, was 18,000 men. 3 Since there was no way of getting to Ullais from a flank because of the two rivers, Khalid had no option but to cross the Khaseef and approach his objective frontally.

The annihilation of the army of Andarzaghar, following close upon the heels of Kazima and the River, shook the Empire of the Chosroes to its foundations. There appeared to be an unearthly quality about this Army of Islam which had emerged like an irresistible force from the desert. Any Persian army that opposed its relentless march vanished. For the proud Persian court, accustomed to treating the dwellers of the desert with contempt, this was a bitter pill to swallow. Never before in its long history had the empire suffered such military defeats, in such rapid succession, at the hands of a force so much smaller than its, own armies, so close to its seat of power and glory.

For the first time the Persians found it necessary to revise their opinions about the Arabs. It was clear that there was something about Islam which had turned this backward, disorganised and unruly race into a powerful, closely-knit and disciplined force of conquest. And it was clear also that there was something about this man Khalid-whose name was now whispered with fear in Persian homes-that added a touch of genius to the operations of his army. But a grand empire of 12 centuries is not beaten with three battles. The Persians were a race of conquerors and rulers who had lost battles before and risen again. The mood of dismay which had gripped Ctesiphon at the first reports of Walaja passed, and was replaced by a single-minded determination to crush this invading army and fling it back into the desert whence it came. Persia picked herself up, dusted herself, and prepared for another round.

1. Ibn Kathir, Al-Bidayah wan-Nihayah, Dar Abi Hayyan, Cairo, 1st ed. 1416/1996, Vol. 6 P. 425.
2. According to Tabari (Vol. 2, P. 560), Ullais was at a junction of the Euphrates. Musil (p. 193) places it at Ash-Shasi, which is now known as Al Asi and is 4 miles west-north-west of Shinafiya. Even now the place can only be approached from between the two rivers, unless one uses a boat to cross one of them.
3. Tabari: Vol. 2, p. 562. There is no record of reinforcements, but the Muslim losses must have been made up by either reinforcements from Arabia or local volunteers from Iraq.

Meanwhile messengers from the surviving Christian Arabs of the Bani Bakr arrived at Ctesiphon and informed the Emperor of their situation. They had sought the help of their fellow-Arabs inhabiting the region between Ullais and Hira; in response thousands of Arabs were even now marching to join the Bani Bakr at Ullais where they would fight a do-or-die battle with Khalid. Would not the Emperor help by sending another army of Persian warriors to join hands with his loyal Arabs subjects and save the Empire?

The Emperor would. He sent orders to Bahman who was still north of the Euphrates. On hearing of Walaja, Bahman had stopped in his tracks and decided not to move until he received further instructions. Now he got the Emperor's order to proceed with his army to Ullais, take under his command the Arab contingents assembled there, and bar Khalid's way to Hira.

But Bahman did not himself go to Ullais. He sent the army under his next senior general, one named Jaban, to whom he passed on the instructions of the Emperor. And Bahman added, *"Avoid battle until I join you, unless it is forced upon you."* [1] As Jaban set off with the army, Bahman returned to Ctesiphon. We do not know the purpose of his journey to the capital, we only know that he wished to discuss certain matters with the Emperor. He arrived at Ctesiphon to find Emperor Ardsheer very ill and remained in attendance on his master.

Jaban moved with his army to Ullais and found a vast gathering of Christian Arabs who had come from the region of Hira and Amghishiya. All had by now realised that Khalid's mission was to take Hira, and felt that Khalid's success would mean more bloodshed and enslavement. To prevent this, they had come to fight Khalid and, if necessary, die fighting. Jaban assumed command of the entire army, the Christian Arab part of which was commanded by a chieftain named Abdul-Aswad, who had lost two sons at Walaja and was burning for revenge. Persian and Arab camped side by side with the Euphrates to their left, the Khaseef to their right and the river junction behind them.

According to the early historians there was a river here which came into prominence as a result of actions taken on conclusion of the Battle of Ullais, as we shall shortly see. This river may once have been a canal, for it was dammed at its junction with the Euphrates just above Ullais, but at the time of the battle the river was dry, or almost dry, because the dam was closed. The Muslims referred to this river as just *the river*. I place this river as the Khaseef (which is now a fair-sized river), for there is no space at Ullais for another river or canal. Since, however, the name *Khaseef* may not have been in use at that time, it is hereafter referred to as *The River*.

Before the arrival of Jaban and the Persians, Muthanna and his light cavalry had appeared at Ullais and made contact with the Christian Arabs. Muthanna informed Khalid of the enemy position, strength and apparent intention to fight. Khalid increased his pace, hoping to catch the Christian Arabs before they were reinforced by other Persian forces. But Jaban beat him to Ullais, perhaps by a few hours; and again Khalid was faced by an enormous army. Again he determined to kill as many enemy warriors as he could lay his hands on, so that fewer would appear against him in the next battle. He also decided to fight the very same day; for the longer battle was delayed the more time the Persians would have to get organised and co-ordinate their plans. It was now the middle of May 633 (end of Safar, 12 Hijri).

Khalid stopped just long enough on the march to array his army in battle formation, appointing Adi bin Hatim and Asim bin Amr once again as the commanders of his wings, before he started the advance towards Ullais. This time no outflanking movements were possible, and he would rely for victory on the speed and violence of his attack rather than on manoeuvre. The Muslim advance to battle continued for some time before Jaban came to know that he was about to be attacked.

1. Tabari: Vol. 2, p. 560.

This information reached Jaban a little before midday, when it was mealtime for the Persian army. The cooks had prepared the soldiers' food, and the Persian soldier, like soldiers of all races and all ages, preferred a hot meal to a cold one and was reluctant to fight on any empty stomach. The Arab auxiliaries, however, were ready for battle. Jaban looked at his soldiers and the tempting pots of food being brought from the kitchens. Then he looked in the direction from which the Muslims were rapidly approaching in battle array. The soldiers also saw the Muslim army. They were brave men; but they were also hungry men. *"Let us eat now"*, they said to Jaban. *"We will fight later."*

"I fear", replied Jaban, *"that the enemy will not let you eat in peace."* [1]

"No!" said the Persians, disobeying their commander. *"Eat now; fight later!"* The meal-cloths were spread on the ground and steaming dishes were laid out upon them. The soldiers sat down to eat. They thought they had time. Meanwhile the Arab auxiliaries, less sophisticated in their eating habits, had formed up for action.

The Persians had eaten but one or two mouthfuls when it became evident that the Muslims were about to assault. If they delayed battle any longer, a full belly would be of no use to them, for they would be slaughtered anyway. Hurriedly they left their dishes; and as hurriedly Jaban deployed them on the battlefield along with the Arabs. He was not a minute too soon. He used the Christian Arabs to form the wings of his army, under the chiefs Abdul-Aswad and Abjar, and massed his Persian troops in the centre.

The battlefield ran south-east of Ullais between the Euphrates and *The River*. The Persian army was deployed with its back to Ullais, while in front of it was arrayed the army of Islam. The northern flank of both armies rested on the Euphrates and their southern flank on the river. The battle front was about 2 miles from river to river.

It was a very hard battle. The Battle of Walaja had been the fiercest battle of the campaign so far, but his was fiercer still. This became a battle that Khalid would never forget.

We do not know the details of the manoeuvres and other actions which took place in the battle. We know that Khalid killed the Arab commander, Abdul-Aswad, in personal combat. We know that the imperial army, though losing heavily in men, would not yield before the assaults of the Muslims. If ever an army meant to fight it out to the last, it was the imperial army of Ullais. The Arab auxiliaries were indeed fighting a do-or-die for if this battle were lost then nothing could save Hira. The Persians fought to vindicate the honour of Persian arms.

For a couple of hours the slogging continued. The fighting was heaviest on the bank of the river, where a large number of Persians fell in combat. The Muslims-tired, angry, frustrated-could see no opening, no weakening of the Persian and Arab resistance. Then Khalid raised his hands in supplication and prayed to Allah:

*"O Lord! If You give us victory, I shall see that
no enemy warrior is left alive until their river runs with
their blood!"* [2]

The Muslims renewed their assaults with greater fury; and Allah gave them victory. Early in the afternoon the imperial army was shattered and its soldiers fled from the battlefield. Thousands lay dead, especially in, and on the bank of, the river whose sandy bed was red with their blood.

As the Persian army fled from the battlefield, Khalid launched his cavalry after it. *"Do not kill them"*, he ordered the cavalry. *"Bring them back alive."* [3] The bed of the river was soaked with blood ... but the river was not "running with blood" as Khalid had pledged!

1. Tabari: Vol. 2, p. 561.
2. Tabari: Vol. 2, p. 561.
3. *Ibid.*

The Muslim cavalry broke up into several groups and galloped out in pursuit of the fugitives who had crossed the Khaseef and were fleeing in the direction of Hira. Parties of desperate Persians and Arabs were isolated from one another, surrounded, overpowered, disarmed and driven back to the battlefield like flocks of sheep. As each group was brought back, it was herded to the river, and every man was beheaded in the river bed or on the bank whence his blood ran into the river. The pursuit by the Muslim cavalry, the capture and return of the Persian and Arab warriors, and their killing in the river went on for the rest of that day and the whole of that night and the whole of the next day and part of the next. 1 Every vanquished warrior who fell into the victors' hands was decapitated. Khalid was keeping his pledge! Not till sometime on the third day was the last man killed.

Once the killing had stopped, a group of officers gathered around Khalid on the river bank. They looked upon a messy sight. Qaqa turned to Khalid and said, *"If you kill all the people of the earth their blood will not flow as long as this river is dammed. The earth will not absorb all the blood. Let the water run in the river. Thus you shall keep your pledge."* 2

Others added, *"We have heard that when the earth absorbs some of the blood of the sons of Adam, it refuses to accept more."* 3

Khalid ordered that the dam be opened. As it was opened the water rushed over the bed of the river and the blood lying in pools on the bed flowed with the water. This river then became known as the River of Blood.

As night fell after the day on which the battle was fought, while the Muslim cavalry was out bringing in the fugitives, the army of Khalid sat down to eat the food of the Persians, laid out upon the meal?cloths. The desert Arab marvelled at the fine fare on which the Persian soldier was fed.

The Battle of Ullais was over. An enormous amount of booty fell into Muslim hands and included the families of the defeated imperial warriors. According to Tabari, 70,000 Persians, and Christian Arabs were killed by the Muslims including those beheaded in the river. 4 But Jaban escaped.

On the following day Khalid entered into a pact with the local inhabitants of the district. They would pay the Jizya and come under Muslim protection; but this time another clause was added to the pact: the local inhabitants would act as spies and guides for the Muslims.

The episode of the River of Blood has been twisted and exaggerated beyond all limits by certain writers who have been unable to resist the temptation of resorting to sensationalism. This has led to some misconceptions which it would be well to correct.

These writers tell us that the river actually ran with blood; that there was a mill downstream of the battlefield powered by the water of this river; that so much blood flowed in the river that for three days the mill was grinding not with water but with blood!

This is a fantastic untruth. Balazuri makes no mention at all of any mill. Tabari, coming to the end of his account of this battle, mentions the mill, "...*as related by Shuaib, who heard it from Saif, who heard it from Talha, who heard it from Mugheerah.*" According to Mugheerah, there was a mill down-stream, powered by the water of this river; this mill was used for grinding corn for the army of Khalid for three days; and *the water was red*. [5]

In so far as this report may be correct, it still says nothing about the mill being run by blood. And there is no other mention in the early accounts of the mill. The facts are as they have been narrated above. When the dam was opened, on Qaqa's advice, the water naturally turned red and remained so for quite some time. But to run a mill with *whole blood* for three days would require the lives of millions of men. The story of the river running with blood for three days can be accepted as something from the Arabian Nights; it is not history.

Furthermore, to call what happened a "killing of prisoners" is an oversimplification. Normally they would have been killed in the pursuit, as had happened before and would happen again, with no questions asked. In this battle Khalid had pledged to make the river run with blood, so those thousands of men, instead of being killed in the pursuit, were brought to the river and killed. And that is all that there is to the episode of *the River of Blood*.

Of the battles which he had fought in the time of the Holy Prophet, the Battle of Mutah had a special place in the memory of Khalid. Nowhere else had he had to take command of so disastrous a situation and save the Muslims from the jaws of death. Of the battles fought in Iraq, the Battle of Ullais was similarly engraved upon his memory.
One day, after the campaign had been fought to a successful conclusion, Khalid sat chatting with some friends. He said, *"At Mutah I broke nine swords in my hand. But I have never met an enemy like the Persians. And among the Persians I have never met an enemy like the army of Ullais."* [6]

Coming from a man like Khalid, there could be no finer tribute to the valour of Persian arms. But the Persian court was now down and out. Ardsheer lay dying, and the empire would send no more armies to face the Sword of Allah. Ullais was the swansong of Ardsheer, great-great-grandson of Anushirwan the Just.

1. Tabari: Vol. 2, p. 561.
2. Ibid: Vol. 2, pp. 561-2.
3. *Ibid.*
4. *Ibid.*
5. Tabari: Vol. 2, p. 562.
6. Tabari: Vol. 2, p. 569.

Chapter 23

(The Conquest of Hira)

"And the day we surrounded the citadels
One after another, at calm Hirah.
We forced them down from their thrones,
Where they had acted as cowardly opponents."
[Al-Qa'qa' bin Amr, commander in Khalid's army]1

In the middle of May 633 (beginning of Rabi-ul-Awwal, 12 Hijri) Khalid marched from Ullais towards Amghishiya. This place was very near Ullais; in fact Ullais acted as an out post of Amghishiya! 2 The same morning the army reached Amghishiya, and found it a silent city.

Amghishiya was one of the great cities of Iraq-a rival the richness to Hira in size, in the affluence of its citizens and in find the and splendour of its markets. The Muslims arrived to city intact, and its markets and buildings abundantly stocked with wealth and merchandise of every kind; but of human beings there was no sign. The flower of Amghishiya's manhood had fallen at Ullais. Those who remained-mainly women and children and the aged-had left the city in haste on hearing of the approach of Khalid and had taken shelter in the neighbouring countryside, away from the route of the Muslim army. The fear which the name of Khalid now evoked had become a psychological factor of the highest importance in the operations of his army.

The Muslims took Amghishiya as part of the legitimate spoils of war. They stripped it of everything that could be lifted and transported, and in doing so accumulated wealth that dazzled the simple warriors of the desert. After it had been thoroughly ransacked, Khalid destroyed the city. 3 It is believed that the spoils taken here were equal to all the booty that had been gained from the four preceding battles in Iraq; and as usual, four-fifths of the spoils were distributed among the men while one-fifth was sent to Madinah as the share of the State.

By now the Caliph had become accustomed to receiving tidings of victory from the Iraq front. Every such message was followed by spoils of war which enriched the state and gladdened the hearts of the Faithful. But even Abu Bakr was amazed by the spoils of Amghishiya. He summoned the Muslims to the mosque and addressed them as follows:

"O Quraish! Your lion has attacked another lion and overpowered him. Women can no longer bear sons like Khalid!" 3

This was one of the finest compliments ever paid to Khalid bin Al Waleed.

These were difficult days for Azazbeh, governor of Hira. He had heard of the disaster that had befallen the Persian army, at Kazima, at *the River*, at Walaja and at Ullais; and it was obvious that Khalid was marching on Hira. If those large armies, commanded by distinguished generals, had crumbled before the onslaught of Khalid, could he with his small army hope to resist? There were no instructions from the ailing Emperor.

Azazbeh. was the administrator of Hira as well as the commander of the garrison. He was a high official of the realm-a 50,000 dirham-man. The Arab king of Hira, Iyas bin Qubaisa who has been mentioned earlier, was a king in name only. Other chieftains who were like princes of the realm also had no governmental authority except in purely Arab or tribal matters. It fell to Azazbeh to defend Hira; and as a true son of Persia, he resolved to do his best.

He got the army garrison out of its quarters and established a camp on the outskirts of Hira. From here he sent his son forward with a cavalry group to hold the advance of Khalid, and advised him to dam the Euphrates in case Khalid should think, of moving up in boats. This young officer rode out to a place where the River Ateeq joined the Euphrates, 12 miles downstream from Hira. Here he formed a base, from which he sent a cavalry detachment forward as an outpost to another river junction a few miles ahead, where the Badqala flowed into the Euphrates, a little above Amghishiya. 5

1. Ibn Kathir, Al-Bidayah wan-Nihayah, Dar Abi Hayyan, Cairo, 1st ed. 1416/1996, Vol. 6 P. 425.
2. Tabari: Vol. 2, p. 563; Amghishiya was also known as Manishiya.
3. Tabari: Vol. 2, p. 563.
4. *Ibid*.
5. The River Ateeq still exists. It is a small river, hardly more than a large stream, and may have been a canal in those days. Taking off from the area of Abu Sukhair, the Ateeq flows west of Euphrates, going up to 5 miles away from the main river, and rejoins the Euphrates a mile above modern Qadisiya (which is 8 miles south-east of the old, historical Qadisiya). In the latter part of its journey, this stream is also known as Dujaij. The Badqala was a canal or channel which joined the Euphrates near Amghishiya (Tabari: Vol. 2, p. 563). In his account of this operation, Tabari is both confusing and confused, and has got the two river junctions mixed up.

Khalid had now resumed his march on what was to be the last leg of his journey to Hira. He decided to use the river for transport and had all the heavy loads of the army placed in boats. As the army advanced on camels and horses, the convoy of boats, manned and piloted by local Arabs, moved alongside. Khalid had not gone far, however, when the water level fell and the boats were grounded. The son of Azazbeh had dammed the river.

Leaving the army stranded at the bank of the Euphrates, Khalid took a detachment of cavalry and dashed off at a fast pace along the road to Hira. Before long he arrived at Badqala, to encounter the Persian horse sent forward by the son of Azazbeh as an outpost. These green Persians were no match for the Muslim veterans; and before they could organise themselves for defence, Khalid's horsemen bore down upon them and slaughtered them down to the last man. Next Khalid opened the dam so that the water flowed once again in the right channel; and the army resumed its advance by river.

The son of Azazbeh also was not as wakeful as, the situation demanded. In the belief that his outpost at Badqala was sufficient precaution against surprise by the Muslims-not for a moment doubting that the outpost would inform him of the approach of danger-he had relaxed his vigilance. Then suddenly he was hit by Khalid. Most of the Persians in this group were killed, including the young commander; but a few fast riders managed to get away to carry the sad news to Azazbeh.

From these riders Azazbeh heard of the loss of the cavalry group and the death of his son. From couriers who came from Ctesiphon he heard of the death of Ardsheer. Heartbroken at the loss of his son and staggered by the news of the Emperor's death, he found the burden of his responsibilities too heavy for his shoulders. He abandoned all intentions of defending Hira against Khalid; and crossing the Euphrates with his army, withdrew to Ctesiphon. Hira was left to the Arabs.

Khalid continued his advance towards his objective. It is not known when he abandoned the boats and took to the road, but this must have happened a few miles downstream of Hira. Expecting stiff opposition at Hira, Khalid decided not to approach it frontally. Moving his army round the left, he bypassed Hira from the west and appeared at Khawarnaq, which was a thriving town 3 miles north-north-west of Hira. [1] He passed through Khawarnaq and approached Hira from the rear. There was no opposition to his columns as they entered the city. The inhabitants were all there. They neither fled nor offered any resistance, and were left unmolested by the Muslim soldiers as they entered deeper into the city.

Soon the situation became clearer; it was a mixed situation of peace and war. Hira was an open city; the Muslims could have it. But the four citadels of Hira, each manned by strong garrisons of Christian Arabs and commanded by Arab chieftains, were prepared for defence and would fight it out. If Khalid wanted any of these citadels, he would have to fight for it.

Each of the four citadels had a palace in which the commanding chieftain lived; and each citadel was known after its palace. The citadels were: the White Palace commanded by Iyas bin Qubaisa ('King' of Iraq); the Palace of Al Adassiyin commanded by Adi bin Adi; the Palace of Bani Mazin commanded by Ibn Akal; and the Palace of Ibn Buqaila commanded by Abdul Masih bin Amr bin Buqaila.

Against each citadel Khalid sent a part of his army under a subordinate general. These generals, besieging the citadels in the order in which they have been mentioned above, were: Dhiraar bin Al Azwar, Dhiraar bin Al Khattab (no relation of Umar), Dhiraar bin Al Muqarrin and Muthanna. All the generals were ordered to storm the citadels; but before doing so they would offer the garrisons the usual alternatives-Islam, the Jizya or the sword. The garrisons would have one day in which to think it over. The generals moved out with their forces and surrounded the citadels. The ultimatum was issued. The following day it was rejected by the Christian Arabs and hostilities began.

1. Nothing remains of Khawarnaq but a mound 600 Yards west of the Nejef road.

The first to launch his attack was Dhiraar bin Al Azwar against the White Palace. The defenders stood on the battlements and in addition to shooting arrows at the Muslims, used a catapult to hurl balls of clay at their assailants. Dhiraar decided to knock out the catapult. Working his way forward with a picked group of archers, he got to within bow-range of the catapult and ordered a single, powerful volley of arrows. The entire crew of the catapult was killed, and many of the enemy archers too. The rest hastily withdrew from the battlements

Similar exchanges of archery were taking place at the other citadels, though none of the others had a catapult. It was not long before the four chieftains asked for terms. They agreed to nominate one from amongst themselves who would speak for all, to negotiate directly with Khalid. The man chosen was the chieftain of the Palace of Ibn Buqaila-Abdul Masih bin Amr bin Buqaila.

Abdul Masih came out of his citadel and walked towards the Muslims. He walked slowly, for he was a very, very old man, "whose eyebrows had fallen over his eyes." [1]

Abdul Masih was in his time the most illustrious son of Arab Iraq. He was a prince. Known as the wisest and oldest of men, he enjoyed no official authority from the Persian court, but was held in reverence by the Iraqis and wielded considerable influence in their affairs. He also had a sparkling, if impish sense of humour. He had become a noted figure as early as the time of Anushirwan the Just. Meeting Anushirwan shortly before the latter's death, Abdul Masih had warned him that after him his empire would decay.

Slowly the old sage approached Khalid. When he stopped, there began one of the most unusual dialogues ever recorded by historians.

"How many years have come upon you?" asked Khalid.

"Two hundred", replied the sage.

Awed by the great age of the man, Khalid asked, *"What is the most wonderful thing that you have seen?"*

"The most wonderful thing that I have seen is a village between Hira and Damascus to which a woman travels from Hira, with nothing more than a loaf of bread."

He was alluding to the incomparable order and system which existed in the time of Anushirwan. The meaning of his words, however, was lost on Khalid, who, concluded that the man must be stupid. Without raising his voice Khalid remarked, *"Have you gained nothing from your great age but senility? I had heard that the people of Hira were cunning, deceitful scoundrels. Yet they send me a man who does not know from where he comes."*

"O Commander!" protested the sage. *"Truly do I know from where I come."*

"Where do you come from?"

"From the spine of my father!"

"Where do you come from?" Khalid repeated.

"From the womb of my mother!"

"Where are you going?"

"To my front."

1. Abu Yusuf: p. 143.

"What is to your front?"

"The end."

"Woe to you!" exclaimed Khalid. *"Where do you stand?"*

"On the earth."

"Woe to you! In what are you?"

"In my clothes."

Khalid was now losing his patience. But he continued his questioning.

"Do you understand me?"

"Yes."

"I only want to ask a few questions."

"And I only want to give you the answers."

Exasperated with this dialogue, Khalid muttered: *"The earth destroys its fools, but the intelligent destroy the earth. I suppose your people know you better than I do."*

"0 Commander," replied Abdul Masih with humility, *"it is the ant, not the camel, that knows what is in its hole!"*

It suddenly struck Khalid that he was face-to-face with an unusual mind. Everything that the sage had said fell into place; every answer had meaning and humour. His tone was more respectful as he said, *"Tell me something that you remember."*

An absent look came into the eyes of Abdul Masih. For a few moments he looked wistfully at the towers of the citadels which rose above the rooftops of the city. Then he said, *"I remember a time when ships of China sailed behind these citadels."* He was mentally again in the golden age of Anushirwan.

The preamble was over. Khalid now came to the point. *"I call you to Allah and to Islam"*, he said. *"If you accept, you will be Muslims. You will gain what we gain, and you will bear what we bear. If you refuse, then the Jizya. And if you refuse to pay the Jizya, then I bring a people who desire death more ardently than you desire life."*

"We have no wish to fight you," replied Abdul Masih, *"but we shall stick to our faith. We shall pay the Jizya."*

The talks were over. Agreement had been reached. Khalid was about to dismiss the man when he noticed a small pouch hanging from the belt of a servant who had accompanied the sage and stood a few paces behind him. Khalid walked up to the servant, snatched away the pouch and emptied its contents into the palm of his hand. *"What is this?"* he asked the sage.

"This is a poison that works instantaneously."

"But why the poison?"

"I feared", replied Abdul Masih, *"that this meeting might turnout otherwise than it has. I have reached my appointed time. I would prefer death to seeing horrors befall my people and my land."* 1

1. This dialogue has been taken from Balazuri: (p. 244) and Tabari (Vol. 2, pp. 564-6).

In the end of May 633 (middle of Rabi-ul-Awwal, 12 Hijri) the terms of surrender were drawn up. A treaty was signed. The citadels opened their gates and peace returned to Hira. The objective given by the Caliph had been taken after four bloody battles and several smaller engagements. Khalid led a mass victory prayer of eight *rakats* . 1

According to the treaty, the people of Hira would pay the Muslim State 190,000 dirhams every year. The pact included certain supplementary clauses: Hira would give the Muslim army one saddle (the army was one saddle short!), 2 the people of Hira would act as spies and guides for the Muslims. And then there was the clause about an Arab princess!

One day at Madinah the Holy Prophet was sitting in the company of some of his followers, talking of this and that. The subject turned to foreign lands, and the Prophet remarked that soon the Muslims would conquer Hira. Thereupon one Muslim, a simple, unlettered man by the name of Shuwail, 3 said eagerly, *"O Messenger of Allah! When we have conquered Hira may I have Kiramah bint Abdul Masih?"*

Kiramah, the daughter of Abdul Masih, was a princess. The people of Arabia had heard of her as a breathtaking beauty-a woman more beautiful than any other in existence. The Prophet laughed as he replied, **"She shall be yours?"** 4

Hira was now conquered. As Khalid's troops came to hear of his talks with Abdul Masih and the preparations to draw up the terms of surrender, Shuwail, who was serving under Khalid, approached the Sword of Allah. *"O Commander!"* he said. *"When Hira surrenders may I have Kiramah bint Abdul Masih? She was promised to me by the Messenger of Allah."*

Khalid found it difficult to believe that the Prophet had promised a princess of the house of Abdul Masih to this simple fellow. *"Have you any witnesses?"* he asked. *"Yes, by Allah!"* replied Shuwail, and brought witnesses whose testimony proved the veracity of the man's statement. Khalid then included this point as a clause in the pact: Kiramah bint Abdul Masih would be given to Shuwail!

The women of the house of Abdul Masih wailed in distress when they were given the devastating news. Was a princess who had lived all her life in splendour and refinement to be handed over to a crude Arab of the desert? What made the situation ludicrous was that Kiramah was an old woman of 80. She had once been the leading beauty of the day, but that was a long time ago.

The princess herself solved the problem. *"Take me to him"*, she said. *"This fool must have heard of my beauty when I was young, and thinks that youth is eternal."* 5 Accompanied by a maid, she left the Palace of Ibn Buqaila.

Excited by visions of amorous delight, Shuwail awaited his prize. Then she stood before him. The poor man's shock and dismay made a pathetic sight as he looked at the lined face. He was left speechless.

The princess broke the embarrassed silence. *"Of what use is an old woman to you? Let me go!"*

Now Shuwail saw his chance of making her pay for her freedom. *"No,"* he replied, *"not except on my terms."*
"And what are your terms? State your price."

"I am not the son of the mother of Shuwail if I let you go for less than a thousand dirhams."

The shrewd old woman assumed a look of alarm. *"A thousand dirhams!"* she exclaimed.

1. A unit of prayer.
2. Balazuri: p. 246.
3. Tabari: Vol. 2, p. 569. According to Balazuri, however, this man's name was Khuraim bin Aus. (p. 245).
4. *Ibid.*
5. *Ibid.*

"Yes, not a dirham less."

Quickly the princess handed over 1,000 dirhams to the exulting Arab and returned to her family.

Shuwail rejoined his comrades, many of whom were more knowledgeable than he. Bursting with pride he told them the story: how he had released Kiramah, but made her pay through *the nose*-1,000 dirhams!

He was quite unprepared for the laughter which greeted his boastful account. *"1,000 dirhams!"* his friends exclaimed. *"For Kiramah bint Abdul Masih you could have got much, much more."*

Bewildered by this remark, the simple Arab replied, *"I did not know that there was a sum higher than a thousand!"* When Khalid heard the story he laughed heartily, and observed, *"Man intends one thing, but Allah intends another."* [1]

Once Hira was his, Khalid turned to the subjugation of other parts of Iraq, starting with the nearer districts. He wrote identical letters to the mayors and elders of the towns, offering them the usual alternatives-Islam, the Jizya or the sword. All the districts in the vicinity of Hira had the good sense to submit; and pacts were drawn up with the chiefs and mayors, laying down the rate of Jizya and assuring the inhabitants of Muslim protection. These pacts were witnessed by several Muslim officers, including Khalid's brother, Hisham, who served under him in this campaign.

Meanwhile the affairs of Persia were going from bad, to worse. The Persians were split over the question of the succession to the throne. In opposition to Khalid, they were

united, but this was a sterile unity, offering no positive results. With the military affairs of the Empire in disarray, Bahman had assumed the role of Commander-in-Chief, and was working feverishly to put the defences of Ctesiphon in order against a Muslim attack which he was certain would come. Bahman aimed at nothing more ambitious than the defence of Ctesiphon; and in this he was being realistic, for over the rest of the region west of the Lower Tigris the Persians had no control.

Over this region the Arab horse was now supreme. Khalid, having crushed four large Persian armies, knew that there was no further threat of a counter-offensive from Ctesiphon, and that he could venture into Central Iraq in strength. He made Hira his base of operations and flung his cavalry across the Euphrates. His mounted columns galloped over Central Iraq up to the Tigris, killing and plundering those who resisted and making peace with those who agreed to pay the Jizya. For the command of these fast-moving columns he used his most dashing generals Dhiraar bin Al Azwar, Qaqa, Muthanna. By the end of June 633 (middle of Rabi-ul-Akhir 12 Hijri) the region between the rivers was all his. There was no one to challenge his political and military authority.

Along with military conquest Khalid organised the administration of the conquered territories. He appointed officers over all the districts to see that the Jizya was promptly paid and that the local inhabitants provided intelligence about the Persians and guides for the movement of Muslim units. Khalid also sent two letters to Ctesiphon, one addressed to the court and the other to the people. The letter to the Persian court read as follows:

In the name of Allah, the Beneficent, the Merciful. From Khalid bin Al Waleed to the kings of Persia.
Praise be to Allah who has disrupted your system and thwarted your designs. And if He had not done so it would have been worse for you. Submit to our orders and we shall leave you and your land in peace; else you shall suffer subjugation at the hands of a people who love death as you love life. 2

The letter addressed to the people was in much the same words, with the added promise of Muslim protection in return for the payment of the Jizya. Both the letters were carried by local Arabs of Hira and delivered at Ctesiphon. There was no reply!

1. Tabari: Vol. 2, p. 569; Balazuri: p. 245.
2. Tabari: Vol. 2, p. 572.

Chapter 24

(Anbar and Ain-ut-Tamr)

After Ain at-Tamr, when Ayadh wrote to Khalid requesting reinforcements, Khalid wrote back,
"Wait a while: there will come to you mounts
Carrying lions in shining armour,
*Battalions followed by battalions."*1

The portion of Central Iraq lying between the Euphrates and the Tigris, below Ctesiphon, was now under Muslim control. The inactivity of the Persians confirmed Khalid's belief that Ctesiphon was no longer in a position to interfere with his operations, let alone pose a threat to his base at Hira or his communications with the desert. Hence Khalid turned his attention to the north, where his forces had not yet ventured. There were two places which offered a likelihood of opposition-Anbar and Ain-ut-Tamr, both manned by sizable Persian garrisons and Arab warriors who would resist the advance of the Muslims. Both were governed by Persian officers.

Khalid decided to take Anbar first. This was an ancient fortified town and commercial centre to which trade caravans came from Syria and Persia. It was also famous for its large granaries. At the end of June 633 (middle of Rabi-ul-Akhir, 12 Hijri) Khalid marched from Hira with half his army (about 9,000 men), leaving behind a strong garrison at Hira and several detachments in Central Iraq. Moving along the west bank of the Euphrates, he crossed the river somewhere below Anbar. As his scouts moved out eastwards to keep the approaches from Ctesiphon under observation, he moved the army to Anbar and laid siege to the town. The Muslims found that the town was protected not only by the walls of the fort, but also by a deep moat filled with water. The moat was within close bow-range of the wall so that those attempting to cross it would have to face accurate fire from archers on the walls. The bridges over the moat had been destroyed at the approach of the Muslims. 2

Anbar was the chief town of the district of Sabat, which lay between the two rivers west of Ctesiphon. In Anbar resided the governor of Sabat, a man named Sheerzad who was known more for his intellect and learning than his military ability. Sheerzad was now faced with the task of defending the fort against a Muslim army with the forces under his command-the Persian garrison and a large number of Arab auxiliaries in whom apparently he had little faith.

The day after his arrival Khalid moved up to examine the defences of the fort. On top of the wall he saw thousands of Persians and Arabs standing around carelessly in groups, looking at the Muslims as if watching a tournament. Amazed at this sight, Khalid remarked, *"I see that these people know nothing about war."* 2

He collected 1,000 archers-the best of his marksmen-and explained his plan. They would move up casually to the edge of the moat with bows ready, but arrows not fitted. At his command they would instantly fit arrows to their bows and fire salvo after salvo at the garrison. *"Aim at the eyes"*, Khalid told the archers. *"Nothing but the eyes!"* 4

The detachment of archers moved towards the fort. The crowds standing on the wall gaped at the archers, wondering what they would do next. When the archers had got to the moat, Khalid gave the order, and 1,000 swift missiles flew across the moat, followed by another 1,000 and yet another. In a few seconds the garrison had lost 1,000 eyes. A clamour went up in the town: *"The eyes of the people of Anbar are lost!"* As a result of this action the Battle of Anbar is also known as the Battle of the Eyes. 5

When Sheerzad heard of the misfortune that had befallen the garrison, he sent Khalid an offer to surrender the fort if suitable terms were agreed upon. Khalid rejected the offer; the surrender would have to be unconditional. Sheerzad half-heartedly decided to continue resistance.

Khalid resolved to storm the fort. The wall would have to, be scaled, but this was not too difficult a task. The chief problem was crossing the moat, which was deep and wide. There were no boats available nor material with which to make boats or rafts; and the Arab of the desert was no swimmer. Khalid decided to make a bridge of flesh and bone.

1. Ibn Kathir, Al-Bidayah wan-Nihayah, Dar Abi Hayyan, Cairo, 1st ed. 1416/1996, Vol. 6 P. 428.
2. Nothing remains of Anbar except some mounds 3 miles north-west of the present Faluja and about a mile from the Euphrates. One can still pick up pieces of old pottery on the mounds which cover an area half a mile square. According to Yaqut (Vol. 1, p. 367), the Persians called this town Fairoz Sabur.
3. Tabari: Vol. 2, p. 575.
4. *Ibid*.
5. *Ibid*.

For the assault he selected a point where the moat was, narrowest, near the main gate of the fort. He placed his archers in a position from which they could shoot at enemy archers on that part of the wall which overlooked the crossing site, and gave them the task of preventing the enemy archers from shooting at the moat.

Khalid then ordered the collection of all the old and weak camels of the army. These jaded animals were led forward to the edge of the moat and under the covering fire provided by the Muslim archers, were slaughtered in twos and threes and thrown into the moat. Rapidly the pile of carcasses rose until it formed a firm though uneven bridge above the level of the water. Then a group of Khalid's warriors, on receiving his command, rushed on to the bridge of flesh and bone and crossed over to the far side of the moat.

As these warriors prepared to scale the wall, the gate of the fort opened and a body of Persians sallied out to drive the Muslims into the moat. There was some vicious fighting between the two groups, but the Muslims succeeded in repulsing this counterattack; and the Persians, fearing that the Muslims might get into the fort by the gate, withdrew hastily and closed the gate behind them. All this while the Muslim archers kept shooting at the Persian and Arab archers on the wall, making it impossible for them to interfere with the bridge-building and the crossing operation.

Khalid was about to order the scaling of the wall when an emissary of Sheerzad appeared on the gate and delivered another offer from the governor: he would surrender the fort if the Muslims would let him and the Persians depart in safety. Khalid took another look at the wall. He could see that it's scaling and the subsequent fighting inside the fort would not be easy. So he told the envoy that he would agree to the terms provided the Persians left all their possessions behind.

Sheerzad was only too glad to be allowed to get away, and accepted Khalid's terms with relief. The next day the Persian soldiers and their families departed for Ctesiphon and the Muslims entered the fort. The Christian Arabs laid down their arms and agreed to pay the Jizya. This happened in the second week of July 633 (end of Rabi-ul-Akhir, 12 Hijri). Over the next few days, Khalid received the submission of all the clans living in the neighbourhood of Anbar.

Sheerzad journeyed with the Persian garrison to Ctesiphon, where he was severely rebuked by Bahman. Like any ineffective commander, Sheerzad blamed his troops-in this case the Christian Arabs. *"I was among a people who have no sense,"* he lamented, *"and whose roots are among the Arabs."* 1

Khalid appointed an administrator over Anbar, and then once again set out with the army. He recrossed the Euphrates and marched south. As he neared Ain-ut-Tamr, he found a purely Arab army deployed across his path in battle array.

Ain-ut-Tamr was a large town surrounded by date plantations, and is believed to have been named after its dates: Ain-ut-Tamr means Spring of Dates. [2] Garrisoned by Persian soldiers and Arab auxiliaries, this town was in a much stronger position than Anbar to oppose the advance of Khalid. The Persian commander of Ain-ut-Tamr was Mahran bin Bahram Jabeen who was not only an able general but also a wily politician. The Persian garrison of Ain-ut-Tamr was larger, and the Arabs here belonged to the proud, fierce tribe of Namr which considered itself second to none. And there were Christian Arab clans which joined the Namr to put up a united front against the Muslims. The commander of all the Arabs was a renowned chief, Aqqa bin Abi Aqqa.

When Arab scouts brought word of the Muslims marching from Anbar in the direction of Ain-ut-Tamr, Aqqa went to the Persian commander. *"Arabs know best how to fight Arabs."* he said. *"Let me deal with Khalid."*

Mahran nodded agreement. *"True"*, he observed wisely. *"You know better how to fight Arabs. And when it comes to fighting non-Arabs you are like us."* [3]

1. Tabari: Vol. 2, p. 575.
2. Ain-ut-Tamr, of which nothing remains but a spring, was located 10 miles west-north-west of the present Shisasa. Shisasa is also called Ain-ut-Tamr these days, but the original Spring of Dates was situated as indicated above.
3. Tabari: Vol. 2, p. 576.

Aqqa was flattered by the compliment. Seeing that his words were having the desired effect, Mahran continued: *"You go and fight Khalid. And if you should need help, we shall be waiting here to come to your assistance."* [1]

A number of Persian officers were standing beside Mahran during this exchange. When Aqqa had left, they questioned their commander: *"What made you talk like that to this dog?"*

"Leave this matter to me", Mahran replied. *"I plan what is best for you and worst for them. If these Arabs win, the victory shall be ours too. If they lose, they will at least have weakened the army of Khalid, and we shall then fight when our enemy is tired and we are fresh."* [2]

The Persians remained at Ain-ut-Tamr while the Arabs, moved up about 10 miles on the road to Anbar. There, Aqqa deployed his Arab army for battle.

When Khalid arrived to face Aqqa, he was surprised to find an exclusively Arab force arrayed against him; for so far all his battles in Iraq had been fought against mixed forces of Persians and Arabs. However, he deployed his army with the usual centre and wings and placed himself in front of the centre, accompanied by a strong bodyguard. Across the battlefield, in front of the Arab centre, stood Aqqa. Khalid decided that he would take Aqqa alive.

When forming up the Muslims, Khalid had instructed the commanders of the wings to engage the enemy wings on his signal but not to attack with any great violence-only enough to tie down the enemy wings before he launched the attack of the centre. Now Khalid gave the signal, and the Muslim wings moved forward and engaged the opposing wings. For some time this action continued. Aqqa was left perplexed about why the Muslim centre was not attacking. Then Khalid, followed by his bodyguard, charged at Aqqa.

The bodyguard engaged the Arab warriors who stood near Aqqa, while Khalid and Aqqa began to duel. Aqqa was a brave and skilful fighter, prepared to give as good as he took; but to his dismay he soon found himself overpowered and captured by Khalid. When the soldiers in the Arab centre saw heir commander captive, many of them surrendered and the rest of the centre turned and fled. Its example was followed by the wings; and the Arab army, leaving many of its officers in Muslim hands, retreated in haste to Ain-ut-Tamr.

The Arabs arrived at the fort to find the Persians gone. Mahran had sent a few scouts to watch the battle and report its progress. As soon as they saw the Arabs turn their backs to Khalid, these scouts galloped back to inform Mahran of the Arab defeat. Without wasting a moment Mahran led his army out of Ain-ut-Tamr and marched off to Ctesiphon. Discovering that they had been abandoned, the Arabs rushed into the fort, closed the gates, and prepared rather uncertainly for a siege.

The Muslims arrived and besieged the fort. Aqqa and the prisoners were paraded outside the fort, so that the defenders could see that their commander and comrades were helpless captives. This had an unnerving effect on the defenders, who called for a surrender on terms, but Khalid rejected the call. There would be no terms; they could surrender unconditionally and place themselves at his mercy. The Arab elders debated the situation for a while, and then decided that an unconditional surrender involved less risk than fighting on; for in the latter case their chances of survival would be slim indeed. In the end of July 633 (middle of Jamadi-ul-Awwal, 12 Hijri) the defenders of Ain-ut-Tamr surrendered to Khalid.

On the orders of Khalid, warriors who had defended the fort and those who had fought the Muslims on the road to Anbar were beheaded. [3] These included the chief Aqqa bin Abi Aqqa. The remainder were made captive, and the wealth of Ain-ut-Tamr was taken and distributed as spoils of war.

In Ain-ut-Tamr there was a monastery in which the Muslims found 40 boys-mainly Arabs-who were being trained for the priesthood. They were all taken captive. Among these captives there was a boy called Nusair, who was later to have a son called Musa, and Musa would become famous as the Muslim governor of North Africa and the man who launched Tariq bin Ziyad into Spain.

After a few days spent in dealing with problems of organisation and administration, Khalid prepared to return to Hira. He was about to set out when he received a call for help from Northern Arabia. After a brief consideration of this request, Khalid changed the direction of his march and gave his men a new destination-Daumat-ul-Jandal.

1. Tabari: Vol. 2, p. 576.
2. *Ibid*: p. 577.
3. Tabari: Vol. 2, p. 577.

Chapter 25

(Daumat-al-Jandal Again)

After the episode of Daumat-ul-Jandal, Khalid returned to Hirah, whose inhabitants received him with singing and amusement. He heard one of them say to his companion, ***"Pass by us, for this is a day when evil is happy."*** [1]

Daumat-ul-Jandal was one of the large commercial towns of Arabia, widely known for its rich and much-frequented market. It was also an important communication centre, a meeting point of routes from Central Arabia, Iraq and Syria. In Part 1 of this book, I have described how Khalid came to Daumat-ul-Jandal during the Prophet's expedition to Tabuk and captured Ukaidar bin Abdul Malik, the master of the fort. Ukaidar had then submitted and sworn allegiance to the Prophet, but subsequent to the operations of Amr bin Al Aas and Shurahbil bin Hasanah in the apostasy, he had broken his oath and decided to have nothing more to do with Madinah. Now he ruled over a principality of Christians and pagans.

At about the time when Khalid set off from Yamamah for the invasion of Iraq, Abu Bakr had sent Ayadh bin Ghanam to capture Daumat-ul-Jandal and once again bring the northern tribes into submission. The Caliph probably intended to send Ayadh to Iraq, to assist Khalid, after this task had been completed. Ayadh arrived at Daumat-ul-Jandal to find it strongly defended by the Kalb-a large Christian Arab tribe inhabiting this region and the eastern fringe of Syria. He deployed his force against the southern face of the fort, and the situation that now developed was, from the military point of view, absurd. The Christian Arabs considered themselves to be under siege, but the routes from the northern side of the fort were open. The Muslims, engaged closely against the fort, considered themselves so heavily committed that they could not break contact. According to early historians both sides were under siege! The operations considered mainly of archery and sallies by the garrison of the fort, which were invariably repulsed by the Muslims. This state of affairs continued for several weeks until both sides felt equally tired and equally hurt by the stalemate.

Then one day a Muslim officer said to Ayadh, *"In certain circumstances wisdom is better than a large army. Send to Khalid for help."* [2] Ayadh agreed. He wrote Khalid a letter explaining the situation at Daumat-ul-Jandal and seeking his help. This letter reached Khalid as he was about to leave Ain-ut-Tamr for Hira.

It did not take Khalid long to make up his mind. The situation on the Iraq front was now stable and he had able lieutenants to deal with the Persians, should they decide to launch a counter-offensive from Ctesiphon. He sent a letter to Qaqa at Hira telling him that he would act as Khalid's deputy and command the front in his absence. He left a garrison at Ain-ut-Tamr. And with an army of about 6,000 men, he left Ain-ut-Tamr the following day to join Ayadh. Ahead of him sped Ayadh's messenger, carrying Khalid's letter, which contained nothing more than the following in verse:

> *Wait a while for the horses come racing.*
> *On their backs are lions brandishing polished swords;*
> *Regiments in the wake of regiments.*

197

The movement of Khalid was discovered by the defenders of Daumat-ul-Jandal a good many days before his arrival, and there was alarm in the fort. With their present strength they could hold off the Muslim force under Ayadh, but they would not have a chance if Khalid's army also took the field against them. In desperate haste they sent couriers racing to neighbouring tribes. The Christian Arab tribes responded spiritedly to the appeal for help. Contingents from several clans of the Ghassan and the Kalb joined the defenders of the fort, many of them camping under the fort walls because of the insufficient room within This put Ayadh in a delicate situation, and he prayed for the early arrival of Khalid.

The Christian Arab forces were led by two great chiefs: Judi bin' Rabi'a and Ukaidar. The only chief who had any personal experience of dealing with Khalid was Ukaidar, and this man had been ill at ease ever since he heard of the march of Khalid from Ain-ut-Tamr. When the clans gathered at Daumat-ul-Jandal, Ukaidar called a conference of the tribal chiefs. *"I know more about Khalid than anyone else"*, he said. *"No man is luckier than he. No man is his equal in war. No people face Khalid in battle, be they strong or weak, but are defeated. Take my advice and make peace with him."* 3

1. Ibn Kathir, Al-Bidayah wan-Nihayah, Dar Abi Hayyan, Cairo, 1st ed. 1416/1996, Vol. 6 P. 429.
2. Tabari: Vol. 2, p. 578.
3. *Ibid*

But they spurned his advice and determined to fight it out with Khalid. Ukaidar, however, had by now completely lost his nerve. He could not bring himself to face another encounter with the Sword of Allah, and one night he slipped out of the fort and set off on the road to Jordan. But it was too late. Khalid's army had just arrived and one of his mounted detachments, under Asim bin Amr, intercepted and captured the fleeing chief.

Again Ukaidar stood before Khalid. If he hoped that memories of the peaceful ending of their last encounter would kindle a spark of kindness in the heart of Khalid, he was mistaken. In Khalid's mind the situation could not be clearer: Ukaidar had broken his oath of allegiance; he was a rebel. Khalid ordered the execution of Ukaidar, and the sentence was carried out without delay. This was the end of Ukaidar bin Abdul Malik, prince of the Kinda, master of Daumat-ul-Jandal.

The following day Khalid took Ayadh under command and incorporated his detachment into his own army. He deployed Ayadh's men on the south of the fort to block the Arabian route; positioned part of his army of Iraq to the east, the north and the west of the fort, covering the routes to Iraq and Jordan; and kept the remainder back as a strong reserve. Khalid appreciated that at present the fort was strongly manned and to storm it in its present state would prove a costly operation. He therefore decided to wait, in the hope that the defenders, tiring of the siege, would sally out to fight him in the open. Then he could inflict the maximum damage upon them and storm the fort after the garrison had been weakened. He accordingly held his forces some distance back from the fort.

With the departure of Ukaidar the entire Christian Arab army had come under the command of Judi bin Rabi'a. Judi waited for the Muslims to make the first move, but the Muslims remained inactive. When some time had passed and Judi saw that the besiegers were making no attempt to close up on the fort, he became impatient for a clash with

Khalid. Consequently he ordered two sallies. One group would attack Ayadh on the Arabian route while the other, a large group comprising his own clan, the Wadi'a, operating under his direct command, would attack Khalid's camp to the north.

Ayadh drove back the Arabs who came out to attack him. Leaving behind many dead, they hastily returned to the fort and closed the gate. This group was lucky. It had only had to face an inexperienced general like Ayadh bin Ghanam and men who were not of the calibre of the hardened veterans of Khalid.

The other and larger group-the clan of Wadi'a operating under Judi-came out at the same time as the group against Ayadh, and made for Khalid, who stood back from the fort and deployed his army for battle. Seeing no move from Khalid's side, Judi became bolder. He formed up his clan for battle and advanced to meet Khalid. The two forces were now very close, and Judi imagined that he would send the Muslims, reeling from the battlefield. Then suddenly Khalid struck at Judi with the utmost violence and speed.

The Arabs never knew just what hit them. In minutes they had collapsed like a house of cards. Judi was captured along with hundreds of his clansmen, while the rest, losing all cohesion and order, fled in panic towards the fort. The Muslims were not just pursuing them; they were with them, among them, all over them. If the first to reach the gate of the fort was a Christian Arab, the second was a Muslim. The Arabs who had remained in the fort saw a horde rushing towards the gate of which at least half was Muslim. They closed the gate in the face of their comrades, and the clan of Wadi'a which had sallied out with Judi was locked out. Hundreds were made prisoner by the Muslims. The rest perished-some in the short violent battle and the rest in the pursuit to and the fighting at the gate. It was with bitterness that they recollected the counsel of Ukaidar. Such indeed was Khalid! But now it was too late.

The first part of Khalid's plan had been accomplished. He next moved the army close to the fort to let the defenders see that there was no possibility of escape, and then called upon the garrison to surrender, but the garrison refused to comply.

Khalid had Judi and his captive clansmen paraded near the fort for all to see. Then, under the horrified gaze of the defenders, Judi and the captives were beheaded. But this, instead of breaking the spirit of the defenders of Daumat-ul-Jandal, as Khalid had hoped, hardened their determination to fight to the last.

The siege continued for a number of days. Then one day Khalid stormed the fort. The defenders put up such resistance as they could, but against the superb, battle-conditioned troops of Khalid they never had a chance. Most of the garrison was slaughtered, but women and children and many youths were taken captive. This happened in about the last week of August 633 (middle-of Jamadi-ul-Akhir, 12 Hijri).

Khalid had always been attracted by beautiful women. And he appears to have had an especial fondness for the womenfolk of the chiefs who fought him. He purchased the lovely daughter of Judi and kept her as a slave!
Khalid spent the next few days in settling the affairs of Daumat-ul-Jandal. Then he set off

for Hira, taking Ayadh with him as a subordinate general. He would return to find the situation in Iraq somewhat altered, for the Persians were on the warpath again.

> *"We attacked them with mounted troops, and they saw*
> *The darkness of death around those leafy gardens.*
> *By morning they said we were a people who had swarmed*
> *Over the fertile country from rugged Arabia."*
> [Al-Qa'qa' bin Amr, commander in Khalid's army][1]

Khalid had not gone from Ain-ut-Tamr many days when word of his departure arrived at the Persian court. It was believed that Khalid had returned to Arabia with a large part of his army; and Ctesiphon breathed more easily. After a few days, this mood of relief passed and was replaced by an angry desire to throw the Muslims back into the desert and regain the territories and the prestige which the Empire had lost. The Persians had resolved not to fight Khalid again; but they were quite prepared to fight the Muslims without Khalid.

Bahman set to work. By now he had organised a new army, made up partly of the survivors of Ullais, partly of veterans drawn from garrisons in other parts of the Empire, and partly of fresh recruits. This army was now ready for battle. With its numerous raw recruits, however, it was not of the same quality as the armies which had fought Khalid south of the Euphrates. Bahman decided not to commit this army to battle until its strength had been augmented by the large forces of Christian Arabs who remained loyal to the Empire. He therefore initiated parleys with the Arabs.

The Christian Arabs responded willingly and eagerly to the overtures of the Persian court. Apart from the defeat at Ain-ut-Tamr, the incensed Arabs of this area also sought revenge for the killing of their great chief, Aqqa. They were anxious, too, to regain the lands which they had lost to the Muslims, and to free the comrades who had been captured by the invaders. A large number of clans began to prepare for war.

Bahman divided the Persian forces into two field armies and sent them off from Ctesiphon. One, under Ruzbeh, moved to Husaid, and the other, under Zarmahr, moved to Khanafis. For the moment these two armies were located in separate areas for ease of movement and administration, but they were not to proceed beyond these locations until the Christian Arabs were ready for battle. Bahman planned to concentrate the entire imperial army to either await a Muslim attack or march south to fight the Muslims at Hira.

But the Christian Arabs were not yet ready. They were forming into two groups: the first, under a chief named Huzail bin Imran, was concentrating at Muzayyah; the second, under the chief Rabi'a bin Bujair, was gathering at two places close to each other-Saniyy and Zumail (which was also known as Bashar). These two groups, when ready, would join the Persians and form one large, powerful army.

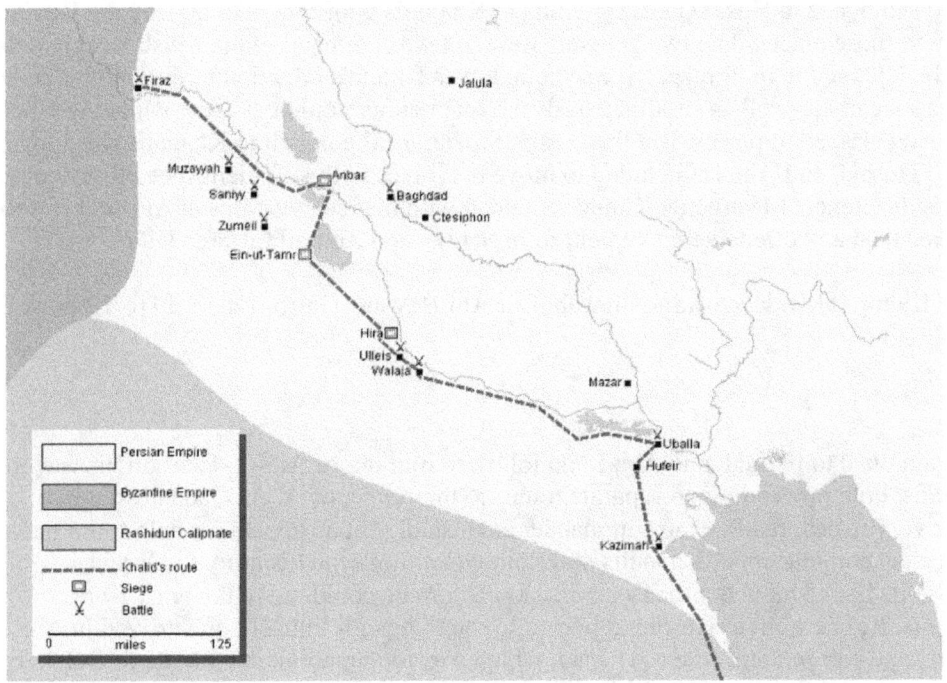

It was while these preparations were in progress that Qaqa, commanding the Iraq front in the absence of Khalid, took counter-measures. He pulled back some of the detachments which Khalid had sent across the Euphrates and concentrated them at Hira. And he sent two regiments forward-one to Husaid and the other to Khanafis. The commanders of these regiments were ordered to remain in contact with the Persian forces at these places, to delay the advance of the Persians, should they decide to push forward, and to keep Qaqa informed of Persian strengths and movements. These regiments moved to their respective objectives and made contact with the Persians. In the mean time, Qaqa kept the rest of the army in readiness to take the field.

This was the situation that greeted Khalid on his arrival at Hira in the fourth week of September 633 (middle of Rajab, 12 Hijri). The situation could assume dangerous proportions, but only if the four imperial forces succeeded in uniting and took offensive action against Hira. Any plan that the Muslims adopted would have to cater for two strategical requirements: (a) to prevent the concentration of the imperial forces into one great, invincible army, and (b) to guard Hira against the enemy in one sector while the Muslims operated against the enemy in the other.

Khalid decided to fight the operation in a way which had now become typical of him. He would take the offensive and destroy each imperial force separately in situ. With this strategy in mind, he divided the Muslim garrison of Hira into two corps, one of which he placed under Qaqa and the other under Abu Laila. Khalid sent them both to Ain-ut-Tamr, where he would join them a little later, after the troops who had fought at Daumat-ul-Jandal had been rested.

A few days later the entire Muslim army was concentrated at Ain-ut-Tamr, except for a small garrison left under Ayadh bin Ghanam to look after Hira. The army was now organised in three corps of about 5,000 men each, one of which was kept in reserve. Khalid sent Qaqa to Husaid and Abu Laila to Khanafis with orders to destroy the Persian armies at those places. The two generals were to take command of the Muslim regiments already deployed in their respective sectors. It was Khalid's intention to fight both Persian armies speedily as well as simultaneously, so that neither could get away while the other was being slashed to pieces. But this was not to be; for the march to Khanafis was longer than to Husaid, and Abu Laila failed to move his forces with sufficient speed to make up for this difference. Meanwhile Khalid remained with his reserve corps at Ain-ut-Tamr to guard against any offensive movement from Saniyy and Zumail towards Hira.

1. Ibn Kathir, Al-Bidayah wan-Nihayah, Dar Abi Hayyan, Cairo, 1st ed. 1416/1996, vol. 6 p. 426.

Qaqa marched to Husaid, and Abu Laila followed him out of Ain-ut-Tamr on his way to Khanafis, both proceeding on separate routes to their objectives. As Qaqa neared his objective, Ruzbeh, the Persian commander at Husaid, sent an appeal for help to Zarmahr, the Persian commander at Khanafis. Zarmahr would not send his army to Husaid, because he had to have Bahman's permission before he could move the army from Khanafis. But he went to Husaid in person to see things for himself, and arrived just in time to take part in the Battle of Husaid, which was fought about the middle of October 633 (first week of Shaban, 12 Hijri).

As soon as Qaqa arrived at Husaid, he deployed his corps and launched it against the Persian army, which was much larger in strength. Ruzbeh was slain by Qaqa. Zarmahr also stepped forward with a challenge which was accepted by a Muslim officer who killed him. There was no dearth of courage among the Persians, but they were nevertheless roundly defeated by Qaqa and driven from the battlefield. Leaving behind a large number of dead, the Persians retreated in haste to Khanafis, where they joined the other Persian army, now under the command of another general, named Mahbuzan.

The Persian survivors of Husaid arrived at Khanafis only a short while before the corps of Abu Laila. Reports of the Muslims' approach had been received. Being a sensible general, Mahbuzan drew the right lesson from the defeat at Husaid and decided to avoid battle with the Muslims. Setting off at once from Khanafis, he moved to Muzayyah where he joined the Arab force gathered under the command of Huzail bin Imran. So Abu Laila arrived at Khanafis to find the Persians gone. He occupied Khanafis and informed Khalid of the departure of the Persians for Muzayyah.

At Ain-ut-Tamr Khalid heard of the defeat of the Persian army at Husaid. He next heard of the movement of the second Persian army, along with the remnants of the first, from Khanafis to Muzayyah. This move left Ctesiphon uncovered and vulnerable to attack, though it would no doubt have a garrison for local defence. Muzayyah now contained the strongest concentration of imperial forces. The Arab concentrations at Saniyy and Zumail, on the other hand, ceased to be a threat to Hira, as with the reverses suffered by the Persians at Husaid and Khanafis, these Arabs were not likely to venture out of their camps with aggressive intentions.

Khalid now had a choice of three objectives: the imperial capital, the imperial army at Muzayyah, and the Arab force at Saniyy and Zumail. He considered the possibility of attacking Ctesiphon, but discarded it for two reasons. Firstly, according to Tabari, he feared the displeasure of the Caliph which he would earn by an attack on Ctesiphon. [1] Abu Bakr apparently did not wish it. Secondly, and this was a purely military consideration, by advancing to Ctesiphon he would expose his flank and rear to the strong forces at Muzayyah. These forces could then either attack him in the rear while he was engaged with Ctesiphon, or advance and capture his base at Hira, severing his communications with the desert.

Of the two remaining objectives, Khalid selected Muzayyah. The other was a smaller objective and could be dealt with later without difficulty. By now the exact location of the imperial camp at Muzayyah had been established by Khalid's agents, and to deal with this objective he designed a manoeuvre which, seldom practised in history, is one of the most difficult to control and co-ordinate-a simultaneous converging attack from three directions made *at night*.

Khalid first issued orders for the move. The three corps would march from their respective locations at Husaid, Khanafis and Ain-ut-Tamr along separate routes he had specified between the Euphrates and the Saniyy-Zumail line, and meet on a given night and at a given hour at a place a few miles short of Muzayyah. This move was carried out as planned, and the three corps concentrated at the appointed place. Here Khalid gave orders for the attack. He laid down the time of the attack and the three separate directions from which the three corps would fall upon the unsuspecting enemy. He was putting his army to a severe test of precision; only a highly efficient military machine could carry out such a finely timed manoeuvre at night.

1. For Khalid's mission is Iraq, see Note 4 on Appendix B.

And so this manoeuvre was carried out. The Persians and the Arabs slept peacefully, for the last reported locations of the Muslim corps showed them at a considerable distance and there was no apparent danger of a surprise attack. This proved to be their last night in Muzayyah. The imperial army knew of the attack only when three roaring masses of Muslim warriors hurled themselves at the camp.

In the confusion of the night and the panic of the moment the imperial army never found its feet. Terror became the mood of the camp as soldiers fleeing from one Muslim corps ran into another. Thousands were slaughtered. The Muslims struck to finish this army as completely as they had finished the army of Andarzaghar at Walaja; but large numbers of Persians and Arabs nevertheless managed to get away, helped by the very darkness that had cloaked the surprise attack.

By the time the sun rose over the eastern horizon no, living warrior of the imperial army remained at Muzayyah.. We do not know the fate of the Persian general, Mahbuzan, but the Arab commander, Huzail bin Imran, made good his escape and joined the Arab force at Zumail.

This action took place in the first week of November 633 (fourth week of Shaban, 12 Hijri). The manoeuvre had worked beautifully; the timing was perfect!

Among the Arabs who lost their lives at Muzayyah were two Muslims. These men had travelled to Madinah a short while before the invasion of Iraq and had met Abu Bakr, accepted, Islam and returned to live among their Christian clansmen. When Madinah heard of the death of these two Muslims at the hands of Khalid's army, Umar walked up to the Caliph and angrily denounced what he called the tyranny of Khalid; but Abu Bakr shrugged it off with the remark: *"This happens to those who live among infidels."* [1] Nevertheless, he ordered that blood-money be paid to their families. As for Khalid, the Caliph repeated his now famous words: *"I shall not sheathe the sword that Allah has drawn against the infidels."*

From Muzayyah, Khalid turned to Saniyy and Zumail-Saniyy was closer and thus became the first objective, for which Khalid decided to repeat the manoeuvre of Muzayyah. His army would operate in three corps as before. From Muzayyah the corps would march on separate axes and converge for the attack on Saniyy on a predetermined night and time. Khalid advanced on the direct route from Muzayyah while the other corps moved wide on his flanks. On the appointed night and at the appointed time-in the second week of November 633 (first week of Ramazan, 12 Hijri)-the three corps fell upon the Arab camp at Saniyy. This time even fewer Arabs survived the slaughter. The women and children and many youths, however, were spared, and taken captive. The Arab commander, Rabi'a bin Bujair, also met his death, and his beautiful daughter was captured; but she was not taken by Khalid. She was sent to Madinah, where she became the wife of Ali. [2]

Khalid was now manoeuvring his army with the effortlessness with which one might move pieces on a chessboard. Two or three nights after Saniyy he did the same to Zumail - three corps attacking from different directions-and the Arabs at Zumail too were swallowed up by the earthquake which hit Muzayyah and Saniyy. [3]

Once he had disposed of the captives and the booty taken at Zumail, Khalid turned his steps towards Ruzab, where Hilal, the son of Aqqa, was gathering more Arab clans to avenge his father's death. But when the Muslims arrived at Ruzab not a soul was to be seen. At the last moment these Arabs had decided that further resistance was futile and had melted away into the desert.

1. Tabari: Vol. 2, p. 581.
2. Tabari: Vol. 2, p. 582.
3. There is uncertainty about the location of these four battlefields. For an explanation see Note 6 in Appendix B.

Khalid could now sit back and rejoice over his victories. In less than a month he had crushed large imperial forces in four separate battles covering an operational area whose length measured 100 miles. He had done this by exploiting the tremendous mobility of his mounted army, by the use of audacity and surprise, and by violent offensive action. He had accomplished the mission given by the Caliph; there was no opposition left for him to crush. The Persians had ventured out of the imperial capital on hearing of Khalid's departure from Ain-ut-Tamr, but Khalid had returned and done it again. Ctesiphon withdrew into its shell.

Several raids were launched by Khalid into the region between the rivers. Places which had so far not felt the heavy hand of war now echoed to the tread of Muslim cavalry and the call of *'Allah is Great!'* But the humble masses of Iraq were left unmolested. These people considered the arrival of the Muslims a blessing; for they brought order and stability such as had not been known since the golden years of Anushirwan the Just.

But it was not in Khalid's nature to sit back and take his ease. It was in his nature to be discontented with past achievements, ever seeking fresh glory and striving towards distant horizons. The Persian capital seemed reluctant to slake his thirst for battle by sending more armies against him so it was a pleasure for Khalid to be reminded that a strong Persian garrison still existed on the Euphrates at Firaz, which marked the frontier between the empires of Persia and Eastern Rome. This was the only Persian garrison left west of Ctesiphon; and since he had been instructed by the Caliph to "fight the Persians", Khalid decided to eliminate this force also. He marched to Firaz. On arrival here in the first week of December 633 (end of Ramadhan, 12 Hijri), Khalid found two garrisons-a Persian and a Roman. These garrisons, representing empires which in the preceding two decades had fought each other in a long and costly war, now united to battle the Muslims, and were joined in this purpose by many local Christian Arab clans.

For more than six weeks nothing happened. The two armies stood and glared at each other across the Euphrates, the Muslims on the south bank and the Romans and Persians on the north bank, neither side willing to cross the river. Then, on January 21, 634 (the 15th of Dhul Qad, 12 Hijri) Khalid was able to entice the allies across the Euphrates onto his side; and their crossing was hardly complete when he attacked them with his usual speed and violence. Thousands of them were slain before the rest found safety in flight.

This was neither a great nor a decisive battle; nor was the enemy force a very large one, as some early historians have stated. (No Persian strategist in his senses would leave a powerful garrison in a peaceful frontier town like Firaz while Central and Western Iraq was being lost and Ctesiphon itself was threatened.) Its importance lies only in the fact that it was the last battle in a brilliant campaign.

Khalid spent the next 10 days at Firaz, then, on January 31 634, the army left Firaz on its way to Hira. For this march it was formed into an advance guard, a main body and a rear guard; and Khalid let it be known that he would travel with the rear guard. But as the rear guard filed out of Firaz, Khalid and a few close friends struck out on their own in a southerly direction. They were off to Makkah, to perform the Pilgrimage which was due in a fortnight. This was to be a peaceful adventure; almost an escapade!

The actual route taken by Khalid is not known. All that is known is that he and his comrades traversed a trackless waste-a difficult and inhospitable region which no guides knew and into which even bandits feared to enter. 1 But they made it. At Makkah they performed the pilgrimage inconspicuously to avoid being recognised. Then they rushed back to Iraq. The speed at which Khalid and his wild, adventurous comrades travelled can be judged by the fact that the Muslim rear guard had not yet entered Hira when Khalid rejoined it. He rode into Hira with the rear guard as if he had been there all the time! Only the commander of the rear guard had known the secret; but the men did wonder why Khalid and a few others had shaven heads! 2

Shortly after this adventure, Khalid went out on another. Tiring of the peace and quiet which now prevailed in Iraq, he decided to lead a raid in person in the area close to Ctesiphon. Along with Muthanna he raided the prosperous market of Baghdad and returned laden with spoils.

1. Tabari: Vol. 2, p. 583.
2. It is traditional for Muslims to shave the head when they perform the pilgrimage.

If Khalid had hoped that he would not be recognised in Makkah, he was mistaken. He had hardly got back from the raid on Baghdad when he received a letter from Abu Bakr warning him "not to do it again!" The warning was accompanied by another great mission: Khalid was to *proceed to Syria*. The Campaign in Iraq was over. 1

The invasion of Iraq was a splendid success. The Muslims had fought several bloody battles with Persian armies much larger in size, and they not only won every battle but also inflicted crushing defeats on the Persians and their Arab auxiliaries. And the Persian Army was the most fearsome military machine of the time!

Khalid's strategy in this campaign, and it was one from which he never deviated, was to fight his battles close to the desert, with his routes to the desert open in case he should suffer a reverse. The desert was not only a haven of security into which the Persians would not venture but also a region of free, fast movement in which he could move easily and rapidly to any objective that he chose. He did not enter deep into Iraq until the Persian Army had lost its ability to threaten his routes to the desert.

The Persian military strategy was conditioned by the political necessity of defending the imperial borders, and this led to their fighting their battles with the Muslims on the boundary between the desert and the sown, as Khalid wished. But within this political limitation, they followed a sound course and planned a massive concentration of strength for battle. Qarin should have joined Hormuz; Bahman should have joined Andarzaghar; and Ruzbeh and Zarmahr should have joined the Arab forces at Muzayyah and Saniyy-Zurmail. Had these combinations taken place, the campaign may have taken an altogether different course. But they did not take place, thanks to Khalid's fast movement and his deliberate design to bring the various armies to battle one by one, separating them from each other in time and space.

The main instruments that Khalid used to make his ambitious manoeuvres successful were the fighting quality of the Muslims and the mobility of the army. These he exploited to the limits of human and animal endurance. Though only part of his army was actual cavalry, the entire army was camel mounted for movement and could strike at the decisive place and the decisive time as its commander wished. It could move fast enough to fight a battle at A, and then be present at B for another battle before the enemy could react.

There is no record of the strength of the Persian forces which faced Khalid in the various battles, or of the casualties suffered by either side. Certain casualty figures given for the Persians are probably exaggerated. What is certain is that they were very large armies and suffered staggering losses, especially at Walaja, Ullais, Muzayyah and Saniyy-Zumail, where they ceased to exist as effective fighting forces. The Persian armies that faced Khalid at Kazima, Maqil, Walaja and Ullais probably numbered between 30,000 and 50,000 men. An enemy force up to two or three times their strength would not worry Khalid and his stalwarts. They would take it in their stride. Nor would armies of this size be too large by Persian standards. (At the Battle of Qadissiyah, fought three years later, the Persians fielded an army of 60,000 men!) As for Muslim casualties, considering that the army remained at a high level of effectiveness throughout the campaign, they must have been light.

Above all, it was the personality of Khalid that made the invasion of Iraq possible and successful against such staggering odds. He was the first of the illustrious Muslim commanders who set out to conquer foreign lands and redraw the political and religious map of the world. He imposed no hardship upon his men which he did not bear himself. It was the limitless faith which his warriors had in the Sword of Allah that made it possible for them to brave such dangers.

Khalid swept across Iraq like a violent storm. Like a violent storm he would now dash to Syria and strike the armies of another proud empire-Eastern Rome.

1. For an explanation of the dates of the battles in this campaign see Note 7 in Appendix B.

Chapter 27

(The Perilous March)

"Allah bless the eyes of Rafi, how did he succeed
In finding the way from Qaraqir to Nawa?
Five days it had marched, when the army wept;
No human ever made such a journey before!"
[A soldier who took part in the march]1

At Hira, in late May 634, Khalid opened the Caliph's letter and read:

In the name of Allah, the Beneficent, the Merciful. From the slave of Allah, Atiq, son of Abu Quhafa, 2 *to Khalid, son of Al Waleed. Peace be upon you.*
I render praise unto Allah save whom there is no Allah, and invoke blessings on His Prophet, Muhammad, on whom be the blessings of Allah and peace.
March until you reach the gathering of the Muslims in Syria, who are in a state of great anxiety ...

Khalid stopped reading, fearing that this meant demotion and that at last the pressure of Umar against him had borne fruit. And what bitter fruit! Khalid muttered, *"This must be the work of that left-handed one. He is jealous of me for conquering Iraq."* 3 But his fears turned to joy as he read on:
I appoint you commander over the armies of the Muslims and direct you to fight the Romans. You shall be commander over Abu Ubaidah and those with him.

Go with speed and high purpose, O Father of Sulaiman, and complete your task with the help of Allah, exalted be He. Be among those who strive for Allah.

Divide your army into two and leave half with Muthanna who shall be commander in Iraq. Let not more go with you than stay with him. After victory you shall return to Iraq and resume command.

Let not pride enter your mind, for it will deceive and mislead you. And let there be no delay. Lo, to Allah belongs all bounty and He is the dispenser of rewards. 4

Thus was Khalid appointed Commander-in-Chief of the Muslim forces in Syria. 5

Khalid now set about the preparations for his march. He explained the instructions of the Caliph to Muthanna, divided his army into two and handed over one half of it to Muthanna. But in the division of the army, Khalid tried to keep all the Companions of the Prophet-the Emigrants and the Ansars, men held in special esteem by the soldiers. To this Muthanna objected vehemently. *"I insist on a total execution of Abu Bakr's orders"*, he said. *"I shall have half the Companions also, for it is by their presence that I hope to win victories."* 6

Khalid saw the justice of Muthanna's claim. He revised the division to leave Muthanna a satisfactory share of the Companions, particularly as these included many of the finest officers of the army. This done, Khalid was ready for the march to Syria.

It was Abu Bakr's way to give his generals their mission, the geographical area in which that mission would be carried out, and the resources that, could be made available for that purpose. He would then leave it to his generals to accomplish their mission in whatever manner they chose. This is how he had launched Khalid into Iraq, and this is how he was now launching Khalid into Syria. The mission given to Khalid was clear: he was to move with all speed to Syria, take command of the Muslim forces and fight the Romans until victory was achieved. What route Khalid should take to get to Syria was left to him, and this was the most important immediate decision that Khalid had to take. The detailed locations of the Muslim forces in Syria were not known to him. He knew, however, that they were in the general area of Busra and Jabiya, and he had to get there fast.

1. Ibn Kathir, Al-Bidayah wan-Nihayah, Dar Abi Hayyan, Cairo, 1st ed. 1416/1996, Vol. 7 P. 10.
2. Although the Caliph is known to history as Abu Bakr, his actual name was Abdullah, and he had also been given the name of Atiq by the Holy Prophet.
3. Tabari. Vol. 2, p. 608.
4. Ibid. Vol. 2, pp. 600, 605; Waqidi: *Futuh*, p. 14 (All references to Waqidi in the remainder of this book are from his *Futuh-ush-Sham*).
5. Other versions of how Khalid assumed command in Syria suggest that he himself prevailed upon the other generals to let him command the army, or that the generals themselves appointed him commander on account of his military stature. These versions are not correct. Khalid was expressly appointed Commander-in-Chief in Syria by Caliph Abu Bakr.
6. Tabari: Vol. 2, p. 605.

There were two known routes available to Khalid for his march. The first was the southern route via Daumat-ul-Jandal whence the army could move along the normal caravan track into Syria. This was the easiest and simplest approach, with ample water on the way and no enemy to interfere with his movement. But it was also the longest route and the movement would take considerable time to complete. The Caliph had emphasised speed, as the situation of the Muslims was apparently serious. So after due consideration Khalid rejected this route.

The other route was the northern one along the Euphrates to North-Eastern Syria. This too was a well-travelled route, but it would take Khalid away from the Muslim armies, and Roman garrisons on the Euphrates would bar his way. He could, no doubt, overcome this opposition, but again there would be delay. He had to find another way of getting to the Muslim forces in Syria.

Khalid called a council of war and explained the situation to his officers. *"How can we find a route to Syria"*, he asked, *"by which we avoid the front of the Romans? They will certainly try to prevent us from going to the aid of the Muslims."* His reference was to Roman garrisons along the northern route.

"We know of no way", the officers replied, *"that could take an army, though a single man might take such a route. Beware of leading the army astray!"* [1]

But Khalid was determined to find a new route, and asked his question again. None responded except one noted warrior by the name of Raafe bin Umaira. Raafe explained that there was indeed a route through the Land of Samawa. The army could proceed from Hira to Quraqir via Ain-ut-Tamr and Muzayyah, and this would be an easy march. Quraqir was a well-watered oasis in the west of Iraq. Thence to Suwa there was a little known route which led through a barren, waterless desert. At Suwa again there was ample water, and one day's journey before Suwa there was a spring which he knew would provide sufficient water for the army. The most dangerous part of the journey was from Quraqir to this spring, about 120 miles.

But Raafe cautioned: *"You cannot take this route with an army. By Allah, even a lone traveller would attempt it at the peril of his life. It involves five days of extreme hardship without a. drop of water and the ever-present danger of losing the way."* [2]

The officers present nodded agreement. To take the army on such a route, where the entire force could get lost and die of thirst, was something that no man in his right senses would consider.

In a quiet voice Khalid said, *"We shall take this route!"* Seeing the look of alarm on the faces of his officers, he added, *"Let not your resolve be weakened. Know that the help of Allah comes according to your deserts. Let not the Muslims fear anything so long as they have the help of Allah."* [3]

The effect of his words was instantaneous. With one voice his officers replied, *"You are a man on whom Allah has bestowed His goodwill. Do as you wish."* [4] And with cheerful enthusiasm the army of Khalid set about its preparations for the march to Syria on a route that no army had travelled before and which was known only to one man, Raafe bin Umaira.

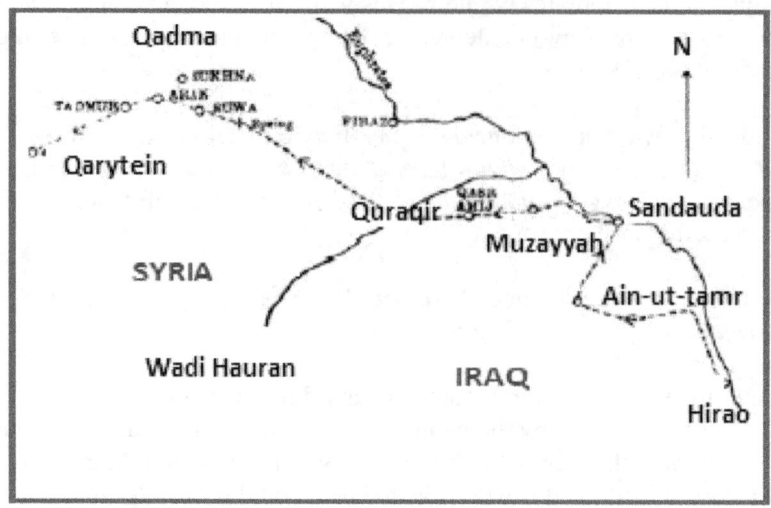

In early June 634 (beginning of Rabi-ul-Akhir, 13 Hijri) Khalid marched from Hira with an army of 9,000 men. 5 No women and children accompanied the army; they were left behind under Khalid's orders for despatch to Madinah, where they would remain until it was convenient to have them moved to Syria. The army moved via Ain-ut-Tamr, Sandauda 6 and Muzayyah to Quraqir; and Muthanna accompanied Khalid up to here before returning to Hira to resume watch over the new frontier with Persia. For the night the army camped at Quraqir and filled its water skins and other containers with supplies of water that were expected to last the men and animals five days.

1. Tabari: Vol. 2, p. 603.
2. *Ibid*: Vol. 2, p. 609.
3. *Ibid*: Vol. 2, p. 603.
4. *Ibid*: Vol. 2, p. 609.
5. The strength of the force that took part in this march has been given variously as 500, 700, 800, 6.000 and 9,000; but the last figure is the correct one. It was the strength of half the army as ordered by Abu Bakr; and all the early writers in their accounts of the campaign in Syria, have said that the Muslim forces included 9,000 men who marched with Khalid from Iraq.
6. Sandauda is the ruined Mashaihad which lies a few miles east of the present Ramadi. (Musil, p. 299)

Early next morning, as the perilous march was about to start, Raafe again approached Khalid. He seemed uncertain of himself. *"O Commander,"* he pleaded *"You cannot traverse this desert with an army. By Allah, even a lone traveller would attempt this journey only at the peril of his life."*

Khalid turned on him angrily. *"Woe to you, O Raafe"*, he said. *"By Allah, if I knew of another route to get to Syria quickly I would take it. Proceed as ordered!"* 1

Raafe proceeded as ordered and led Khalid's army of 9,000 men into the desert. As usual the men rode on camels, while the horses were led. It was the month of June when the sun beat mercilessly upon the sands of the desert, destroying all traces of life and daring man to set foot on the tortured, waterless waste. Sensible men would not do this-certainly not at this time of year; certainly not in such large numbers; and certainly not when the fate of the Muslims in Syria hung on their safe arrival. But the greatest glories of man have never been achieved by sensible men. These soldiers were not sensible men. They were the warriors of Khalid, the Sword of Allah, setting out to perform one of the greatest feats of military movement in history.

The first three days passed uneventfully. The men were oppressed by the intense heat and glare, but they were inured to hardship and as long as there was water, all was well. But the water, which was meant to last five days, finished at the end of the third day. They had another two days' journey ahead of them with not a drop of water. 2

Silently the column resumed the march on the fourth day. The heat now appeared to become more intense. There was no conversation on the march, for the men could think only of water and the horrors of getting lost in the desert and dying of thirst. They shuddered to think of what would happen if Raafe lost the way or was otherwise incapacitated. That night the men camped as usual, but there was no sleep. With the agony of fire in their throats and their tongues swollen in their mouths, they could only repeat in their minds the prayer: **Sufficient for us is Allah, and what a good protector He is! [Quran: 3-173]**

On the fifth morning began the last stage of the march which would, by Allah's will, get them to the spring which Raafe knew. Mile after weary mile the column trudged in silence. Hour after painful hour the men struggled through sandy wastes, tortured by the pitiless glare and heat. The day's march was completed and the men still lived, though most of them had reached the limits of human endurance. The column was no longer a neat, orderly formation as it had been at the start of the march. Many of the warriors were straggling in the rear of the column, hoping against hope that they would not fall by the wayside.

As the head of the column reached the area where the spring was supposed to be, Raafe the guide could no longer see. He had been suffering from opthalmia and the blinding glare of the sun had worsened the condition of his eyes. He now wrapped part of his turban over his eyes and halted his camel. The men following him were horrified to see this, and called to him piteously, *"O Raafe! We are on the point of death. Have you not found the water?"* But Raafe could no longer see. In a voice which was little more than a hoarse whisper, he said, *"Look for two hillocks like the breasts of a woman."* The column moved on, and soon after the two hillocks were identified and the guide informed accordingly.

"Look for a thorn tree shaped like a man in a sitting posture", ordered Raafe. A few scouts rode out to look for the tree, but returned a few minutes later to say that no such tree could be found.

1. Tabari: Vol. 2, p. 609.
2. For the romantic legend of the camel carrying water in a pouch in its belly, as was supposedly done on this march, see Note 7 in Appendix B.

"Lo! We belong to Allah and indeed to Him we shall return", said Raafe, quoting a Quranic verse. *"Then we all perish. But look once again."* The men looked again, and this time found the trunk of a thorn tree of which the remainder had vanished. *"Dig under its roots"*, [1] instructed Raafe. The men dug under the roots, and, in the words of Waqidi, *"water flowed out of the earth like a river!"* [2]

The men drank their fill, all the while praising Allah and invoking His blessings on Raafe. Then the animals were watered, and there was still water to spare. Hundreds of men filled their water skins and set off back on the route which they had travelled, looking for stragglers, of whom there were many. All were found and brought in alive.

The perilous march was over. They had made it. It had never been done before, and would never be done again. Khalid had reached the border of Syria, leaving behind the Roman frontier and its garrisons facing Iraq. They were now only a day's march from Suwa, where the desert ended and habitation began.

Khalid had no doubt that he and his army had gone through hell and come very near annihilation. But the real extent of the peril which they had faced was not known to him until Raafe, now smiling, came to him and said, *"O Commander, I have only alighted at this spring once, and that was 30 years ago, when as a boy I travelled hither with my father!"* [3]

In later years a certain caliph wrote to an eminent scholar and asked him for a description of the lands under Muslim rule. The scholar wrote back and gave the required description. When he came to Syria, he said, *"Know, O Commander of the Faithful, that Syria is a land of clouds and hills and winds and abundance upon abundance. It freshens the body and clears the skin, especially the land of Emessa, which beautifies the body and creates understanding and forbearance. Its waters are pure and sharpen the senses. Syria, O Commander of the Faithful, is a land of pleasant verdure and large forests. Its rivers run in the right courses, and in it camels have plenty to drink."* [4]

Indeed, Syria was a beautiful land-the fairest province of the Byzantine Empire. Its temperate climate, conditioned by the Mediterranean, provided relief from the heat of the desert and the cold of northern climes. Antioch, now in Turkey, was the capital of the Asian region of the Byzantine Empire, and second only to Constantinople in glory and political importance. The great cities of Syria-Aleppo, Emessa, Damascus-not only contined immense commercial wealth, but were also seats of culture and civilisation. Its thriving ports on the Mediterranean-Latakia, Tripolis, Beirut, Tyre, Acre, Jaffa-saw ships of the entire known world and bustled with trade and commerce.

Politically, the Syrian region consisted of two provinces. Syria proper stretched from Antioch and Aleppo in the north to the top of the Dead Sea. West and south of the Dead Sea lay the province of Palestine, which included the holy places of three great faiths and cities no less rich and sophisticated than any in the world. The Arabs of the time also spoke of the Province of Jordan, lying between Syria and Palestine; but this was more of a geographical expression that a term denoting a political and administrative unit. And all this was part of the Eastern Roman, or Byzantine Empire. To invade Syria was to invade Rome, and this was not an action to be undertaken light-heartedly.

The Eastern Roman Empire too was declining, and this decline had been going on for a much longer period than that of the Persian Empire. The latter still enjoyed a degree of stability and strength, which was due, among other factors, to the powerful Sasanid

Dynasty that had ruled in unbroken succession for the past four centuries. The Romans, on the other hand, had no such ruling dynasty, nor did they subscribe to the concept of a royal house to which the privilege of rule was confined. On the death of a ruler, the Empire fell to the most successful general or politician or intriguer.

1. Tabari: Vol 2. p. 609.
2. Waqidi: p. 14.
3. Tabari: Vol. 2, pp. 604, 609. For other versions of Khalid's route, which are mistaken, see Note 9 in Appendix B.
4. Masudi: *Muruj*, Vol. 2, pp. 61-2.

But the army of Eastern Rome was still a powerful instrument for the waging of imperial wars and, after the Persian Army, the most efficient and formidable military machine in the world. Its legions were well-equipped and ably led, and could still strike terror into the hearts of the peoples over whose lands they marched. Like any great imperial army, it was not one national unit but a heterogeneous collection of contingents from many peoples inhabiting many lands. In its ranks served Romans, Slavs, Franks, Greeks, Georgians, Armenians, Arabs and tribes from far-flung regions. These soldiers manned garrisons in the cities of Syria, most of which were fortified.

Syria, like Iraq, was partly an Arab land, especially in its eastern and southern parts. The Arabs had been there since pre-Roman times; and when Emperor Constantine made Christianity the State religion of the Empire in the early part of the fourth century A.D., these Arabs also embraced Christianity. But, the Arabs of Syria were people of no consequence until the migration of the powerful Ghassan tribe from the Yemen to Syria, which occurred a few centuries before Islam. For some time the Ghassan fought the Roman garrisons in Eastern Syria. Then as the Romans came to realise and value their martial spirit and warlike traits, they made peace with them and agreed to their living in Syria as a semi-autonomous people with their own king. The Ghassan Dynasty became one of the honoured princely dynasties of the Empire, with the Ghassan king ruling over the Arabs in Jordan and Southern Syria from his capital at Busra. The last of the Ghassan kings, who ruled at the time of Khalid's invasion, was Jabla bin Al Aiham. This man shared with Adi bin Hatim, who has been mentioned earlier in this book, the distinction of being the tallest Arab in history. His feet too touched the ground when he rode his horse! [1]

This then was the Syria, and this its political and military condition, that greeted the Muslim army in the early weeks of the thirteenth year of the Hijra.

The man who commanded the first serious military venture into Syria was a namesake of Khalid, *viz.* Khalid bin Saeed-a man whose military ability was just the opposite of Khalid's! Towards the end of 12 Hijri (beginning of 634) Abu Bakr placed him at Taima, some distance north of Madinah, with a detachment which was to act as a general reserve.

While at Taima, it occurred to Khalid bin Saeed to invade Syria; and for this project he sought the Caliph's permission. Abu Bakr had no intention of attempting the conquest of Syria with a small body of men, especially under an indifferent and untried general. But the Muslims knew little about the detailed military situation in Syria and Abu Bakr decided to let this operation proceed as a reconnaissance in force. He therefore wrote and gave Khalid bin Saeed permission to enter Syria; but cautioned him against getting involved in any serious hostilities which might threaten his withdrawal into the safety of Arabia.

Khalid bin Saeed set out with his small force, entered Syria and ran headlong into some Roman forces. The Roman commander in contact with the Muslims-a skilful tactician by the name of Bahan-lured the unwary Muslims into a trap and executed a pincer movement to encircle them. At this, Khalid bin Saeed lost his nerve and fled, leaving most of his men behind. Luckily for the Muslims, Ikrimah bin Abi Jahl was present at this action; and taking command of the situation, he extricated the Muslims from a blunder that was about to turn into a major tragedy. Ikrimah was able to save the Muslims, but inevitably the expedition bore the stigma of defeat. Khalid bin Saeed was now in disgrace, and Abu Bakr made no secret of his contempt for the man's pusillanimity and lack of skill. (Later, however, this man was allowed to join the Muslims in Syria, and he retrieved his honour by dying in battle.)

The exact location of this action is disputed. Some historians suggest that it took place at Marj-us-Suffar, south of Damascus, but it is unlikely that the expedition could have got that far before being seriously engaged by the Roman army. The benefit of this abortive venture to the Muslims, however, was that it made it clear to the Caliph that the invasion of Syria was not a matter to be taken lightly.

On return from the annual pilgrimage at Makkah in February 634, Abu Bakr issued a call to arms for the invasion of Syria. All was now quiet on the Iraq front. Khalid's campaign in Iraq had proved an unqualified success: it not only expanded the political boundaries of the Muslim State but also filled the coffers of Madinah. The Muslims therefore came to feel that if they could win against the formidable and much-feared Persians, why not also against the Romans who were not so fearsome as an imperial military power? Moreover, the promise of the new religious movement had to be fulfilled and its destiny achieved. Islam had come as a blessing for all mankind; and the message had to be conveyed to all mankind.

1. Ibn Qutaiba: p. 644.

Tribal contingents responded eagerly to the call from Madinah. They came in thousands from all over the peninsula. From as far away as Oman and the Yemen. They came mounted and armed for battle, but also brought their women and children with them. Only those who had apostatised were excluded from the summons. The concentration of the able-bodied manhood of Muslim Arabia was both begun and completed in March 634 (Muharram, 13 Hijri).

Abu Bakr now organised the available manpower into four corps, each of about 7,000 men. The commanders of these corps and the objectives given to them were as follows:

a. *Amr bin Al Aas*: Objective Palestine. Move on Eila route, then across Valley of Araba.
b. *Yazeed bin Abi Sufyan*: Objective Damascus. Move on Tabuk route.
c. *Shurahbil bin Hasanah*: Objective Jordan. Move on Tabuk route after Yazeed.
(Shurahbil had fought in the Iraq Campaign under Khalid, and had recently been sent as a messenger to Madinah, where the Caliph detained him and gave him the command of a

corps for the Syrian Campaign).
d. *Abu Ubaidah bin Al Jarrah*: Objective Emessa. Move on Tabuk route after Shurahbil.

Abu Bakr's intention was to invade Syria and take as much of it as possible. Not being aware of the size and detailed dispositions of the Roman army, he did not strengthen any one corps at the expense of the others. But he realised that the Romans could concentrate a very large army in any sector of the theatre of operations, and consequently ordered that the corps commanders would, keep in touch with each other and that any one of them could seek the help of his comrades if a serious clash with Roman forces appeared imminent on his front. In case the corps had to concentrate for one major battle, the command of the entire Muslim army would be taken by Abu Ubaidah.

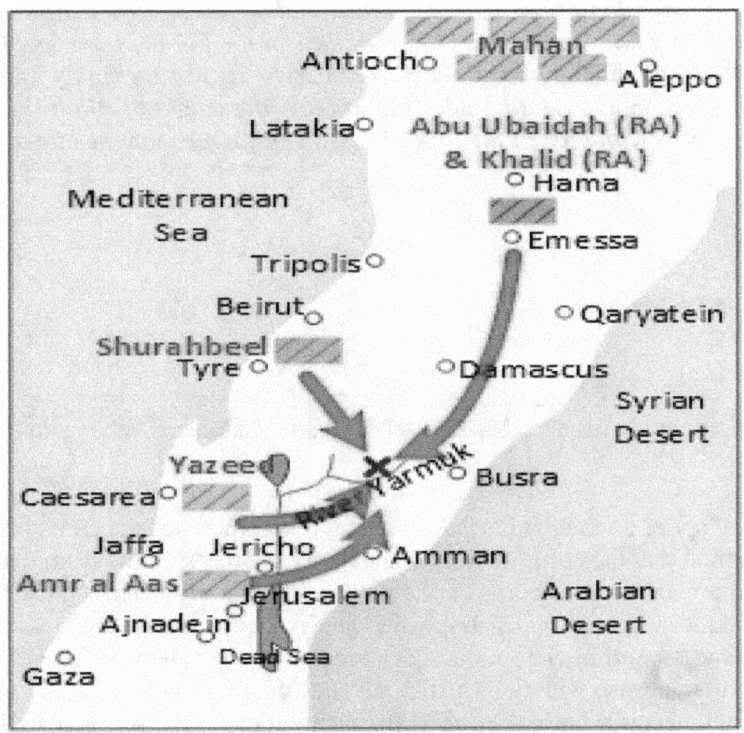

In the first week of April 634 (beginning of Safar, 13 Hijri), the Muslim forces began to move. The first to leave Madinah was Yazeed; and as his column started from its camp outside Madinah, Abu Bakr walked for a short distance by his side. His parting words to Yazeed, which he repeated to the other corps commanders, were as follows:

In your march be not hard on yourself or your army. Be not harsh with your men or your officers, whom you should consult in all matters.

Be just and abjure evil and tyranny, for no nation which is unjust prospers or achieves victory over its enemies.
When you meet the enemy turn not your back on him; for whoever turns his back, except to manoeuvre for battle or to regroup, earns the wrath of Allah. His abode shall be hell, and what a terrible place it is!

And when you have won a victory over your enemies, kill not women or children or the aged and slaughter not beasts except for eating. And break not the pacts which you make.

You will come upon a people who live like hermits in monasteries, believing that they have given up all for Allah. Let them be and destroy not their monasteries. And you will meet other people who are partisans of Satan and worshippers of the Cross, who shave the centre of their heads so that you can see the scalp. Assail them with your swords until they submit to Islam or pay the Jizya.

I entrust you to the care of Allah 1

In making this speech Abu Bakr was following the example of the Holy Prophet, who, when despatching a military expedition, would instruct its commander: **"Fight in the name of Allah: fight but do not exceed the bounds; and do not be treacherous; and do not mutilate; and do not kill women and children; and do not kill the inmates of monasteries."** 2 It is reported that Abu Bakr walked beside Yazeed for nearly 2 miles, and when Yazeed asked him to return, said, *"I heard the Messenger of Allah say that the feet that get covered with dust in the way of Allah shall not be touched by the fire of hell."* 3

1. Waqidi: p. 4.
2. Abu Yusuf: pp. 193-5.
3. *Ibid*.

With these words ringing in his ears, Yazeed set off from Madinah. The invasion of Syria had been launched.

Yazeed made good speed on the road to Tabuk. Behind him marched the corps of Shurahbil, and behind that the corps of Abu Ubaidah, each a day's march from the other. Amr bin Al Aas marched with his corps on the western route to Eila. Yazeed had advanced two or three stages beyond Tabuk when he first contacted the enemy-a force of Christian Arabs sent forward by the Romans as a reconnaissance element. These Arabs withdrew hastily after a brush with the Muslim advance guard. Following their withdrawal, Yazeed made for the Valley of Araba where it meets the southern end of the Dead Sea.

Yazeed arrived at the Valley of Araba at about the same time as Amr bin Al Aas reached Eila. Both corps now made contact with Roman forces of about equal strength which had been sent forward from the main Roman army to prevent the Muslims from entering Palestine. Both Yazeed and Amr bin Al Aas fought the Roman detachments facing them and drove them back with heavy losses. When the Romans defeated by Yazeed withdrew in precipitous haste, Yazeed sent a fast column which overtook the retreating detachment at Dasin, some distance short of Gaza, and caused it considerable damage before rejoining Yazeed at the Valley of Araba. Meanwhile Amr bin Al Aas was moving north along this valley. These engagements took place within a fortnight of the start of the Muslim march from Madinah.

While these actions were being fought by the corps of Yazeed-which had strayed from the objective given by the Caliph-Shurahbil and Abu Ubaidah continued their march northwards on the main route: Ma'an-Mutah-Amman. They were followed a little later by Yazeed. By the end of the month of Safar (early May) Shurahbil and Abu Ubaidah had got to the region between Busra and Jabiya; [1] Yazeed was camped somewhere in North-Eastern Jordan; and Amr waited by the Valley of Araba. It was at this stage that the Muslims came to realise that the Roman eagle was stirring. Indeed the Roman eagle was already on the wing!

The Emperor Heraclius was in Emessa, planning countermeasures against the Muslims. When he first heard of the crushing defeats suffered by the Persian Army at the hands of Khalid, he was not a little surprised, for he had had no higher opinion of the Arabs than did the Persian court once have. But he was not unduly worried. Then came news of the fiasco of Khalid bin Saeed, and Heraclius felt reassured. However, as a precaution, he ordered the positioning of several Roman legions at Ajnadein, whence they could operate against any Muslim force entering Palestine or Jordan.

As the Muslim corps set off from Madinah, the Roman army received intelligence of the move from Christian Arabs. Apprised of the latest situation and the direction of the Muslim movement, Heraclius realized that this was a serious attempt at the invasion of his domain. Soon after this he heard of the defeat of the Roman covering forces sent from Ajnadein at the hands of the leading corps of the Muslim army. He decided to punish these rough intruders and throw them back into the desert whence they had come. On his orders, large detachments of the Roman army began preparations for a move to Ajnadein from garrisons in Palestine and Syria.

By now the Muslim commanders had established contacts with the local population and laid the foundations of an intelligence network. They had already come to know of the existence of a Roman army at Ajnadein. A few days later they received intelligence of the movement of more Roman legions in the direction of Ajnadein; and all corps commanders sent messages to Abu Ubaidah informing him of these moves. Three corps of the Muslim army were in more or less the same region-i.e., Eastern Jordan and Southern Syria-and Abu Ubaidah at once took these corps under his command. Amr bin Al Aas was more isolated from the others and felt that the Roman preparations were being made against his corps. He therefore sought help from Abu Ubaidah.

1. Masudi (*Muruj*, Vol. 4, p. 66) gives the location of Jabiya as 2 miles from Jasim. It was a little to the west of the present Jasim-Nawa line, and after the arrival of the Muslims, became known as a military cantonment.

Some time in the middle of Rabi-ul-Awwal (third week of May), the Caliph received a message from Abu Ubaidah giving a fairly clear picture of the situation in Syria and Palestine. Muslim estimates suggested that presently the Romans would have an army of 100,000 men at Ajnadein, from where it could either strike frontally against Amr bin Al Aas, or manoeuvre against the flanks and rear of the other three Muslim corps. This estimate of Roman strength was not far from the mark, as we shall see later.

The situation had taken a turn for the worse. The Romans were in much larger strength than had been anticipated by the Muslims when the invasion was launched; and it was clear that the Romans were not going to sit in their fortified cities and await attack. They were concentrating into one great army to fight a grand offensive battle in the field. The Muslims would either have to fight a general set-piece battle with the Imperial Roman

Army or withdraw hastily into Arabia, neither of which alternatives was pleasant to contemplate. The Caliph rejected the second one outright. There was no question of returning to Arabia in face of the Roman threat. The invasion of Syria had been launched; it must be sustained. But what caused Abu Bakr the greatest anxiety was the question of who should command the Muslim army? Abu Bakr had ordered that Abu Ubaidah would take command of the army whenever the corps were united for battle. Abu Ubaidah was a wise, intelligent man, and a widely esteemed and venerated Muslim. He was also a man of unquestionable personal courage. But knowing his mild and gentle nature and his lack of experience in the command of military forces in major operations, Abu Bakr had serious misgivings about his ability to lead the entire Muslim army in a serious clash with the powerful and sophisticated army of Eastern Rome.

Abu Bakr reached the best conclusion which was possible under the circumstances: he would send Khalid bin Al Waleed to command the Muslim army in Syria! Khalid had recently shattered the Persian army in several bloody battles. Khalid would know what to do. This decision made Abu Bakr feel lighter, as if a heavy burden had been lifted off his shoulders. *"By Allah,"* he said, *"I shall destroy the Romans and the friends of Satan with Khalid bin Al Waleed!"* [1] He consequently despatched a fast rider to Hira with instructions for Khalid to move with half his army to Syria, take command of the Muslim forces and fight the Romans.

The next chapter takes up the thread of events which constituted Khalid's conquest of Syria. This subject is taken up with the full realization of the possibility of error in the account of this campaign, because of the confusion and the contradictions that exist in the narratives of the early historians. There is disagreement about many important aspects of this military history-in the dates of the great battles; in the strengths of the forces deployed in these battles, in the order in which these battles were fought; even, in the case of the odd battle, about who commanded the army at the time. The only writer who has described the campaign in meticulous detail is Waqidi; but his account also contains errors, as it is based on narratives passed down orally from the Syrian veterans, which sometimes conflict.

In this book has been prepared, from all the accounts available, a sequence of events and a version of these events which makes the most military sense and leaves the least room for contradiction. The reader has been spared copious footnotes, explaining each alternative version and each deviation from the commonly accepted version of this campaign; but he will find footnotes in the case of the more important issues, so that he may form his own opinion. And Allah knows best!

1. Tabari: Vol. 2, p. 603.

Chapter 28

(Deeper into Syria)

"O joy to Syria! O joy to Syria! O joy to Syria! For the angels of the Ever-Merciful spread their wings over it."
[Prophet Muhammad (SAWS)][1]

If the soldiers hoped that they would have a day of rest after the harrowing experience of the five days' march-which had brought them closer to annihilation than any battle could have done-they were mistaken. The very next morning Khalid set his army in motion towards Suwa. The men could not complain, for their commander himself took no rest nor looked as if he needed it. In fact as the march began and Khalid rode up and down, the column to see that all was well, the sight of their commander put fresh vigour into the soldiers, and they forgot the horrible memories of the perilous march. This day they would draw their first blood in the Syrian Campaign. They had to draw blood, for Khalid had arrived!

Khalid started his Syrian Campaign wearing a coat of, chain mail which had belonged to Musailima the Liar. At his broad leather belt hung a magnificent sword which had also belonged to Musailima the Liar. These two were trophies of the Battle of Yamamah. Over his chain helmet he wore a red turban, and under the helmet, a red cap. In this cap, if examined carefully, could be seen a few black lines; and in the eyes of Khalid this cap was more precious than all his weapons and armour. Its story shall be told at another time. In his hand Khalid carried a black standard which had been given him by the Holy Prophet. It had once belonged to the Prophet and was known as the Eagle.

With Khalid travelled 9,000 fearless fighters, veterans of many victorious battles, not one of whom would think twice before laying down his life on the orders of his beloved commander. In this army also travelled some of the bravest young officers of the time, who would perform prodigies of valour and laugh at death. There was Khalid's own son, Abdur-Rahman-just turned 18. There was the Caliph's son, also named Abdur-Rahman. There was Raafe bin Umaira, the guide on the Perilous March, who was Khalid's son-in-law and a redoubtable warrior. There was Qaqa bin Amr, the one-man-reinforcement sent to Khalid by the Caliph. And there was one young man of whom we shall hear a great deal in this campaign, Zarrar bin Al Azwar, a slim, sinewy youth whose cheerful countenance and bubbling enthusiasm could make exhausted men want to get up and fight again. Dhiraar was to become Khalid's right-hand man. He would be given the most daring missions and would show both a reckless disregard for danger and a most uncanny knack of survival.

In the early afternoon the column reached Suwa. This was the first settlement near, the border, of Syria and was an oasis surrounded by a grassy area of land used to graze large flocks of sheep and herds of cattle. Moving through this settlement, Khalid put down all resistance and commandeered the grazing flock to stock up the army's food supply for the campaign.

Next day the army arrived at Arak, which was a fortified town defended by a garrison of Christian Arabs under the command of a Roman officer. As the garrison had retired to the safety of the fort on sighting them, the Muslims laid siege to Arak. It was here that Khalid first came to know that his fame had spread beyond the lands in which he had fought. His reputation proved sufficient to bring about a peaceful surrender.

In Arak lived an old scholar who kept himself informed of the affairs of the world. When he was told of the arrival of a hostile army across the desert, he asked, *"Is the standard of this army a black one? Is the commander of this army a tall, powerfully built, broad shouldered man with a large beard and a few pock marks on his face?"* [2]Those who had seen the approach of Khalid and brought the news to Arak confirmed that it was indeed so. *"Then beware of fighting this army"*, warned the sage.

The Roman garrison commander made an offer to surrender the fort, and was astonished at the generous terms offered by the Muslims. Beyond the payment of the Jizya, the people of Arak would pay or suffer nothing. The pact was signed, the fort was surrendered, and the Muslim army camped outside for the night.

The next morning Khalid despatched two columns to subdue Sukhna and Qadma (now known as Qudaim). At the same time, he sent a camel rider to find Abu Ubaidah in the area of Jabiya and tell him to remain at his position until the arrival of Khalid or the receipt of further instructions. Then, with the main body of his army, Khalid marched to Tadmur (Palmyra).

1. Tirmidhi, Ahmad and Hakim from Zayd bin Thabit. Sahih Al-Jami' Al-Saghir No. 3920.
2. Waqidi.

When the columns sent by Khalid arrived at Sukhna and Qadma, they were received joyfully by the inhabitants, who had heard of the generous terms given the day before to Arak. They were only too willing to make friends. There was no trouble at these places and the columns returned to the army without any bloodshed.

At Tadmur, the garrison locked itself in the fort, but hardly had the Muslims arrived and surrounded the fort, when parleys were started for a peaceful surrender. Soon after a surrender was negotiated in which the inhabitants of Tadmur agreed to pay the Jizya and feed and shelter any Muslim warrior passing by their town. The Arab chief of Tadmur also presented Khalid with a prize horse, which he used in several battles of this campaign.

From Tadmur the army marched to Qaryatain, the inhabitants of which resisted the Muslims. They were fought, defeated and plundered.

The next stop was Huwareen (about 10 miles beyond Qaryatain) which contained large herds of cattle. As the Muslims started gathering in the cattle, they were attacked by thousands of Arabs. These were the local inhabitants reinforced by a contingent of the Ghassan from Busra, which had hastened to help their comrades in Huwareen. They too were defeated and plundered.

The following morning the advance was resumed in the direction of Damascus, and after three days of marching the army arrived at a pass about 20 miles from Damascus. This pass lies between the present Azra and Qutaifa and crosses a gently sloping ridge which rises gradually to a height of over 2,000 feet above the level of the surrounding countryside. The ridge is part of the range known as Jabal-ush-Sharq, which is an offshoot of the Anti-Lebanon Range and runs in a north-easterly direction to Tadmur. The pass itself, not a formidable one, is quite long. Khalid stopped at the highest part of it, and here he planted his standard. As a result of this action the pass became known as Saniyyat-ul-Uqab, i.e. the Pass of the Eagle, after the name of Khalid's standard, but is sometimes referred to as just Al Saniyya. [1] At this pass Khalid stayed an hour with his standard fluttering in the breeze, and gazed at the *Ghuta* of Damascus. From where he stood, he could not see the city itself, because it was concealed from view by a rise of ground which stretches east-west, north of the city, but he marvelled at the richness and beauty of the *Ghuta*.[2]

From the Pass of the Eagle, Khalid moved to Marj Rahit, a large Ghassan town near the present Azra on the road to Damascus. 3 The Muslims arrived in time to participate in a joyous festival of the Ghassan, which participation took the form of a violent raid! At Marj Rahit had gathered a large number of refugees from the region over which Khalid had recently operated, and these refugees mingled with the crowds celebrating the festival. The Ghassan were not unmindful of the danger which Khalid's entry into Syria posed for them. They had positioned a strong screen of warriors on the route from Tadmur, below the pass; but this screen was scattered in a few minutes by a swift charge of the Muslim cavalry. Although some Ghassan resistance continued as the Muslims advanced, it ceased once the town was reached. The Muslims raided Marj Rahit. After a little while having collected a large amount of booty and a certain number of captives, Khalid pulled out of the town and camped outside.

The following morning he sent a strong mounted column towards Damascus with the task of raiding the *Ghuta*. Then, having sent a messenger to Abu Ubaidah with instructions to report to him at Busra, Khalid himself set off for Busra with the main body of the army, by-passing Damascus. The mounted column sent to Damascus reached the neighbourhood of the city, picked up more booty and captives, and rejoined Khalid while he was still on the march.

The minor operations following Khalid's entry into Syria were now over.

1. Yaqut: (Vol. 1, p. 936) gives the location of this pass as above the *Ghuta* of Damascus, on the Emessa Road.
2. The Ghuta was, and still is, a green, fertile, well-watered plain, covered with crops, orchards and villages, lying all round Damascus, except to the west and north?west, where stand the foothills of the Anti-Lebanon Range. It formed an irregular D with its base on the foothills, and stretched up to about 10 miles from Damascus.
3. Marj Rahit, which was also a meadow, has been placed by Masudi (*Muruj*, Vol. 3, p.12: he calls it Marj Azra) 12 miles from Damascus. This would be about the centre of the meadow and the location of the town.

Abu Ubaidah had already occupied the District of Hauran which lay north-east of the river Yarmuk. Under his command he had three corps of the Muslim army-his own, Yazeed's and Shurahbil's, but he had fought no battles and captured no towns. One place which worried him a great deal was Busra, a large town which was the capital of the Ghassan Kingdom. It was garrisoned by a strong force of Romans and Christian Arabs under the command of Roman officers.

While Khalid was clearing the region of Eastern Syria, Abu Ubaidah came to know that he would come under Khalid's command upon the latter's arrival. He decided to take Busra quickly, so that Khalid would not have to worry about this problem. He therefore sent Shurahbil with 4,000 men to capture Busra. Shurahbil marched to Busra, the garrison of which withdrew into the fortified town as soon as the Muslims appeared in sight. This garrison consisted of 12,000 soldiers, but expecting that more Muslim forces would soon arrive and that Shurahbil's detachment was only an advance guard, it remained within the walls of the fort. Shurahbil camped on the western side of the town, and positioned groups of his men all round the fort.

For two days nothing happened. The following day, as Khalid set out on the last day of his march to Busra, the garrison of the town came out to give battle to the Muslims outside the city. Both forces formed up for battle; but first there were talks between Shurahbil and the Roman commander, at which the Muslim offered the usual alternatives, Islam, the Jizya, or the sword. The Romans chose the sword, and in the middle of the morning the battle began.

For the first two hours or so the fighting continued at a steady pace with neither side making any headway; but soon after midday, the superior strength of the Romans began to tell and the battle turned in their favour. The Romans were able to move forces around both Muslim flanks, and the fighting increased in intensity. The temper of the Muslims became suicidal as the real danger of their position became evident and they fought ferociously to avoid encirclement, which appeared to be the Roman design. By early afternoon the Roman wings had moved further forward, and the encirclement of Shurahbil's force became a virtual certainty. Then suddenly the combatants became aware of a powerful force of cavalry galloping in mass towards the battlefield from the northwest.

Khalid was about a mile from Busra when the wind carried the sounds of battle to him. He immediately ordered the men to horse, and as soon as the cavalry was ready, led it a gallop towards the battlefield. Beside him rode Abdur-Rahman bin Abi Bakr. But Khalid and the Romans never met. As soon as the Romans discovered the arrival of the Muslim cavalry, they broke contact from Shurahbil and withdrew hastily into the fort. The Muslims under Shurahbil came to regard this occurrence as a miracle: the Sword of Allah had been sent to save them from destruction!

Shurahbil was a brave and pious Muslim in his mid-sixties. A close Companion of the Prophet, he was one of those who used to write down the revelations of the Prophet, and consequently became know as a scribe of the Messenger of Allah. As often as not, he was addressed by this title. As a general, he was competent and sound, having learnt a great deal about the art of war from Khalid, under whom he had fought at Yamamah and in the Iraq Campaign.

It took only a glance for Khalid to assess the relative strengths of the Muslims and the Romans and he wondered why Shurahbil had not awaited his arrival before engaging the garrison at Busra. As soon as the two met and greeted each other, Khalid said, *"O Shurahbil! Do you not know that this is an important frontier town of the Romans and contains a large garrison commanded by a distinguished general? Why did you go into battle with such a small force?"*

"By the order of Abu Ubaidah", replied Shurahbil. Thereupon, Khalid remarked, *"Abu Ubaidah is a man of the purest character, but he does not know the stratagems of war."* [1]

1.

Next morning, the Roman garrison again came out of the fort to give battle. The shock of Khalid's arrival the previous day ad now worn off, and seeing that the combined strength of the Muslims was about the same as their own, the Romans decided to try their luck again. They also hoped to fight and defeat the Muslims before they could get a rest after their march. They did not know that Khalid's warriors were not used to resting!

The two armies formed up for battle on the plain outside the town. Khalid kept his center under his own command, appointing Raafe bin Umaira as the commander of the right wing and Dhiraar bin Al Azwar as the commander of the left wing. In front of the center, he placed a thin screen under Abdur-Rahman bin Abi Bakr. At the very start of the battle, Abdur-Rahman dueled with the Roman army commander and defeated him. As the Roman general fled to the safety of the Roman ranks, Khalid launched a general attack along the entire front. For some time the Romans resisted bravely, while the commanders of the Muslim wings played havoc with the opposing wings, especially Dhiraar, who now established a personal tradition which would make him famous in Syria - adored by the Muslims, and dreaded by the Romans. Because of the heat of the day, he took off his coat of mail; and this made him feel lighter and happier. Then he took off his shirt and became naked above the waist. This made him feel even lighter and even happier. In this half naked condition Dhiraar launched his assaults against the Romans and slaughtered all who faced him in single combat. Within a week, stories of the Naked Champion would spread over Syria, and only the bravest of Romans would feel inclined to face him in combat.

After some fighting, the Roman army broke contact and withdrew into the fort. At this time Khalid was fighting on foot in front of his centre. As he turned to give orders for the commencement of the siege, he saw a horseman approaching through the ranks of the Muslims. This horseman was to achieve fame and glory in the Syrian Campaign that would be second only to Khalid's.

A man in his early fifties, he was tall, slim and wiry with a slight stoop. His lean and clear-cut face was attractive, and his eyes showed understanding and gentleness. His thin beard was dyed. In his hand he held a standard such as only generals carried. This was a yellow standard and is believed to have been the standard of the Holy Prophet at the Battle of Khaibar. 1 His coat of mail did not conceal the simple and inexpensive appearance of the clothes that he wore. As he smiled at Khalid, he revealed a gap in his front teeth; and this gap was the envy of all Muslims. This was Abu Ubaidah, Son of the Surgeon, the One Without Incisors. He had lost his front teeth while pulling out the two links of the Prophet's helmet that had dug into the Prophet's cheek at the Battle of Uhud, and it is said that Abu Ubaidah was the handsomest of "those without incisors"! 2

Though called Abu Ubaidah bin Al Jarrah, his actual name was Amir bin Abdullah bin Al Jarrah. It was Abu Ubaidah's grandfather who was the surgeon (Al Jarrah), but like some Arabs he was known after his grandfather rather than his father. As a Muslim, he belonged to the topmost strata and had been very dear to the Prophet, who had once said, *"Every nation has, its trusted one; and the trusted one of this nation is Abu Ubaidah."* 3 Thereafter Abu Ubaidah had become known as the Trusted One of the Nation-Ameen-ul-Ummat. He was one of the Blessed Ten.

This was the man who had been placed under the command of Khalid, and the new army commander looked with some apprehension at the approach of the old army commander. Khalid had known Abu Ubaidah well at Madinah, and liked and respected him for his great virtue and his devout piety. Abu Ubaidah liked Khalid because of the Prophet's fondness for him and saw in him a military instrument that Allah had chosen to crush disbelief. Khalid was reassured by Abu Ubaidah's smile. As he got near, Abu Ubaidah started to dismount, for Khalid was still on foot. *"Stay on your horse"*, Khalid called to him, and he remained mounted. Khalid walked up to him, and the two top generals in Syria shook hands.

"O Father of Sulaiman," began Abu Ubaidah, *"I have received with gladness the letter of Abu Bakr appointing you commander over me. There is no resentment in my heart, for I know your skill in matters of war."*

1. *Ibid*: p. 138.
2. Ibn Qutaiba: p. 248.
3. *lbid*: p. 247.

"By Allah," replied Khalid, *"but for the necessity of obeying the orders of the Caliph, I would never have accepted this command over you. You are much higher than me in Islam. I am a Companion of the Prophet, but you are one whom the Messenger of Allah had called 'the trusted one of this nation."* [1] And on this happy note Abu Ubaidah came under the command of Khalid.

The Muslims now laid siege to Busra. The Roman commander lost hope, for he knew that most of the available reserves had either moved or were moving to Ajnadein, and doubted that any help would be forthcoming. After a few days of inactivity, he surrendered the fort peacefully. The only condition Khalid imposed on Busra was the payment of the Jizya. This surrender took place in about the middle of July 634 (middle of Jamadi-ul-Awwal, 13 Hijri).

Busra was the first important town to be captured by the Muslims in Syria. The Muslims lost 130 men in the two days of fighting that preceded this victory. The casualties suffered by the Romans and the Christian Arabs are not on record. Khalid now wrote to Abu Bakr, informing him of the progress of his operations since his entry into Syria, and sent one-fifth of the spoils which had been won during the past few weeks. Hardly had Busra surrendered when an agent sent by Shurahbil to the region of Ajnadein returned to inform the Muslims that the concentration of Roman legions was proceeding apace. Soon they would have a vast army of 90,000 imperial soldiers at Ajnadein. This acted as a reminder to Khalid that there was no time to waste.

At this time Yazeed was still south of the River Yarmuk; Amr bin Al Aas was still at the Valley of Araba; and several detachments of the corps of Abu Ubaidah and Shurahbil were spread over the District of Hauran. Khalid wrote to all commanders to march at once and concentrate at Ajnadein; and the Muslims marched, taking with them their wives and children and vast herds of sheep which served as a moving supply depot. At Ajnadein would be fought the first of the mighty battles between Islam and Christendom.

1. Waqidi: p. 23.

Chapter 29

(The Battle of Ajnadein)

"How often has a small force vanquished a large force by the permission of Allah? And Allah is with those who steadfastly persevere."
[Quran 2:249]

In the third week of July 634, the Muslim army marched from Busra; and the march of this army was an amazing sight-one that would earn the immediate disapproval of any regular, disciplined soldier. It had none of the appearance of a normal army. Its advance was more like the movement of a caravan than the march of a military force.

The soldiers of this army had no uniform of any kind, and there was no similarity in the dress that they wore. The men could wear anything they chose, including captured Persian and Roman robes. There were no badges of rank and no insignia to distinguish the commander from the commanded. In fact there were no officers so far as rank was concerned; officership was an appointment and not a rank. Any Muslim could join this army, and regardless of his tribal status would consider it an honour to serve in the ranks. The man fighting as a simple soldier one day could next day find himself appointed the commander of a regiment, or even a larger force. Officers were appointed to command for the battle or the campaign; and once the operation was concluded, they could well find themselves in the ranks again. The army was organised on the decimal system-a system started by the Holy Prophet at Madinah. [1] There were commanders of 10, 100 and 1,000 men, the latter corresponding to regiments. The grouping of regiments to form larger forces was flexible, varying with the situation.

Even in weapons and equipment there was no standard scale for this army. Men fought with whatever weapons they possessed, and had to find their own weapons either by purchase or by taking them from fallen foes. They could have any or all of the normally used weapons of the time-the lance, the javelin, the spear, the sword, the dagger and the bow. For armour they wore coats of mail and chain helmets. And these could be of any colour or design; in fact many of them had been taken from the Persians and the Romans. Most of the men mounted camels; those who possessed horses formed the cavalry.

One remarkable feature of the movement of this great army was that it was independent of lines of communication. Behind it stretched no line of supply, since it had no logistical base. Its food trotted along with the army; and if it ran out of meat, the men, women and children could live for weeks on a simple ration of dates and water. This army could not be cut off from its supplies, for it had no supply depots. It needed no roads for its movement, for it had no wagons and everything was carried on camels. Thus this army could go anywhere and traverse any terrain so long as there was a path over which men and animals could move. This ease of movement gave the Muslims a tremendous edge on the Romans in mobility and speed.

Although this army moved like a caravan and gave the impression of an undrilled horde, from the point of view of military security it was virtually invulnerable. The advance was led by a mobile advance guard consisting of a regiment or more. Then came the main body of the army, and this was followed by the women and children and the baggage loaded on camels. At the end of the column moved the rear guard. On long marches the horses were led; but if there was any danger of enemy interference on the march, the horses were mounted, and the cavalry thus formed would act either as the advance guard or the rear guard or move wide on a flank, depending on the direction from which the greatest danger threatened. In case of need, the entire army could vanish in an hour or so and be safe at a distance beyond terrain which no other large army could traverse. In this fashion the Muslims marched from Busra.

The route of the army has not been recorded; but it undoubtedly lay north of the Dead Sea, for the army arrived at Ajnadein before the corps of Amr bin Al Aas, who joined the army at Ajnadein. Had the army travelled south of the Dead Sea, Amr bin Al Aas, who was still at the Valley of Araba, would have been picked up *en route*. The army probably marched via Jarash and Jericho, then by-passed Jerusalem, which was strongly garrisoned by the Romans, and crossed the Judea Hills stretching south of Jerusalem. Beyond this range it descended into the plain of Ajnadein, arriving there on July 24. The following day Amr bin Al Aas, moving up from the Valley of Araba on the orders of Khalid, arrived at Ajnadein, and his joy knew no bounds. He had been in a state of anxiety for several weeks, expecting the Roman storm gathering at Ajnadein to break over his head any day.

1. Tabari: Vol. 3, p. 8.

The Muslims now established a camp which was a vast affair in view of the strength of the army-32,000 men, the largest Muslim force yet assembled for battle. The camp stood about a mile away from the Roman camp, which was even larger and lined the road from Jerusalem to Bait Jibreen. The opposing camps ran like two parallel lines, so laid out as to enable the armies to take the field at a moment's notice without unnecessary movement.

The Muslims had taken a week to concentrate their army at Ajnadein, a task which took the Romans more than two months. The Roman army, like any regular, sophisticated military force, needed time for its movement, and had to spend weeks in preparation-in collecting supplies, wagons and horses, and in issuing weapons and equipment. Since it travelled with thousands of wagons and carriages, it needed good roads for its movement. But over these two months the Romans had successfully concentrated an army of 90,000 men at Ajnadein under the command of Wardan, Governor of Emessa. Another general, one named Qubuqlar, acted as the Chief of Staff or the Deputy Commander-in-Chief.

The Muslims had marched to Ajnadein as a matter of choice. So long as the Roman army remained at Ajnadein, it posed no immediate threat to the Muslim corps. Only if a forward movement were undertaken by the Romans could a threat to the Muslims arise; and then the normal Arab strategy would be to pull back to the eastern or southern part of Jordan and fight a battle with their backs to the desert, into which they could withdraw in case of a reverse. The Muslims could have waited for the Romans to start the first move.

In this case, why did the Muslim army move away from the desert, and enter deep into a fertile, inhabited region towards a Roman army three times its size? The answer lies in the character of Khalid. It was his destiny to fight battles, and the promise of battle drew him like a magnet. Twelve centuries later another illustrious general, Napoleon, would say, *"Nothing pleases me more than a great battle."* So it was with Khalid. If anyone else had been the commander of the Muslim army, it is doubtful that the Muslims would have moved to Ajnadein.

In the long run, Khalid's decision was the right one. With a large Roman army poised at Ajnadein, the Muslims would have remained tied down to the area occupied by them, which in itself was of little importance. This Roman threat, cleverly engineered by Heraclius, had to be eliminated before the invasion could proceed deeper into Syria. So it came about that the Romans and the Muslims faced each other in their respective camps at Ajnadein. Guards and outposts were positioned by both armies to prevent surprise. The officers rode across the land, carrying out reconnaissances, while the men made their preparations for battle.

The sight of the gigantic Roman camp had a somewhat disturbing impact on the Muslims. Everyone knew the strength of the Roman army-a staggering figure of 90,000. The majority of the Muslims had never taken part in a great battle. The only men who were left unmoved by the sight of the Roman camp were Khalid's 9,000 veterans, who had fought regular battles with large armies in Iraq; but even they had never before faced an army of this size.

Khalid went round visiting the various units in the camp and spoke to their commanders and men. He said, *"Know, O Muslims, that you have never seen an army of Rome as you see now. If Allah defeats them by your hand, they shall never again stand against you. So be steadfast in battle and defend your faith. Beware of turning your backs on the enemy, for then your punishment will be the Fire. Be watchful and steady in your ranks, and do not attack until I give the order."* [1] The personality of their commander and the supreme confidence which emanated from him had a marvellously steadying effect on the Muslims.

In the opposing camp, Wardan called a council of war and spoke to his generals. *"O Romans,"* he said, *"Caesar has placed his trust in you. If you are defeated, you will never again be able to make a stand against the Arabs; and they shall conquer your land and ravish your women. So be steadfast. When you attack, attack as one man-do not disperse your efforts. Seek the help of the Cross; and remember that you are three to each one of them."* [2]

1. Waqidi: p. 35.
2. *Ibid*.

As part of his preparations for battle, which in fact did not take place until some days later, Khalid decided to send a brave scout to carry out a close reconnaissance of the Roman camp. Dhiraar volunteered for the job and was sent forward accordingly. The youth stripped to the waist and rode up to a little hillock not far from the centre of the Roman camp. Here he was seen, and a body of 30 Romans rode out to catch him. As they approached, Dhiraar began to canter back to the Muslim camp; and when they drew nearer, Dhiraar increased his pace. His purpose was to draw these Romans away from their camp, so that others should not be able to come to their assistance. When he had reached a spot between the two armies, Dhiraar turned on his pursuers and attacked the one nearest him with his lance. After bringing him down, Dhiraar assaulted a second and a third and a fourth and so he continued, throughout the combat manoeuvring his horse in such a way that he should not have to tackle more than one man at a time. Against some he used his sword also; and it is believed that he killed 19 of the Romans before the remainder turned and galloped back to their camp. That night the Roman camp was full of stories of the dreaded Naked Champion.

On his return Dhiraar was greeted with joy by the Muslims; but Khalid looked at him sternly and rebuked him for engaging in combat when the task given to him was reconnaissance. To this Dhiraar replied that he was conscious of the possible disapproval of his commander, and that but for this he would have pursued the fleeing Romans to kill every one of them!

Following this incident, Qubuqlar, the Roman deputy commander, sent a Christian Arab to enter the Muslim camp, spend a day and a night with the Muslims and gather all possible information about the strength and quality of the Muslim army. This Christian

Arab had no difficulty in entering the Muslim camp, as he was taken for a Muslim. The following day he slipped out and returned to Qubuqlar, who questioned him about the Muslims, *"By night they are like monks, by day like warriors"*, said the spy. *"If the son of their ruler were to commit theft, they would cut off his hand; and if he were to commit adultery, they would stone him to death. Thus they establish righteousness among themselves."*

"If what you say be true", remarked Qubuqlar, *"it would be better to be in the belly of the earth than to meet such a people upon its surface. I wish it were my portion from Allah to stay away from them, so that He would not have to help either me against them or them against me."* 1

Wardan, the Commander-in-Chief, was full of fight; but Qubuqlar had lost his nerve.

Early in the morning of July 30, 634 (the 28th of Jamadi-ul-Awwal, 13 Hijri), as the men finished their morning prayers, Khalid ordered the move to battle positions, detailed instructions for which had been given the day before. The Muslims moved forward and formed up for battle on the plain a few hundred yards ahead of the camp. Khalid deployed his army facing west on a front of about 5 miles, stretched sufficiently to prevent the more numerous Roman army from overlapping his flanks. The army was deployed with a centre and two strong wings. On either side of the army, next to the wing, as an extension of the front, was positioned a flank guard to counter any Roman attempt to envelop the Muslim flanks or to outflank their position entirely.

The centre was placed under Muadh bin Jabal, the left wing under Saeed bin Amir and the right wing under Abdur-Rahman, the Caliph's son. We also know that the left flank guard was commanded by Shurahbil, but the name of the commander of the right flank guard is not recorded. Behind the centre, Khalid placed 4,000 men under Yazeed, as a reserve and for the close protection of the Muslim camp in which the women and children stayed. Khalid's place was near the centre, where he kept a number of officers near him to be used as champions or as commanders of groups needed for any specific task in battle. These included Amr bin Al Aas, Dhiraar, Raafe and Umar's son, Abdullah.

When the Romans saw the Muslims moving, they also rushed out and began to form up in their battle positions about half a mile from the Muslim front line. They formed up on about the same frontage, but had much greater depth in their dispositions, the detailed layout of which is not known. Wardan and Qubuqlar stood surrounded by their bodyguards in the centre. The massive formations of the Romans, carrying large crosses and banners, were an awe-inspiring sight.

1. Tabari: Vol. 2, p. 610.

When his men had been formed up for battle, Khalid rode along the front, checking units and urging his warriors to fight in the way of Allah. In the few words that he said to each unit, he laid emphasis on concentrating their efforts in time and attacking as one man. *"When you use your bows,"* he said, *"let the arrows fly from your bows as if shot by a single bow to land like a swarm of locusts on the enemy."* He even spoke to the women in the camp and told them to be prepared to defend themselves against any Romans who might break through the Muslim front. They assured him that this was the least that they could do, considering that they had not been allowed to fight in the forefront of the Muslim army!

Forming their positions took the two armies a couple of hours. When all was in readiness, an old bishop wearing a black hat emerged from the Roman centre, walked up half-way towards the Muslim army and called out in perfect Arabic, *"Which of you will come forth and talk with me?"*

Muslims have no priests; and in those days the commander himself acted as the *Imam* [1] of the army. Hence Khalid rode forward, and the bishop asked, *"Are you the commander of this army?"* Khalid replied, *"So they regard me as long as I obey Allah and follow the example of His Prophet; but if I fail in this, I have no command over them and no right to be obeyed."* The bishop thought for a moment, then remarked, *"It is thus that you conquer us."*

He then continued: *"Know, O Arab, that you have invaded a land which no king dares to enter. The Persians entered it and returned dismayed. Others also came and fought with their lives, but could not attain what they sought. You have won over us up till now, but victory does not belong permanently to you"*

"My master, Wardan, is inclined to be generous with you. He has sent me to tell you that if you take your army away from this land, he will give each of your men a dinar, a robe and a turban; and for you there will be a hundred dinars and a hundred robes and a hundred turbans."

"Lo, We have an army numerous as the atoms, and it is not like the armies that you have met before. With this army Caesar has sent his mightiest generals and his most illustrious bishops." [2]

In reply Khalid offered the usual three alternatives; Islam, the Jizya or the sword. Without the satisfaction of one of these alternatives the Muslims would not leave Syria. As for the dinars and the fine clothes, Khalid pointed out that the Muslims would soon possess them anyway, by right of conquest!

With this reply, the bishop returned and informed Wardan of his talks. The Roman commander was furious and swore that he would crush the Muslims with one all-destroying attack.

Wardan now ordered a line of archers and slingers to be positioned ahead of the Roman front within range of the Muslim army. As this line formed up, Muadh the commander of the Muslim centre, began to order his men to attack, but was stopped by Khalid who stood nearby. *"Not till I give the order"* said Khalid. *"And not till the sun has passed its zenith."* [3]

Muadh had wished to attack because the Roman archers, with their better bows, outranged the Muslim bows and to the slingers the Muslims had no effective counter. The only way to deal with the situation would be to get closer to the Romans-to come to grips. But Khalid did not wish to risk a reverse by launching a premature attack against the well-formed legions of the Romans. Thus a couple of hours before noon, the battle began with the action of the Roman archers and slingers.

1. One who leads the prayer.
2. Waqidi: p. 36.
3. *Ibid*.

This phase of the battle went against the Muslims, several of whom were killed while many were wounded. This suited the Romans very well; and for some time the missiles continued to fly from their bows and slings. The Muslims, unable to do anything to offset this Roman advantage, became impatient to attack with sword and lance, but still Khalid restrained them. Finally the impetuous Dhiraar came to Khalid and said, *"Why are we waiting when Allah, the Most High, is on our side? By Allah, our enemies will think that we are afraid of them. Order the attack, and we shall attack with you."* Khalid decided to let individual champions go into combat against Roman champions.

In this duelling the Muslims would have the advantage, and it would be useful to eliminate as many of the Roman officers as possible, as this would in turn reduce the effectiveness of the Roman army. ***"You** may attack, Dhiraar"*, he said. [1] And the delighted Dhiraar urged his horse forward.

Because of the Roman archers, Dhiraar kept on his coat of mail and helmet, and in his hand carried a shield made of elephant hide, which had once belonged to a Roman. Having gone halfway to the Roman line, he stopped and raising his head, gave his personal battle cry:

> *I am the death of the Pale Ones;*
> *I am the killer of the Romans;*
> *I am a scourge sent upon you;*
> *I am Dhiraar bin Al Azwar!* [2]

As a few of the Roman champions advanced to answer his challenge, Dhiraar quickly disrobed; and the Romans knew him at once as the Naked Champion. In the next few minutes, Dhiraar killed several Romans, including two generals, one of whom was the governor of Amman and the other the governor of Tiberius.

Then a group of 10 officers emerged from the Roman army and moved towards Dhiraar. At this move, Khalid picked 10 of his stalwarts, and riding up, intercepted and killed the Romans. Now more champions came forward from both sides, some individually, others in groups. Gradually, the duelling increased in extent and intensity, and continued for about two hours, during which the Roman archers and slingers remained inactive. This phase more than restored the balance in favour of the Muslims, for most of the Roman champions were killed in combat.

While this duelling was still in progress-and it was now past midday-Khalid ordered a general attack; and the entire Muslim front moved forward and hurled itself at the Roman army. The main battle was now on with sword and shield.

This was a frontal struggle with no fine manoeuvre and neither side attempting to outflank the other. It was a hard slogging match at close quarters, and continued for some hours. Then in the late afternoon both sides, now very tired, broke contact and fell back to their original lines. No more could be done on this day.

The losses of the Romans were staggering. Wardan was shocked to learn that thousands of his soldiers lay dead on the battlefield, while very few Muslims had been accounted for. He called a council of war, at which he expressed his misgivings about the outcome of the battle, but his generals swore that they would fight to the last. Wardan asked for ideas; and of the various suggestions made, the one that appealed to him most was a plot to kill the Muslim commander. According to this plan, Wardan would personally go forward in the morning, offer peace and ask for Khalid to come forth and discuss the terms with him. When Khalid had approached near enough Wardan would engage him in combat; then, on his signal, 10 men, suitably concealed nearby, would rush up and cut the Muslim commander to pieces. It was as simple as that. Wardan was a brave general and agreed to the plan. The men would be positioned during the night, and would be carefully briefed for their role.

The Roman commander then sent a Christian Arab named David, who was a member of his staff, with instructions to proceed to the Muslim army and seek Khalid. He was to say to the Muslim commander that sufficient blood had been shed; that there should be no more fighting; that they should make peace; and that Khalid should meet Wardan early next morning between the two armies to discuss terms of peace. Both generals would appear alone.

1. Waqidi: p. 36.
2. *Ibid*: p. 37.

David was horrified to hear these instructions, as they appeared to be against the orders of Heraclius to fight the Muslims and throw them back into the desert. He therefore refused to carry out this mission. Wardan then told him the entire plan of the plot in order to convince him that he intended no disobedience of the instructions of the Emperor. And this, as we shall see, was a mistake.

The sun had not yet set when David walked up to the Muslim army, which was still arrayed in battle order, and asked to see Khalid on a matter of peace proposed by Wardan. As soon as Khalid was informed, he came out to David and stood glaring at him.

The sight of Khalid with his 6 feet and more of bone and muscle could have an unnerving effect on any man at whom Khalid glared. His hard, weather-beaten, battle-scarred face and his piercing eyes gave the impression of pitilessness to those whom Khalid regarded as enemies. The effect on poor David was devastating. Wilting under the gaze of the Sword of Allah, he blurted out: *"I am not a man of war! I am only an emissary!"*

Khalid drew closer. *"Speak!"* he ordered. *"If you are truthful you will survive. If you lie you shall perish."*

The Christian Arab spoke: *"Wardan is pained by all this unnecessary bloodshed and wishes to avoid it. He is prepared to sign a pact with you and spare those who still live. There should be no more fighting until the talks are completed. He proposes that you and he meet alone between the two armies in the morning and discuss terms of peace."*

"If what your master intends is deceit," replied Khalid, *"then by Allah, we ourselves are the root of trickery and there is none like us in stratagem, and guile. If he has a secret plot, it will only hasten his own end and the annihilation of the rest of you. If on the other*

hand he is truthful, then we shall not make peace except on the payment of the Jizya. As for any offer of wealth, we shall soon take it from you anyway." [1]

Khalid's words, uttered with unshakeable conviction, had a profound effect on David. Saying that he would go and convey Khalid's message to Wardan, he turned and began to walk away while Khalid stood staring after him and sensing that all was not as it seemed. David had not gone far before it suddenly struck him that Khalid was right; that victory would go to the Muslims and the Romans would perish no matter what tricks they tried. He decided to save himself and his family by confessing the truth. Consequently he retraced his steps and once again stood before Khalid, to whom he revealed the entire Roman plot, including the place at which the 10 Romans would lie concealed - below a hillock a little to the right of the Roman centre. Khalid promised to spare David and his family on condition that he did not tell Wardan that the Muslims now knew of his plot. To this, David agreed.

On his return to the Roman Army, David informed Wardan of the initial talks he had had with Khalid and Khalid's agreement to the rendezvous as planned; but said nothing of the second conversation he had with the Muslim. Wardan was delighted.

At first Khalid thought of going to the hillock alone and killing all 10 Romans himself. His adventurous soul thrilled at the prospect of a glorious fight. But when he discussed the matter with Abu Ubaidah, the latter dissuaded him and suggested that he should detail 10 valiant fighters instead. To this Khalid agreed. The 10 men he chose included Dhiraar, who was appointed leader of the party. He instructed Dhiraar to be prepared to next morning to dash out from the front rank of the Muslims and intercept and kill the 10 Romans when they appeared. But Dhiraar was no less adventurous in spirit that Khalid and insisted that he and his men be allowed to use the hours of darkness to find the Romans in their place of concealment and kill then in their lair. Knowing Dhiraar as he did, Khalid acceded to his request. Shortly before midnight Dhiraar and his nine comrades set off from the camp.

1. Waqidi: p. 39.

Soon after sunrise, Wardan came forward in full imperial regalia, wearing bejewelled armour with a bejewelled sword hanging at his side. Khalid walked up from the Muslim centre and stood in front of Wardan. The two armies were already arrayed in battle order as on the previous day.

Wardan started negotiations with an attempt to browbeat the Muslim. He expressed his low opinion of the Arabians; how wretched were the conditions in which they lived, and how miserably starved they were in their homeland. Khalid's response was sharp and aggressive. *"O Christian dog!"* he snapped. *"This is your last chance to accept Islam or pay the Jizya."* [1] At this, Wardan, without drawing his sword, sprang at Khalid and held him, at the same time shouting for the 10 Romans to come to his aid.

From behind the hillock he saw, out of the corner of his eye, 10 Romans, emerge and race towards him. Khalid also saw then and was horrified, for he was expecting to see Muslims emerge from behind the hillock. He has made no other arrangement for his own protection, and he wondered with a sense of deep sorrow, if Dhiraar had at last met his match. As the group of Romans got nearer, however, Wardan noticed that the leader of these 'Romans' was naked to the waist; and then the terrible truth dawned upon him.

During the night Dhiraar and his nine comrades had got to the hillock, killed all 10 Romans noiselessly, and then, such was Dhiraar's impish sense of humour, put on the garments and armour of the Romans. Later, however, Dhiraar discarded the garments and reverted to his normal fighting dress! As the first light of dawn appeared, these 10 Muslims said the prayer of the Morning and then awaited the call of the Roman commander.

Wardan left Khalid and stepped back, looking on helplessly as the 10 Muslims surrounded the pair. Dhiraar now advanced with drawn sword. At this Wardan implored Khalid, *"I beseech you, in the name of whatever you worship, to kill me yourself; do not net this devil come near me"* 2

In reply Khalid nodded to Dhiraar, and Dhiraar's sword flashed in the sun and severed Wardan's head.

It was Khalid's way so to time his attack as to get the maximum benefit from any tactical advantage which he had gained over his enemy. When no other advantage was possible and manoeuvre was restricted, he would exploit the psychological effect of killing the enemy commander-in-chief or some other prominent general, and strike a powerful blow with the entire army while the enemy was stunned by the moral setback of such a loss. Here again Khalid did the same. As soon as Wardan was killed, he ordered a general attack: the centre, the wings and the flank guards swept forward and assaulted the Romans, who were now under the command of Qubuqlar.

As the two armies met, another phase of violent hand-to-hand fighting began. Soon the fighting became vicious, with no quarter given or taken. The Muslims struck fiercely at the Roman formations, and the Romans struggled desperately to hold the assault. Khalid and all his officers fought in front of the men, and so did many of the Roman generals who were prepared to die for the glory of the empire. The battlefield soon turned into a wreckage of human bodies, mostly Roman, as the men struggled mightily without respite.

At last, as the two sides were reaching the point of exhaustion, Khalid threw his reserve of 4,000 men under Yazeed into the centre; and with the added impetus of this reinforcement, the Muslims broke through at several places, driving deep wedges, into the Roman army. In the centre a Muslim group got to where Qubuqlar stood with his head wrapped in a cloth, and killed him. It is believed that Qubuqlar had ordered his head to be so wrapped because he could no longer bear to see such carnage. With the death of Qubuqlar, the Roman resistance weakened, and soon after collapsed entirely. The Romans fled from the field of battle.

It was safer to stand and fight the Muslim Arabs in battle than to run from them. Against a fleeing enemy, the Arab of the desert was in his element. As the Romans sought to escape, they turned in three directions; some fled towards Gaza, others towards Jaffa, but the largest group of fugitives made for Jerusalem. Khalid forthwith launched his cavalry in several regiments to pursue the enemy on all three routes; and at the hands of this cavalry the Romans suffered even more grievous damage than in the two days of fighting on the plain of Ajnadein. The pursuit and the killing of the fugitives continued till sunset, when the pursuing columns returned to camp.

1. MISSING REFERENCE
2. MISSING REFERENCE
3. Tabari: Vol. 2, pp. 610-611.

The Roman army had been torn to pieces.

It was a complete victory. The Romans had been fought in a set-piece battle after the regular imperial fashion, and were not only defeated tactically but also slaughtered mercilessly. The Roman army assembled at Ajnadein had ceased to exist as an army, although a sizable Portion of it managed to get away, especially the part that fled to Jerusalem and found safety within its walls. In the first great encounter between Islam and Byzantium, the followers of Muhammad had conquered.

It had been a full and fierce battle, but without any fine manoeuvres. The Roman army had not attempted any outflanking movement, since it was too large and too cumbersome to do so. The Muslims had not because their army was comparatively small, and manoeuvres against the flanks and rear of the enemy could only have been carried out by weakening the centre-a clearly unjustifiable risk. Hence this had been a frontal clash of massed bodies of men in which Muslim leadership and the courage and skill of the warriors prevailed over the great size of the Roman legions. The only choice of manoeuvre available to Khalid had been to time his assaults to get the maximum benefit from the prevailing situation, which he did as has been described. And of course, when the Roman army broke, Khalid showed his typical drive by organising the pursuit to ensure that as many Romans as possible were brought down before the rest reached a place of safety.

Victory in the Battle of Ajnadein opened the way for the conquest of Syria. This land could not, of course, be conquered with a single battle; for large imperial forces remained in the cities of Syria and Palestine, and the Roman Emperor could draw on the resources of the whole Empire, which stretched from Armenia to the Balkans. But the first great clash With the Romans was over; and the Muslims could now continue their campaign with the confidence that they would have no less success in the mighty battles that undoubtedly lay ahead.

Three days after the battle, according to Waqidi, Khalid wrote to Abu Bakr and informed him of the battle, giving the Roman casualties as 50,000 dead at the cost of only 450 Muslims. [1] The Roman Commander-in-Chief, his deputy, and several top generals of the Roman army had been killed. Khalid also informed the Caliph that he would shortly march on Damascus. At Madinah the news of this victory was received with joy and shouts of *Allah-o-Akbar*, and more volunteers came forward to join the holy war in Syria. These included Abu Sufyan, who, along with his wife, the redoubtable Hind, journeyed to Syria to join the corps of his son, Yazeed. In reply to Khalid's letter, Abu Bakr wrote to him to besiege Damascus until it was conquered, and thereafter attack Emessa and Antioch. Khalid was not, however, to advance beyond the northern frontier of Syria.

Heraclius was at Emessa when the news of the crushing defeat of the Roman army struck him like a bolt from the sky. Heraclius felt devastated. He journeyed to Antioch; and expecting the Muslims to advance on Damascus, ordered the remnants of the Roman army at Jerusalem (but not its local garrison) to intercept the Muslims at Yaqusa and

delay their advance. At the same time he ordered more forces into motion towards Damascus to strengthen that city and prepare for a siege.

A week after the Battle of Ajnadein, Khalid marched with the Muslim army and, again by-passing Jerusalem from the south, moved towards Damascus. At Fahl, which held a strong Roman garrison, he left a mounted detachment under Abul A'war to keep the garrison tied down in the fort; with the rest of the army he moved on and reached the bank of the River Yarmuk at Yaqusa, [2] where he was again faced by Roman forces on the north bank. The Romans were not in a position to offer serious resistance, as they were still shaken by their defeat at Ajnadein; their main purpose here was only a rear-guard action to gain more time for the reinforcement of Damascus. Nevertheless a battle did take place at Yaqusa in mid-August 634 (mid-Jamadi-ul-Akhir, 13 Hijri), and the Romans were again defeated. [3]

The Romans fell back in haste; and Khalid advanced upon Damascus.

1. Waqidi: p. 42.
2. Also known as Waqusa.
3. Some early writers, including Tabari, appear to have confused this action at Yaqusa with the Battle of Yarmuk, which was fought in the same general area, and have given the year of Yarmuk as 13 Hijri, which is incorrect.

Chapter 30

(The Conquest of Damascus)

"Damascus is one of the most blessed cities of al-Sham (Syria, Jordan, Palestine)."
[Prophet Muhammad (SAWS)][1]

Damascus was known as the paradise of Syria. A glittering metropolis which contained everything that makes a city great and famous, it had wealth, culture, temples and troops. It had history. The main part of the city was enclosed by a massive wall, 11 metres high, [2] but outside the battlements lay some suburbs which were not protected. The fortified city was a mile long and half a mile wide and was entered by six gates: the East Gate, the Gate of Thomas, the Jabiya Gate, the Gate of Faradees, the Keisan Gate and the Small Gate. Along the north wall ran the River Barada, which, however, was too small to be of military significance.

At the time of the Syrian campaign, the Roman Commander-in-Chief at Damascus was Thomas, son-in-law of Emperor Heraclius. A deeply religious man and a devout Christian, he was known not only for his courage and skill in the command of troops but also for his intelligence and learning. Under him served, as his deputy, a general by the name of Harbees about whom little is known except that he was there.

The general who was in active command of the garrison, however, was Azazeer, a veteran soldier who had spent a lifetime campaigning in the East and had acquired fame in countless battles against the Persians and the Turks. He was acknowledged as a great champion and was proud of the fact that he had never lost a duel. Having served in Syria for many years, he knew Arabic very well and spoke it fluently.

Azazeer's garrison consisted of no less than 12,000 soldiers, but Damascus as a city had not been prepared for a siege. Although its walls and bastions were in good order, nothing had been done for the storage of food and fodder-a task which, for a garrison and a population so large, would take weeks and months. The Romans can hardly be blamed for this neglect, for ever since the final defeat of the Persians by Heraclius in 628, there had been no threat of any kind to Syria; and it was not until the Battle of Ajnadein had been fought that the Romans realised the full extent of the danger which threatened them.

Heraclius, working from his headquarters at Antioch, now set about the task of putting things right and preparing Damascus for a siege. Having ordered the remnants of the army of Ajnadein to delay the Muslims at Yaqusa, he sent a force of 5,000 soldiers from Antioch to reinforce the garrison of Damascus. This force was placed under a general named Kulus, who promised the Emperor that he would bring the head of Khalid on a lance. [3] Kulus arrived at Damascus at about the time when the battle of Yaqusa was fought. The strength of the garrison at Damascus was thus raised to 17,000 men; but Kulus and Azazeer were professional rivals and there was little love lost between them. Each wished to see the downfall of the other.

Thomas worked feverishly to prepare the city for a siege. Provisions were rapidly gathered from the surrounding countryside to sustain the garrison and the inhabitants in case the lines of supply were severed by the besiegers. However, not enough could be gathered for a long siege. Scouts were sent out to watch and report on the movement of the Muslims; and the bulk of the army, leaving strong guards and a reserve in Damascus, was ordered to prepare to fight a battle outside Damascus. The idea was to defeat and drive back the Muslims before they could invest the city; but it was with mounting anxiety that the Damascenes awaited the arrival of Khalid.

1. Abu Dawud, Ahmad and Hakim from Abu Darda. Sahih Al-Jami' Al-Saghir No.2116.
2. Damascus City has risen 4 metres since then, so that the wall is now only 7 metres above ground level.
3. Waqidi: p. 20.

Khalid had by now organized a military staff-a simple beginning of what later in military history would emerge as the General Staff. He had collected from all the regions in which he had fought-Arabia, Iraq, Syria and Palestine-a small group of keen and intelligent men who acted as his 'staff officers', mainly functioning as an intelligence staff. [1] They would collect information, organize the despatch and questioning of agents, and keep Khalid up to date with the latest military situation. Intelligence was one aspect of war to which Khalid paid special attention. Ever watchful and ever ready to exploit fleeting opportunities, it was said of him that "he neither slept nor let others sleep, and nothing was concealed from him." [2] But this was a personal staff rather than the staff of an army headquarters; wherever Khalid went, this staff went with him.

Khalid had also made a notable change in the organisation of the army. From his army of Iraq, which after Ajnadein numbered about 8,000 men, he had organised a force of 4,000 horsemen, which the early historians refer to as 'the Army of Movement'. For want of a better translation, it shall here be called the Mobile Guard. This force, like the army of Iraq, which now comprised just one corps of the Muslim army, was kept under his personal command by Khalid, and was earmarked as a mobile reserve for use in battle as required. The Mobile Guard was undoubtedly the finest body of men in the army-a *corps d'elite*.

From Yaqusa, Khalid marched with his corps of Iraq in the lead. This was followed by the other corps and the women and children. By now the families of the warriors from Iraq, which had been sent to Madinah before the Perilous March, had also joined the Muslim army in Syria. After three days, of marching along the Jabiya route, the leading elements arrived at Marj-us-Suffar, about 12 miles from Damascus, and discovered a large Roman army barring their way. This Roman force, consisting of about 12,000 soldiers and commanded by Kulus and Azazeer, had been sent forward by Thomas to fight a battle in the open and drive the Muslims away from Damascus, or if that were not possible, delay the Muslim advance and thus gain more time for the provisioning of the city. For the night the leading Muslim corps camped about a mile from the Roman position, while the other corps were still some distance behind.

Marj-us-Suffar (the Yellow Meadow) stretched south from Kiswa, a small town 12 miles from Damascus on the present road to Dar'a. At the southern edge of the town ran a small, wooded wadi and from this wadi stretched southwards the Marj-us-Suffar. Just west of the town rose a low ridge, and the Roman position was in front of this and south of the *wadi*. 3

The following morning, on August 19, 634 (the 19th of Jamadi-ul-Akhir, 13 Hijri), Khalid moved up his corps; and the Muslims and the Romans marshalled their forces for the Battle of Marj-us-Suffar. The rest of the Muslim army was rushing to the battlefield, but it would be another two hours or so before it arrived. The leading corps, which was now deployed for battle, would act as a firm base on which the whole army would form up on arrival. The Romans appeared to remain on the defensive since they made no move to engage the Muslims. In the mean time Khalid started a phase of duels that would keep the Romans occupied until the arrival of the remaining Muslim corps.

This phase was rather like a tournament with gallants displaying their courage and skill, except that a good deal of blood was shed. The Romans played the game sportingly, for they too had champions as gallant as any; and among these the two generals, Kulus and Azazeer, were considered the bravest and the best. The rank and file of the two armies stood by as spectators and cheered their 'players'.

Khalid started this bloody tournament by calling forward a number of his stalwarts, including Dhiraar, Shurahbil and Abdur-Rahman bin Abi Bakr. All these cavaliers rode out from the Muslim front rank, galloped about the space between the two armies and threw their individual challenges. Against each of them a Roman officer emerged, and the champions paired off for combat. Practically every Roman was killed. After killing his opponent the Muslim champion would gallop across the front of the Roman army, taunting and challenging; and on getting a suitable opportunity, would even strike down one or two men in the front rank before retiring to the Muslim army.

1. Waqidi: Vol. 2, p. 47.
2. Tabari: Vol. 2, p. 626.
3. The town, the ridge and the wadi are still there, and the plain is still yellowish in appearance.

As in earlier encounters, Dhiraar, naked above the waist, did the most damage and slew the largest number of Romans, thrilling the spectators with his daredevilry.

When this had gone on for an hour or so, Khalid decided that it was time for the 'heavy-weight bout'! He called back the Muslim officers and rode forward himself. As he got into the centre of the battlefield, he called:

I am the pillar of Islam!
I am the Companion of the Prophet!
I am the noble warrior,
Khalid bin Al Waleed! [1]

Since he was the commander of the Muslim army, his challenge had to be met by a top ranking Roman general. Kulus had by now lost some of his zest for battle, because he had been intimidated by the sad fate of all the Romans who had come forward to duel with the Muslims this morning. It appears that he was unwilling to accept the challenge of Khalid; but egged on by the taunts of his rival, Azazeer, he rode out from the front of the Roman army. On getting near Khalid he indicated that he wished to talk; but Khalid paid no heed to his sign and attacked him with his lance. Kulus parried the thrust, showing uncommon skill in doing so. Khalid charged at him again, but the thrust was parried.

Khalid decided not to use the lance any more. He came near his opponent, dropped his lance and grappled with him with his bare hands. Catching Kulus by the collar he jerked him off his horse, whereupon the Roman fell to the ground and made no effort to rise. At this Khalid signalled for two Muslims to come to him. When they came forward, he ordered them to take Kulus away as a prisoner, which they did.

While the Romans were dismayed by the sight of this encounter, Azazeer was secretly pleased and hoped that the Muslims would kill Kulus. Now he came forward, and regarding himself as a greater fighter than Kulus, had no doubt that he would soon make short work of Khalid. But he would first amuse himself by making fun of the Muslim commander. Azazeer stopped a few paces from Khalid and said in Arabic, *"O Arab brother, come near me so that I can ask you some questions."*

"O enemy of Allah" replied Khalid. *"Come near me yourself or I shall come and take your head."* Azazeer looked surprised, but urged his horse forward and stopped at duelling distance. In a gentle, persuasive tone he continued: *"O Arab, brother, what makes you come to fight in person? Do you not fear that if I kill you, your comrades will be left without a commander?"*

"O enemy of Allah, you have already seen what a few of my comrades have done. If I were to give them permission, they would destroy your entire army with Allah's help. I have with me men who regard death as a blessing and this life as an illusion. Anyway, who are you?"

"Do you not know me?" Azazeer exclaimed. *"I am the champion of Syria! I am the killer of Persians! I am the breaker of Turkish armies!"*

"What is your name?" asked Khalid.

"I am named after the angel of death. I am Israel!"

At this Khalid laughed. *"I fear that he after whom you are named seeks you ardently... to take you to the abyss of hell!"*

Azazeer ignored this remark and went on in an unconcerned way: *"What have you done with your prisoner, Kulus?"*

1. Waqidi: pp. 41, 48.

"He is held in irons."

"What prevents you from killing him? He is the most cunning of the Romans."

"Nothing prevents me except the desire to kill both of you together."

"Listen," said the Roman, *"I shall give you 1,000 pieces of gold, 10 robes of brocade and five horses if you will kill him, and give me his head."*

"That is the price for him. What will you give me to save yourself?"

"What do you want of me?"

"The Jizya!"

This enraged Azazeer, who said, *"As we rise in honour, so you fall in disgrace. Defend yourself, for now I kill you."*

These words were hardly out of the Roman's mouth when Khalid assailed him. He struck several times with his sword, but Azazeer, showing perfect mastery over the art, parried every blow and remained unharmed. A cry of admiration rose from the Muslim ranks at the skill with which the Roman was defending himself against their commander, who had few equals in combat and those only among the Muslims. Khalid also stopped in amazement.

The face of the Roman broke into a smile as he said, *"By the Messiah, I could easily kill you if I wished. But I am determined to take you alive, so that I may then release you on condition that you leave our land."*

Khalid was infuriated by the cool, condescending manner of the Roman general and his success in defending himself. He decided to take the Roman alive and humble him. As he moved forward to attack again, however, to his great surprise, Azazeer turned his horse and began to canter away. Believing that the Roman was fleeing from combat, Khalid pursued him and the spectators saw the remarkable spectacle of two generals galloping, one after the other, in the no-man's-land between the two armies. Several times the riders galloped round the field; and then Khalid began to lag behind, his horse sweating and winded. The Roman was better mounted, and his horse showed no sign of fatigue.

This apparently was a pre-determined plan of Azazeer, for when he saw Khalid's mount exhausted, he reined in his horse and waited for Khalid to catch up. Khalid was now in a most unforgiving mood, since in this race his opponent had got the better of him, and it did not help his temper to hear the Roman mock at him: *"O Arab! Do not think that I fled* in fear. In fact I am being kind to you. Lo, I am the taker of souls! I am the angel of death!"

Khalid's horse was no longer fit for combat. He dismounted and walked towards Azazeer, sword in hand. The Roman gloated at the sight of his opponent approaching on foot while he himself was mounted. Now, he thought, he had Khalid just where he wanted him. As Khalid got within striking distance, Azazeer raised his sword and made a vicious sideways swipe to cut off the Muslim's head; but Khalid ducked to let the blade swish past harmlessly inches above his head. The next instant he struck at the forelegs of the Roman's horse, severing them completely from the body, and horse and rider came tumbling down. Now all courage left Azazeer. He got up and tried to run, but Khalid sprang at him and catching him with both hands, lifted him bodily off the ground and hurled him down. Next he caught Azazeer by the collar, jerked him up and marched him back to the Muslim army, where he joined Kulus in irons. [1]

This grand duel was hardly over when two more Muslim corps, those of Abu Ubaidah and Amr bin Al Aas, arrived at the battlefield. Khalid deployed them as the wings of his army; and as soon as the battle formation was complete, ordered a general attack.

1. The description of these duels and the dialogue are taken from Waqidi: pp. 19-21.

The Romans stood firm for an hour or so, but could not hold the Muslims longer. The loss of a large number of their officers, including the two top generals, had had a depressing effect on their spirits; and the fact that Damascus stood just behind, beckoning to them to come and be safe within its walls acted as a temptation to withdraw. So they retreated, in good order, leaving behind a large number of dead. The Roman army arrived at the city and entered its walls, closing the gates behind it.

The Muslims spent the night on the plain, and the following day marched to the city. Here, on August 20, 634 (the 20th of Jamadi-ul-Akhir, 13 Hijri), Khalid launched the Muslim army into the siege of Damascus. [1]

Khalid had already left behind a mounted detachment at Fahl to keep the Roman garrison occupied and prevent it from coming to the aid of Damascus or interfering with the movement of messengers and reinforcements from Madinah. Now he sent out another detachment on the road to Emessa to take up a position near Bait Lihya, about 10 miles from the city, [2] and instructed its commander to send out scouts to observe and report the arrival of Roman relief columns. If unable to deal with such columns himself, the detachment commander would seek Khalid's help. Having thus arranged a blocking position to isolate Damascus from Northern Syria, which was the most likely region whence relief columns could approach Damascus, Khalid surrounded the city with the rest of the army.

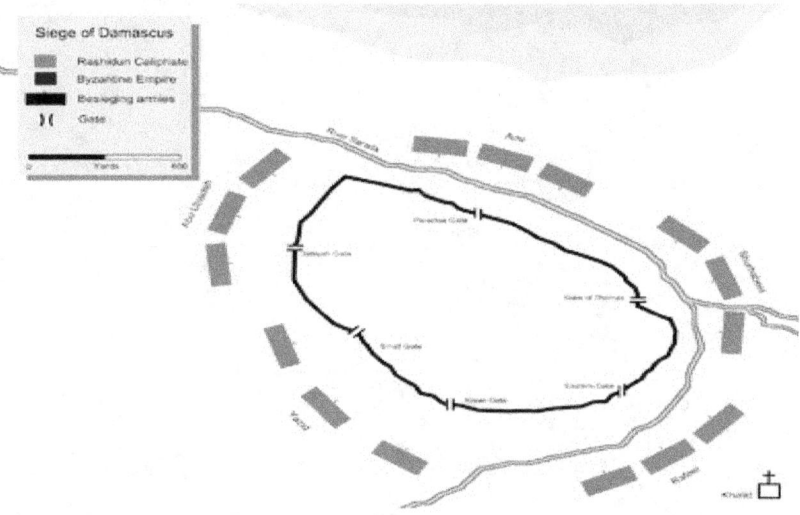

Muslim troop deployment (Red) during the siege of Damascus.

amascus now held a Roman garrison of about 15,000 to 16,000 soldiers, a considerable civil population comprising the permanent inhabitants and a large number of people from the surrounding region who had taken refuge in the city. The Muslim strength at Damascus is not recorded, but must have been quite a bit less than in the preceding month. Muslim dead in the three battles just fought - at Ajnadein, at Yaqusa and at the Marj-us-Suffar - undoubtedly ran into four figures; and thousands more must have been wounded in these battles and rendered unable to participate in the siege. Moreover, a group had been sent out as a blocking force and a detachment left at Fahl. In view of all this, I estimate the Muslim strength at Damascus at about 20,000 men. With this strength Khalid besieged the city.

He positioned the corps of Iraq, which included elements of the Mobile Guard, at the East Gate. He placed the bulk of this corps under Raafe, and himself stayed a short distance away from the East Gate with a reserve of 400 horsemen from the Mobile Guard. He established his headquarters in a monastery which, as a result, became known as Dair Khalid, i.e. *Monastery of Khalid* (and it is believed that the monks living in this monastery helped the Muslims in various ways, including the care of the Muslim wounded). 3 At each of the remaining gates, he deployed a force of 4,000 to 5,000 men whose commanders were as follows:

Gate of Thomas : *Shurahbil*,
Jabiya Gate : *Abu Ubaidah*,
Gate of Faradees : *Amr bin Al Aas*,
Keisan Gate : *Yazeed*,
Small Gate : *Yazeed*.

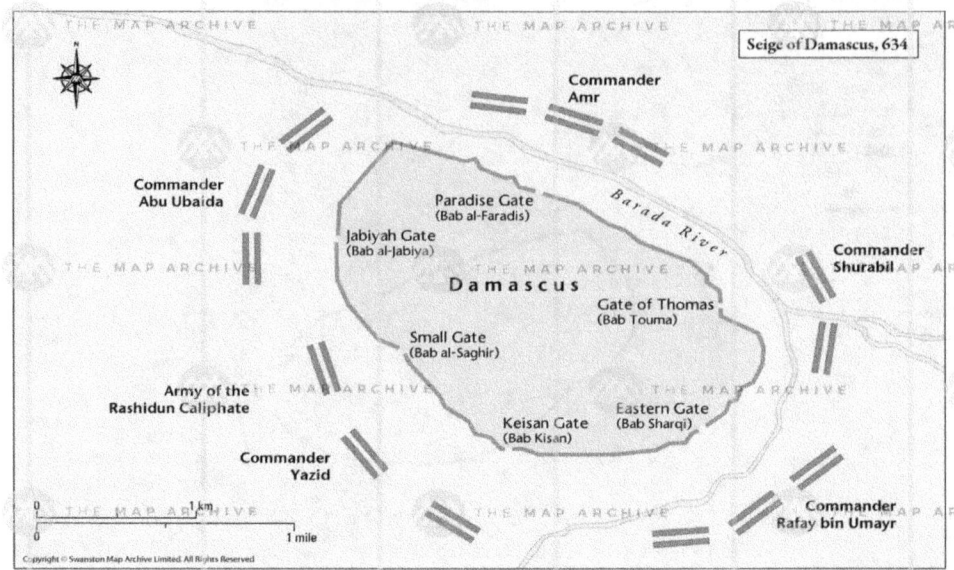

To the corps commanders Khalid gave instructions to the effect that they would: (a) camp outside bow-range of the fort; (b) keep the gate under observation; (c) move archers up to engage any Roman archers who appeared on the battlements; (d) throw back any Roman force which sallied out from the gate; and (e) seek Khalid's help in case of heavy pressure. Dhiraar was placed in command of 2,000 horsemen from the Mobile Guard, and given the task of patrolling the empty spaces between the gates during the night and helping any corps attacked by the Romans.

With these instructions the Muslim corps deployed, and the siege began. Tents were pitched, and Dhiraar started his patrolling. Every main avenue of relief and escape was closed, but this applied only to formed bodies of men. Individuals could still be lowered from the wall at many places during the night, and thus Thomas was able to keep in touch with the outside world and with Heraclius at Antioch.

1. For an explanation of the details of the Battle of Marj-us-Suffar, see Note 10 in Appendix B.
2. Bait Lihya no longer exists, and its exact location is not known. It was a small town in the *Ghuta* (Yaqut: Vol. 1, p. 780), and I have placed it at the outer edge of the *Ghuta* because to position a blocking force nearer the city would be militarily unsound.
3. This monastery, which was also known as Dair-ul-Ahmar (the Red Monastery), no longer exists, but its general location is known. About a quarter of a mile from the East Gate, stretching eastwards, stands a garden. The monastery was in this garden, and according to Waqidi (p. 43), was less than half a mile from the gate.

On the day following the arrival of the Muslims, Khalid had Kulus and Azazeer brought in irons near the East Gate where they could be seen by the Romans on the wall. Here both generals were offered Islam, and both rejected the offer. Then, in full view of the Roman garrison, the two generals were beheaded, the executioner being none other than Dhiraar.

For three weeks the siege continued with no major action except for a few half-hearted Roman sallies which the Muslims, had no difficulty in repulsing. During the day the two sides would keep up a sporadic exchange of archery, though no great damage was suffered by either side. This was to be a siege to the bitter end. Damascus would, if necessary, be starved into submission. 1

Soon after Heraclius heard of the defeat of the Roman army at Marj-us-Suffar by Khalid and the commencement of the siege of the city, he undertook measures to raise fresh forces. The recent blows suffered by the Empire were serious enough; but the successful advance of the Muslims had now created an even more critical situation, and Damascus itself was in danger. If Damascus fell, it would be a staggering blow to the prestige it could not recover without mobilizing the entire military resources of the Empire-a task not to be undertaken except in the direct emergency. And Damascus was in danger of falling not because of insufficient troops in the city but because of insufficient supplies. It had not been provisioned for a long siege.

Within 10 days of the start of the siege, Heraclius had raised a new army of 12,000 men drawn from garrisons in various parts of Northern Syria and the Jazeer. 2 This army was launched from Antioch with a large baggage-train carrying supplies, and the commander was instructed to reach Damascus at any cost and relieve the beleaguered garrison. The relief column marched via Emessa, made contact with Muslim scouts between Emessa and Damascus, and from here onwards was prepared for battle at a moment's notice.

On September 9, 634 (the 10th of Rajab, 13 Hijri), a messenger came galloping into Khalid's camp to inform him that a large Roman army of undetermined strength was advancing rapidly from Emessa, and in a day or so would make contact with the blocking force deployed at Bait Lihya. Khalid was not surprised to hear this, for he had guessed that Heraclius would do everything in his power to relieve Damascus; and it was for this reason that Khalid had placed the blocking force on the main route by which a relief column could approach the city.

He immediately organised a mounted force of 5,000 men and placed it under Dhiraar. He instructed Dhiraar to proceed with all speed to the area of Bait Lihya, take command of the regiment already deployed there and deal with the relief column approaching from Emessa. He cautioned Dhiraar against being rash and instructed him to seek reinforcements before committing his force to battle in case the enemy strength proved too large. Such words of caution, however, were wasted on Dhiraar; if there was one quality which he did not possess it was caution. With Raafe as his second-in-command, Dhiraar rode away from Damascus and picking up the blocking force, moved forward to a low ridge a little short of Saniyyat-ul-Uqab (the Pass of the Eagle) and deployed his force in ambush.

Next morning the Roman army appeared in sight. The Muslims waited. As the head of the Roman column got close to the ambush, Dhiraar ordered the attack. His men rose from their places of concealment, and led by their half-naked commander, rushed at the Romans. But the Romans were prepared for such a contingency. They deployed so quickly in battle formation that the action became a frontal engagement, with the Muslims attacking and the Romans standing firm in defence on higher ground in front of the Pass of the Eagle. The Muslims now realised the full strength of the enemy, which amounted to twice their own. But this did not matter to Dhiraar. Assaulting furiously in front of his men, he got far ahead of his comrades and before long was completely surrounded by the Romans. His enemies recognised him as the Naked Champion; and decided to take him alive and show him as a prize to their Emperor. Dhiraar was wounded by an arrow in the right arm but continued to fight as the Romans closed in. At last, however, after he had suffered several wounds, he was overpowered by the Romans, who then sent him to the rear.

1. According to Tabari (Vol. 2, p. 626) the Muslims also used catapults at this siege; but this is unlikely because the Muslims had no siege equipment, nor did they know much about using it.

2. Jazeera literally means island, and this name was used to designate the region between the rivers Euphrates and Tigris in present day North-Eastern Syria, North-Western Iraq and South-Eastern Turkey.

The loss of Dhiraar had a depressing effect on the Muslims, but Raafe was a worthy successor to the dashing Dhiraar. Taking command, he launched several attacks to get through to Dhiraar and rescue him; but his efforts proved fruitless, and the action turned into a stalemate. Raafe realised that there was nothing that he could do to break the Roman force deployed in front of him; and in the afternoon he sent a message to Khalid telling him about the engagement, about the enemy strength and about the loss of Dhiraar-probably still alive as a prisoner.

The sun was still well above the horizon when Khalid received news of this engagement. He realised that the Roman strength at Bait Lihya was too large for Raafe to tackle on his own. And this placed Khalid in a serious dilemma. The Roman relief column had to be defeated and driven back towards Emessa, and this could be done quickly only if Khalid himself took command at Bait Lihya with a sizable reinforcement from Damascus. Failing this, the Roman relief column would have every chance of breaking through the Muslim blocking force, and this could have a disastrous effect on the Muslim siege of Damascus. But there was also the problem of timing. If an immediate move were made to reinforce Raafe, the Roman garrison would observe the move and sally out to break the grip of the weakened besieging force. The relieving Romans at Bait Lihya had to be beaten; yet the besieged Romans in Damascus had to be kept in the dark about the movement of Muslim reinforcements from Damascus. Khalid decided to risk a delay and carry out on move till the latter part of the night, by when the beleaguered garrison would be less likely to discover the move.

Preparations wore made accordingly. The command at Damascus was taken over by Abu Ubaidah who would see to the siege operations during Khalid's absence. After midnight a detachment of 1,000 Muslim warriors under Maisara bin Masruq took up positions at the East Gate and some other readjustments were made at the other gates. Then, some time between midnight and dawn, Khalid set off with his Mobile Guard of 4,000 horses. The Guard moved swiftly through the remainder of the night and early the following morning arrived at the scene of battle between Raafe and the Romans. The fighting was continuing on this second day of battle with no decision in sight. Indeed the Muslims were now tired of attacking the Romans who stood like a rock against the Muslim assaults.

As Khalid approached the battlefield he suddenly saw a Muslim rider flash past him from behind and gallop off towards the Roman front. Before Khalid could stop him, he was gone. A slim, lightly-built person, dressed in black, this rider wore a breastplate and was armed with a sword and a long lance. He sported a green turban and had a scarf wrapped around his face, acting as a mask, with only his eyes visible. Khalid arrived on the battlefield in time to see this rider throw himself at the Romans with such fury that everyone present thought that he and his horse must both be mad. Raafe saw this rider before he saw Khalid and remarked, *"He attacks like Khalid, but he is clearly not Khalid."* [1] Then Khalid joined Raafe.

Khalid took a little time to organize Raafe's group and his own Mobile Guard into one and deploy it as a combined force for battle. Meanwhile the masked rider treated the Muslims to a thrilling display of horsemanship and attacks with the lance. He would go charging on his own, strike the Roman front atone point and kill a man; then go galloping away to another part of the front, again strike someone in the Roman front line and so on. A few Romans came forward to tackle him but all went down before his terrible lance. Marvelling at this wondrous sight, the Muslims could still see nothing more of the warrior than a youthful figure and a pair of bright eyes shining above the mask. The rider appeared bent on suicide as with his clothes and lance covered with blood, he struck again and again at the Romans. The example of this warrior put fresh courage into the men of Raafe, who forgot their fatigue and went into battle with renewed high spirits as Khalid gave the order to attack.

The masked rider, now joined by many others, continued his personal war against the Romans as the entire Muslim force attacked the Roman front. Soon after the general attack had begun, Khalid got near this rider and called, *"O warrior, show us your face."* A pair of dark eyes flashed at Khalid before the rider turned away and galloped off into another assault at the Romans. Next, a few of Khalid's men caught up with him and said, *"O noble warrior, your commander calls you and you turn away from him! Show us your face and tell us your name so that you may be properly honoured."* Again the rider turned away as if deliberately trying to keep his identity a secret.

1. Waqidi: p. 27.

As the masked rider returned from his charge, he passed by Khalid, who called to him sternly to stop. The rider pulled up his horse, and Khalid continued, *"You have done enough to fill our hearts with admiration. Who are you?"*

Khalid nearly fell off his horse when he heard the reply of the masked rider, for it was the voice of a girl! *"O commander, I only turn away from you out of modesty. You are the glorious commander, and I am of those who stay behind the veil. I fight like this because my heart is on fire."*

"Who are you?"

"I am Khaulah, sister of Dhiraar. My brother has been captured, and I must fight to set him free."

Khalid marvelled at the old man, Al Azwar, who had fathered two such dauntless fighters, a boy and a girl. *"Then come and attack with us"*, he said. [1]

The Muslim attack continued in force and at about midday the Romans began to withdraw from the battlefield in good order. The Muslims followed, keeping up a steady pressure, but there was no sign of Dhiraar, dead or alive. Then, as good luck would have it, some local Arabs came to the Muslims with the information that they had seen 100

Romans riding to Emessa with a half-naked man in their midst, tied to his horse. Khalid at once guessed that Dhiraar had been sent away from the battlefield and ordered Raafe to take 100 picked riders, move wide around the flank of the Romans, get to the Emessa road and intercept the escort taking Dhiraar to Emessa. Raafe at once selected 100 stalwarts and set off, accompanied, of course, by Khaulah bint Al Azwar.

Raafe got to the Emessa road at a point which the escort had not yet reached and waited in ambush. When the 100 Romans arrived at this point, Raafe and his men assailed them, killed most of the soldiers and set Dhiraar free. The Naked Champion and his loving sister were happily reunited. The party again made a wide detour to avoid the Roman army, and rejoined Khalid who was very, very grateful to Raafe for rescuing Dhiraar.

Under the unrelenting pressure of the Muslims, the Romans increased the pace of their retreat. As the Muslims struck with greater ferocity, the retreat turned into a rout, and the Romans took to their heels and fled in the direction of Emessa.

Khalid could not pursue the fleeing enemy because he had to get back to Damascus. The Muslim forces investing the city had been weakened by 9,000 men with the departure of first Raafe's detachment and then the reinforcement of the Mobile Guard. In case the Romans should attack in strength against any Muslim corps, there would be a serious danger of their breaking through. Consequently Khalid sent only a mounted regiment under Samt bin Al Aswad to follow the Romans to Emessa. Samt got there in due course and found that the Romans had withdrawn into the fort. The local inhabitants of Emessa, however, approached Samt and let it be known that they had no desire to fight the Muslims, with whom they would make peace and even feed any soldiers quartered in their city. After a friendly exchange of messages, Samt returned to Damascus.

Meanwhile Khalid had rejoined the Muslim army at Damascus. He resumed command and re-established the Muslim dispositions as they had been before the appearance of the Roman relief column.

The news of the sad fate of the relief column spread among the inhabitants of Damascus, and it was a grievous blow indeed. The Damascenes had pinned their hopes on Heraclius sending such a force to save them. Heraclius had in fact done his best, but the hopes of the city had been shattered by Khalid's action at Bait Lihya. Heraclius could no doubt raise more forces, but that would take time. Meanwhile the supplies were running low and there was no fresh ray of hope to brighten the horizon and give assurance and strength to the people of Damascus.

1. Waqidi: p. 28.

A number of questions were raised wherever people assembled. Even if Heraclius raised a fresh column-and this was unlikely in the near future-what assurance did they have that it would achieve better success than the last one? If the Muslims could do what they did to an army of 90,000 men at Ajnadein, what chance did the relatively small force at Damascus have of avoiding a military defeat and the plunder and captivity which would doubtless follow? How much longer would the supplies last? Would it not be better to make peace with, the Muslims on whatever terms were offered, and in this manner avoid total destruction? Spirits fell and discontent rose in Damascus, especially in the non-

Roman section of the population. The situation was becoming increasingly more desperate, and the tension increasingly more unbearable.

Then a delegation of prominent citizens approached Thomas. They apprised him of their fears and suggested that he consider the possibility of making peace with Khalid; but Thomas assured them that he had sufficient troops to defend the city, and would soon take the offensive to drive the Muslims away. Special services were held in the churches and prayers offered for deliverance from the peril which threatened the city. Thomas decided to attempt a powerful sally from the fort. He was a brave man, and as long as there was some chance of success, he would not surrender.

The following morning, early in the third week of September 634, Thomas drew men from all sectors of the city and formed a strong force to break out through the Gate of Thomas. His immediate opponent here was Shurahbil with his corps of about 5,000 men. Thomas started the operation with a concentrated shower of arrows and stones against the Muslim archers in order to drive them back and get more room for debouching from the gate. The Muslims answered the Roman salvos with their own volleys of arrows. At the very beginning of this exchange several Muslims were killed, one of whom was Aban bin Saeed bin Al Aas-a man who had only recently got married to an unusually brave woman. As soon as she heard that she had become a widow, she took a bow and joined the Muslim archers, seeking revenge. On the wall of the fort, near the Gate of Thomas, stood a priest with a large cross, the sight of which was intended to give added courage to the Romans. Unfortunately for this priest, the young Muslim widow chose him as her target. The arrow she shot at him drove through the man's breast; and priest and cross came tumbling down to the foot of the wall, to the delight of the Muslims and the dismay of the Romans. However, in this exchange the Romans got the better of the Muslims; and after a while the besiegers were driven back to a line out of range of the Roman archers and slingers.

Next the gate was opened and the Roman foot-soldiers covered by the archers and slingers on the wall, rushed through the gate and fanned out into battle formation. As soon as the deployment was complete, Thomas ordered the attack against the corps of Shurahbil, which had also formed up a few hundred yards from the gate. Thomas himself led the assault, sword in hand, and according to the chronicler, 'roared like a camel!' [1]

Very soon there was heavy fighting between the two bodies of men. Shurahbil's corps was outnumbered but held its ground, not yielding an inch, and Roman losses began to mount. Thomas now noticed Shurahbil and guessing that he was the commander of this Muslim force, made for him. Shurahbil saw him coming, and with a blood-covered sword in his hand prepared to meet him. But before Thomas could reach Shurahbil, he was struck in his right eye by an arrow, again fired by the widow, and fell to the ground. He was quickly picked up by his men and carried away, while at the same moment the Romans began to fall back to the fort. Thus, under pressure from swordsmen and under the punishing fire of Muslim archers deployed on the flanks, the Romans returned to the fort, leaving behind a large number of dead, several of whom had fallen to the arrows of the widow of Aban.

Inside the fort the surgeons examined the eye of Thomas. The arrow had not penetrated deep, but they found that it could not be extracted. They therefore cut off the shaft where it entered the eye, and Thomas, instead of being depressed by the loss of his eye and the pain of his wound, showed himself to be a man of extraordinary courage. He swore that he would take a thousand eyes in return; that he would not only defeat these Muslims but would follow them into Arabia, which, after he had finished with it, would be fit only for the habitation of wild beasts. He ordered another great sally to be carried out that night.

1. Waqidi: p. 46.

Meanwhile Shurahbil was not a little worried. He had lost quite a large number of men, killed and wounded, and feared that if another determined sally were made by the Romans, they might succeed in breaking through his corps. He consequently, asked Khalid for reinforcements; but Khalid had no men to spare. He could not weaken the other corps, because the Romans could attack at any gate, and might well choose another gate for their next sally. He instructed Shurahbil to hold on as best he could, and assured him, that Dhiraar with his 2,000 men would get to him in case of heavy pressure. If need be he himself, with his reserve, would come and take over the battle at the Gate of Thomas. Shurahbil prepared for another sally by the Romans, quite determined to hold on to the last man.

For the sally of the night, Thomas again selected the Gate of Thomas as the point of main effort in order to exploit the damage which he had undoubtedly caused to the corps of Shurahbil. But he planned to make sallies from other gates also. The locations of the various Muslim corps and their commanders, were known in detail to the garrison. To keep the Muslim corps at other gates tied down, so that they would not be able to come to the aid of Shurahbil, Thomas ordered sallies form the Jabiya Gate, the Small Gate and the East Gate. For the last he allotted rather more forces than for the others, so that Khalid would be unable to move to Shurahbil's help and take command in that decisive sector. Attacking from several gates also gave more flexibility to the operation. Thus, if success were achieved, in any sector other than the Gate of Thomas, that could be converted into the main sector and the success exploited accordingly.

In his orders Thomas emphasised the need for swift attacks, so that the Muslims would be caught unawares and destroyed in their camps. No quarter would be given. Any Muslim wishing to surrender would be killed on the spot-any, that is, but Khalid, who was to be taken alive. The moon would rise about two hours before midnight. Soon after, on the signal of a gong to be struck on the orders of Thomas, the gates would be flung open and the attacks launched simultaneously.

In the moonlight the Roman attacks began as planned. At the Jabiya Gate there was some hard fighting, and Abu Ubaidah himself entered the fray with drawn sword. The Son of the Surgeon was an accomplished swordsman, and several Romans fell under his blows before the sally was repulsed and the Romans hastened back to the city.

At the Small Gate Yazeed had fewer troops than were positioned at the other gates, and the Romans gained some initial success. But luckily Dhiraar was nearby and joined Yazeed with his 2,000 warriors. Without a moment's delay Dhiraar hurled his men at the enemy, whereupon the Romans reacted as if they had been assailed by demons and hastily withdrew to the fort with Dhiraar close upon their heels.

At the East Gate the situation soon became more serious, for a larger Roman force had been assigned to this sector. From the sounds of battle Khalid was able to judge that the enemy had advanced farther than he should have been allowed to; and fearing the Raafe might not be able to hold the attack, went into battle himself with his reserve of 400 veterans from the Mobile Guard. As he got to the Romans, he gave his personal battle cry: *"I am the noble warrior, Khalid bin Al Waleed."* This battle cry was by now known to all the Romans, and had the effect of imposing caution upon them. In fact it marked the turning point in the sally at the East Gate. Soon the Romans were in full retreat with the Muslims cutting down the stragglers. Most of this force was able to re?enter the city and close the East Gate behind it.

The heaviest fighting, however, took place at the Gate of Thomas, where Shurahbil's corps, having fought a hard action during the day, had to bear the brunt of the fighting of the night. The moonlight helped visibility as the Romans rushed out of the gate and began to form up for battle. In this process they were subjected to withering fire from Shurahbil's archers, but in spite of some losses, they completed their deployment and advanced to battle. For two hours the fighting continued unabated with Shurahbil's men struggling desperately to hold the Roman attack. And hold it they did.

Shortly after midnight Thomas, who was himself fighting in the front rank, singled out Shurahbil. The Muslim commander could be easily identified by the orders that he was shouting to his warriors. The two commanders paired off and began to duel with sword and shield.

For some time while the rest of the soldiers were locked in wild, frenzied combat, the duel of the two champions continued with no success to either. Then Shurahbil, seeing an opening, struck with all his might at the shoulder of Thomas; but his sword landing on the hard metal shoulder-pad of the Roman's breastplate, broke into pieces. Shurahbil was now at the mercy of Thomas. Luckily for him, at that very moment two Muslims, came up beside him and engaged Thomas. Shurahbil pulled back, picked up the sword of a fallen Muslim and again returned to combat. But Thomas was no longer there.

By now the Romans had had enough of battle. Seeing that there was no weakening in the Muslim front, Thomas decided that to continue the attack would be fruitless and would lead to even heavier casualties among his men. He ordered a withdrawal, and the Romans moved back at a steady pace. The Muslims made no attempt to follow, though their archers did a certain amount of damage. Again the young widow used her bow with deadly effect.

This was the last attempt by Thomas to break the siege. The attempt had failed. He had lost thousands of men in these sallies, and could no longer afford to fight outside the walls of the city. His soldiers shared his disillusionment. They would fight to defend the city, but would not venture to engage the Muslims outside the fort. Thomas now gave more authority to his deputy, Harbees, delegating to him several of the functions of command which hitherto he had himself exercised.

After the failure of the nocturnal sally, the despair of the Damascenes knew no bounds. The dark clouds which threatened the great city had no silver lining. There was widespread grumbling among the people who now wished for nothing but peace; and in this desire they were joined by Thomas, who had fought gallantly in defence of the city and answered the call of honour. He was prepared to make peace and surrender the fort on terms, but was Khalid prepared to make peace? He was known as a man of violence who looked upon battle as a sport; and since he undoubtedly knew the internal conditions prevailing in Damascus, would he accept anything less than an unconditional surrender, by which they would all be placed at his mercy?

By now the Romans had come to know the Muslim generals very well. They knew that Abu Ubaidah was next in command after Khalid, and wished he were the first in command. The Son of the Surgeon was essentially a man of peace-gentle, kind, benevolent-and looked upon war as a sacred duty rather than a source of pleasure and excitement. With him they could make peace, and he would doubtless be generous in his terms. But Abu Ubaidah was not the army commander. For two or three days this dilemma continued; and then the matter was taken out of their hands by Jonah the Lover.

Jonah, son of Marcus, was a Greek who was madly in love with a girl, also Greek. Actually she was his wife. Just before the arrival of the Muslims they had been married, but the ceremony of handing over the bride to the husband had not been completed when the Muslims arrived and laid siege to Damascus. Thereafter Jonah asked her people several times to hand over his bride to him but they refused, saying that they were too busy fighting and that this was a matter of survival; and how could Jonah think of such things at a time like this? Actually Jonah could think of little else!

Just after dusk, on or about September 18, 634 (the 19th of Rajab, 13 Hijri), Jonah lowered himself with the aid of a rope near the East Gate, and approaching the nearest Muslim guard, asked to see Khalid. As soon as he was ushered into the presence of the commander, he narrated his sad story and explained the purpose of his visit. Would Khalid help him get his bride if he gave intelligence which would lead quickly to the capture of Damascus? Khalid would. He then informed Khalid that in the city this night the people were celebrating a festival in consequence of which there was revelry and drunkenness everywhere, and few sentries would be found at the gates. If Khalid could scale the wall, he would have no difficulty in opening any gate he chose and forcing an entry into the city.

Khalid felt that he could trust the man. He appeared sincere in what he said. Khalid offered him Islam, and Jonah accepted it. During the past few years he had heard much about Islam and was favourably inclined. At the hands of Khalid, Jonah now accepted the new faith, whereafter Khalid instructed him to return to the city and wait, which Jonah did.

As soon as the Greek had departed, Khalid ordered the procurement of ropes and the preparation of rope ladders. There was no time to make a co-ordinated plan of attack for the whole army; and so Khalid decided that he would storm the fort by the East Gate, with just the corps of Iraq which was positioned there. The moon would rise at about midnight, and soon after that the assault would begin.

According to Khalid's plan, 100 men would scale the wall at a place near the East Gate, where it was known to be the most impregnable. Here certainly there would be no sentries. At first three men would climb up with ropes. Then rope ladders would be fastened to the ropes and hauled up by the three to be used by the rest of the picked hundred to get to the top. Some men would remain at the top, while others would descend into the fort, kill any guards found at the gate and open the gate. Thereupon the entire corps would rush in and start the attack.

The three leaders who were to scale the wall were Khalid, Qaqa and Maz'ur bin Adi. The ropes were thrown up, lassoing the epaulements on the wall, after which these three indomitable souls climbed up hand by hand. There was no guard at the top. The rope ladders were drawn up, and on these others began to climb in silence. When half the group had arrived at the top, Khalid left a few men to assist the remaining climbers, and with the rest descended into the city. A few Roman soldiers were encountered on the way down and put to the sword. Thereafter the party rushed to the gate, where two sentries stood on guard. Khalid killed one while Qaqa killed the other. But by this time the alarm had been raised and parties of Romans began to converge towards the East Gate. Khalid knew that it was now touch and go.

The rest of the Muslim party hastily took up a position to keep the Romans away while Khalid and Qaqa dealt with the gate, which was locked and chained. A few blows shattered both lock and chain, and the gate was flung open. The next instant the corps of Iraq came pouring in. The Roman soldiers who had converged towards the gate never went back; their corpses littered the road to the centre of the city.

All Damascus was now awake. The Roman soldiers rushed to their assigned positions, as per rehearsed drills, and manned the entire circumference of the fort. Only a small reserve remained in the hands of Thomas as Khalid began his last onslaught to get to the centre of Damascus, killing all who stood in his way-the regiments defending the sector of the East Gate.

It was shortly before dawn, and now Thomas played his last card-brilliantly. He knew that Khalid had secured a firm foothold in the city, and it was only a matter of time before the entire city would lie at his feet. From the absence of activity at the other gates, he guessed that Khalid was attacking alone and that other corps were not taking part in the storming of the fort. He hoped-and this was a long shot-that the other corps commanders, especially Abu Ubaidah, would not know of the break-in by Khalid. Thomas acted fast. He threw in his last reserve against Khalid to delay his advance for as long as possible, and at the same time sent envoys to the Jabiya Gate to talk with Abu Ubaidah and offer to surrender the fort peacefully and to pay the Jizya.

Abu Ubaidah received these envoys with courtesy and heard their offer of surrender. He believed that they had come to him because they were afraid to face Khalid. At the distance at which he was placed from the East Gate, if he heard sounds of battle at all, he must have assumed that it was a sally by the Romans; for it could not have occurred to him that Khalid would scale the wall with ropes. Abu Ubaidah had no doubt in his mind that Khalid also would agree to peace to put an end to the bloodshed and ensure a quick occupation of Damascus. Consequently he took upon himself the responsibility of the decision and accepted the terms of surrender. Damascus would be entered peacefully; there would be no bloodshed, no plunder, no enslavement and no destruction of temples; the inhabitants would pay the Jizya; the garrison and any local inhabitants who wished to do so would be free to depart from the city with all their goods. After this the Roman envoys went to the corps commanders at the other gates and informed then that a peace had been arranged with the Muslim commander and that the gates would be opened shortly, through which the Muslims could enter in peace. There would be no resistance.

Soon after dawn Abu Ubaidah, followed by his officers and the rest of his corps, entered Damascus in peace from the Jabiya Gate, and marched towards the centre of the city. He was accompanied by Thomas and Harbees and several dignitaries and bishops of Damascus. Now Abu Ubaidah, walking like an angel of peace, and Khalid advancing like a tornado, arrived simultaneously at the centre of Damascus, at the Church of Mary. Khalid had just broken through the last Roman resistance. The other corps commanders had also entered the city and were moving peacefully towards the centre.

Abu Ubaidah and Khalid stared at each other in amazement. Abu Ubaidah noted that Khalid and his men held dripping swords in their hands, and he guessed that something had happened of which he was not aware. Khalid noticed the peaceful air surrounding Abu Ubaidah and his officers, whose swords were in their sheaths and who were accompanied by Roman nobles and bishops.

For some time there was no movement. Then Abu Ubaidah broke the tense silence. *"O Father of Sulaiman,"* he said, *"Allah has given us this city in peace at my hand, and made it unnecessary for the Muslims to fight for it."*

"What peace!" Khalid bristled. *"I have captured the city by force. Our swords are red with their blood, and we have taken spoils and slaves."*

It was clear that there was now going to be a terrible row between these two generals, which could have serious consequences. Khalid was the commander and had to be obeyed; what is more, he was not a man who would take any nonsense from his subordinates. Furthermore, his towering personality and his unquestioned judgement in military matters made him difficult to argue with, especially on this occasion, when he was determined to regard the conquest of Damascus as a consequence of the use of force and not of peaceful negotiation. Abu Ubaidah, on the other hand, had none of the military stature or operational genius of Khalid, and would be the last person to assert otherwise. But as a Muslim he was in the topmost class, one of the Blessed Ten, the Trusted One of the Nation. He was the Al Asram, the One without the Incisors-and no one could forget how he had lost his front teeth.

Abu Ubaidah was wrong in making peace without Khalid's knowledge and permission, but he was determined to see that the word of a Muslim was honoured and unnecessary bloodshed avoided. He respected Khalid's leadership and knew that he would have to be handled with great care. Abu Ubaidah was in fact the only man in Syria with high enough standing to question any decision of Khalid. Even Khalid would not raise his voice when speaking to Abu Ubaidah, no matter how great his anger. What made the situation less dangerous was the fact that these two men held each other in genuine affection and respect for the various qualities which made them great. Abu Ubaidah also knew that he could silence Khalid with a few words, for he was armed with an authority of which Khalid was unaware. But he decided not to use this authority except as a last resort, when all manner of persuasion had failed. In this he was being kind to Khalid, but more of that later.

"O Commander," said Abu Ubaidah, *"know that I have entered the city peacefully."*

Khalid's eyes flashed with anger, but he restrained himself; in a voice which was not without respect, he replied, *"You continue to be heedless. How can they have peace from you when I have entered the city by force and their resistance is broken?"*

"Fear Allah, O Commander! I have given them a guarantee of peace, and the matter is settled."

"You have no authority to give them peace without my orders. I am commander over you. I shall not sheathe my sword until I have destroyed them to the last man."

"I never believed," pleaded Abu Ubaidah, *"that you would oppose me when I gave a guarantee of peace for every single one of them. I have given them peace in the name of Allah, exalted be He, and of the Prophet, on whom be the blessings of Allah and peace. The Muslims who were with me agreed to this peace, and the breaking of pacts is not one of our traits."*

At this stage some of Khalid's soldiers, tiring of listening to the argument and seeing some Romans standing on one side, began to wave their swords and moved towards the Romans to kill them. Abu Ubaidah saw this movement and rushing past Khalid, ordered the men to desist until the discussion between him and Khalid was over. The men obeyed. Only Abu Ubaidah could have done this; and Khalid could do nothing but try and control his rising anger.

Now the other three corps commanders got together and began to discuss the situation. After a few minutes they reached agreement among themselves and conveyed their opinion, to Khalid: Let there be peace, because if the Romans in Syria heard that the Muslims had given a guarantee of safety and then slaughtered those whose safety had been guaranteed, no other city would ever surrender to the Muslims, and that would make the task of conquering Syria immeasurably more difficult.

The emotions of Khalid never interfered with his reason; and the reason of Khalid saw the military wisdom of the advice tendered by the generals. For a moment he glared at Thomas and Harbees. Then he said, *"Alright, I agree to peace, except for these two accursed ones."*

"These two were the first to enter my peace," Abu Ubaidah said to Khalid. *"My word must not be broken. May Allah have mercy upon you!"*

Khalid gave up. *"By Allah!"* he exclaimed, *"but for your word I would certainly have killed them. Let them get out of the city, both of them, and may Allah's curse follow them wherever they go!"*

Thomas and Harbees were nervously watching the altercation between the two Muslim generals while interpreters were translating their statements. Thus they understood all and breathed a sigh of relief as they came to know of the conclusion of the dialogue. They now moved to Abu Ubaidah with an interpreter and asked for permission to depart on any route they chose.

"Yes," said Abu Ubaidah. *"You may go on any route you choose. But if we conquer any place at which you are residing, you will not then be under a guarantee of peace."*

Thomas, fearing a pursuit by Khalid, then requested, *"Give us three days of peace; then the truce would be ended. Thereafter if you catch up with us, do as you will-kill us or enslave us."*

Here Khalid entered the talks. *"Agreed, except that you may take nothing with you but food for the journey."*

"This again would amount to a breaking of the pact," objected Abu Ubaidah. *"My pact with them allows them to take all their belongings."*

"Even to this I agree," said Khalid, *"but no weapons."*

Now Thomas protested: *"We must have some weapons for our defence against other enemies than you. Otherwise we stay here; and you can do with us as you please."* Thomas understood very well how important it was for these Muslims to honour their pacts, and was exploiting this sense of honour.

Khalid went so far as to agree that every man could take one weapon with him, either a sword or a lance or a bow. The last of the problems was thus settled. [1]

Immediately after this, and it was now shortly after sunrise, a pact was drawn up and signed by Khalid. It read as follows -

"In the name of Allah, the Beneficent, the Merciful. This is given by Khalid bin Al Waleed to the people of Damascus. When the Muslims enter, they (the people) shall have safety for themselves, their property, their temples and the walls of their city, of which nothing shall be destroyed. They have this guarantee on behalf of Allah, the Messenger of Allah, on whom be the blessings of Allah and peace, the Caliph and the Faithful, from whom they shall receive nothing but good so long as they pay the Jizya." [2]

1. This dialogue between Khalid and Abu Ubaidah is taken from Waqidi: pp. 51-52.
2. Balazuri: p. 128.

The rate of Jizya was fixed at one dinar per man and a certain amount of food to be provided to the Muslims, the scale of which was also laid down.

Damascus had been taken. The greatest prize in Syria, with the exception of Antioch, was now in Muslim hands; but those who had conquered the city looked upon their victory with mixed feelings.

The Muslims had fought hard for this prize. While their casualties were much lower than those of the Romans, they had nevertheless paid a heavy prize for the conquest. They had struggled heroically for a month and given their blood and sweat for this victory. They had taken the city by the sword-especially the corps of Iraq, which had stormed it on the last night and crushed all resistance. But the fruits of their labour had been snatched away by the clever diplomacy of Thomas and the simple generosity and large-heartedness of

Abu Ubaidah. The Son of the Surgeon had no business to do this; but he was, after all the Trusted One of the Nation, and not a word of censure was raised against him.

The Muslims gathered in groups to see the Roman convoy march out of the city. The convoy consisted of the garrison and thousands of civilians who preferred not to remain under Muslim rule and moved out of Damascus with their wives and children. Thomas's wife, the daughter of Heraclius, travelled with her husband. With the convoy went hundreds of carriages and wagons carrying all the belongings of the travellers and the merchandise of the city, including 300 bales of the finest brocade belonging to Heraclius. Some Muslims looked in anger, others in sorrow, as they saw Damascus drained of all its wealth. It was a bitter moment for the victors of Damascus.

Khalid stood with some of his officers and men, gazing at the saddening sight. It appeared that the Romans were leaving nothing of value in Damascus. There was pain in the heart of Khalid. He was the commander of the army; he had conquered Damascus by the sword; he had stormed the fort. And Abu Ubaidah had done this!

He looked at the others and saw faces red with anger. All this should have been theirs by right of conquest. All along the route stood groups of Muslims watching in silence. They could easily have pounced upon the convoy and taken what they wished, but such was the discipline of this army, and such its respect for the moral obligation of the given word, that not a single soldier stirred to interfere with the movement of the convoy.

Khalid fought to control his rage. Then he raised his arms, to heaven, and in an anguished voice prayed aloud: *"O Allah! Give all this to us as sustenance for the Muslims!"* [1] But it was hopeless. Or was it?

Khalid heard a respectful cough behind him, and turned to see Jonah the Lover, still as sad as he had looked the night before in Khalid's tent. Jonah, meeting his bride after the surrender, had asked her to come away with him, and at first she was willing enough. But when he had told her that he was now a friend of the Muslims and had accepted their faith, she recoiled from him and swore that she would have nothing more to do with him. She decided to leave Damascus, and was even now travelling in the convoy of Thomas. Jonah, still the distracted lover maddened by his passion for the girl, had come to seek Khalid's help.

Could not the Muslims take the girl by force and deliver her to him? No, they could not. She was covered by the guarantee of safety and could not be touched.

Could the Muslims not pursue and attack the convoy? No, they could not. The guarantee of safety for the convoy would last three days, and during that period no pursuit could be undertaken.

After three days then? It was no use. Going at the terrified pace which it had adopted, the convoy would be so far away after three days that the Muslims would never catch up with it.

1. Waqidi: p. 52.

Oh yes, they could. He, Jonah, knew several short cuts which fast-moving horseman could use to overtake the convoy, while the convoy itself was bound to the roads and could not shorten its route. Still no use. Several Syrian forts-Emessa, Baalbeek, Tripolis-were close enough to reach in three or four days, and the convoy would be safely within the walls of any of these before the Muslims could catch up with it.

Oh no, it would not. The convoy was not going to any of these places. He, Jonah, knew that the convoy was making for Antioch and would take many days to get there. He, Jonah, would be the guide of the Muslims. All he wanted in return was the girl!

Khalid's eyes brightened. The possibilities which the proposal of Jonah opened up were like water to the thirsty. He beckoned to a few of his officers-Dhiraar, Raafe, Abdur-Rahman bin Abi Bakr. They would launch a pursuit after three days! Plans were formulated, orders issued, preparations made. When the three days' grace period was over, the Mobile Guard would dash out in pursuit and go at breakneck speed. On Jonah's suggestion it was decided that all would be dressed like local Arabs, so that any Roman units encountered on the way would mistake them for such and not intercept their movement. Hope stirred in the hearts of the Faithful!

On the morning of the fourth day, shortly after sunrise, at the exact time when the period of grace ended, the Mobile Guard galloped away from Damascus with Khalid and Jonah in the lead. Abu Ubaidah was left as commander at Damascus.

The route taken by the Mobile Guard is not recorded. It is stated by Waqidi that the Muslims caught up with the convoy a short distance from Antioch, not far from the sea, on a plateau beyond a range of hills called Al Abrash by the Arabs and Barda by the Romans. [1] Here there had been a heavy downpour, and the convoy had dispersed on the plateau, seeking shelter from the inclement weather, while the goods lay all over the place. The Romans had not the least suspicion of the thunderbolt that was about to strike them. So many bundles of brocade lay scattered on the ground that this plain became known as Marj-ud-Deebaj, i.e. the Meadow of Brocade, and for this reason the action described has been named the Battle of the Meadow of Brocade.

The weather had now cleared. Jonah and other scouts established the location of the convoy without being spotted, and brought sufficient intelligence for Khalid to plan his attack. He took a few hours to give his orders and position the Mobile Guard for its task. Khalid, the master of movement and surprise, here again showed his superb skill in the application of these military principles.

The Romans received their first indication of the presence of the Muslims when about a regiment of cavalry came charging at them from the south, along the road from Damascus, led by the half-naked Dhiraar. The Romans were surprised that Dhiraar had caught up with them, but seeing that he had only a small force, they decided to make mincemeat of him and then rest again. They formed up to meet the Muslim charge, and began to fight like the brave Romans that they were.

Half an hour later another body of Muslim cavalry, 1,000 horse led by Raafe, appeared from the east; and the Romans now realised the mistake that they had made in believing that only a regiment had caught up with them. The Muslims no doubt had two regiments. The first was a feint to draw the attention of the Romans, while the second delivered the main blow from a flank. But it did not matter; they would make mincemeat of two regiments instead of one. The Romans re-formed and received the charge of Raafe also.

Half an hour later, when another regiment of cavalry made its appearance from the north, i.e. from the direction of Antioch, under the command of Abdur-Rahman, the Romans were seriously alarmed. This was more dangerous than they had imagined. They were cut off from Antioch, and would have to deal quickly with these three regiments in order to break out to the north or retreat to the west, the latter being the only way left open to them. The Romans again re-formed though their spirits now were not so high. The Muslim regiments struck at the massed Romans with sword and lance and played havoc; but the Romans were able to hold their position, and the fighting proceeded fiercely for another hour.

1. This range was probably what is now known as Jabal Ansariya, the northern end of which stretches to the south of Antioch. Travelling across this range from Aleppo to Latakia one sees many stretches of level ground on the higher parts of the range.

Then from the west appeared a fourth Muslim regiment which charged at a gallop at the Roman mass. From the battle cry of its leader, the Romans knew who was the commander of this last group:

I am the noble warrior,
Khalid bin Al Waleed!

There was much slaughter-in the usual manner of Khalid. Khalid himself killed Thomas and Harbees in single combat, and at one time got so deep into the Roman army that he was separated from his comrades and surrounded by his enemies. He would not have come out alive but for Abdur-Rahman, who broke through with a party of horsemen and rescued him.

After some more fighting, Roman resistance collapsed. Since the Muslims were too few to completely surround the Roman army and the fighting had become confused as it increased in violence, thousands of Romans were able to escape and make their way to safety. But all the booty and a large number of captives, both male and female, fell to the Muslims. Jonah found his beloved. He moved towards her to take her by force; but she saw him coming, and drawing a dagger from the folds of her dress, plunged it into her breast. As she lay dying, Jonah sat beside her with silent tears running down his cheeks. He swore that he would remain true to the memory of the bride he was not destined to possess, and would not look at another girl.

When Khalid came to know of the loss suffered by Jonah, he sent for him and offered him another young woman who stood nearby-one who was both beautiful and rich, judging by the clothes and the jewellery which she wore. His first look at the young woman left Jonah dumbfounded. When he found his speech again, he informed Khalid that this woman was none other than the daughter of Heraclius, widow of Thomas. He could not possibly take her, for soon Heraclius would send either an army to get her back by force or envoys to arrange for her ransom.

The Muslims now marched back with spoils and captives enough to delight any conquering army. Their return route also is not recorded, but there was no mishap on the journey. When a day's march from Damascus, they saw a small cloud of dust approaching along the road from Antioch. As this cloud got nearer, it revealed a small party of riders, obviously not intending battle, since they were too few for such a purpose.

From this party a Roman noble rode forth and approached Khalid. *"I am the ambassador of Heraclius"*, he said. *"He says to you, 'I have come to know what you have done to my army. You have killed my son-in-law and captured my daughter. You have won and got away safely. I now ask you for my daughter. Either return her to me on payment of ransom or give her to me as a gift, for honour is a strong element in your character'. This is what Heraclius says."*

Honour was indeed a strong element in the make-up of Khalid. So was gallantry and so was generosity. Throughout his fife he had been generous in giving-a generosity which later would get him into serious trouble. Now he decided to be generous to Emperor of Roman. *"Take her as a gift"*, he said grandly. *"There shall be no, ransom."* The ambassador took the daughter of Heraclius, and with profuse thanks, returned to Antioch.

Jonah remained inconsolable. Nothing would cheer him up. Khalid offered him a large reward from his own share of the spoils, with which he could procure another wife, by purchase if necessary; but Jonah declined. He would remain true to his promise of celibacy. He also remained true to his new faith and fought under the banner of Islam for two years until the Battle of Yarmuk, where he fell a martyr.

The return of the Mobile Guard loaded with spoils was greeted with joy by the Muslims at Damascus. The Sword of Allah had done it again! The force had been absent for about 10 days, and the Muslims had been seriously perturbed; but now all was well. Khalid at once sent off a letter to Madinah, addressed to Abu Bakr, informing him of the conquest of Damascus and how Abu Ubaidah had been 'deceived by the Romans'; of his pursuit of the Roman convoy, the killing of Thomas and Harbees, the capture of the spoils and captives; of the daughter of
Heraclius and her release. This letter was written on October 1, 634 (the 2nd of Shaban, 13 Hijri). 2

The messenger carrying this letter had not gone many hours when Abu Ubaidah called Khalid aside and told him that Abu Bakr was dead and Umar was now Caliph. He held out a letter which the new Caliph had written him (i.e. Abu Ubaidah). Hesitantly Khalid took the letter and began to read. The most important line seemed to stand out mockingly: *"I appoint you commander of the army of Khalid bin Al Waleed..."*

Khalid looked up from the letter ...

1. Waqidi: p. 58.
2. For an explanation of the dates of the siege and conquest of Damascus, see Note 11 in Appendix B.

Chapter 31

(The Unkind Cut)

"Praise be to Allah who decreed death upon Abu Bakr, who was more beloved to me than Umar. Praise be to Allah who gave authority to Umar, who was less beloved to me than Abu Bakr, and compelled me to love him."
[Khalid ibn al-Walid, upon breaking the news of Abu Bakr's death to his army.]

"You have done deeds which no-one has done, but people do nothing, for Allah is the Doer."
[Arabian poet, quoted by Umar to Khalid][1]

In Madinah, as the old Caliph lay dying, the sent for writing materials and wrote an order: After him Umar would be the Caliph and the Believers would swear allegiance to him. This was the last order of Abu Bakr.

On August 22, 634 (22nd Jamadi-ul-Akhir, 13 Hijri), Abu Bakr died and Umar became Caliph. On the same day the new Caliph issued his first order: Khalid was dismissed from the command of the Muslim army in Syria! He wrote to Abu Ubaidah as follows:

In the name of Allah the Beneficent, the Merciful.

I urge upon you the fear of Allah who lives eternally while everything else perishes; who has guided us away from wrongdoing and taken us out of darkness into light.

I appoint you commander of the army of Khalid bin Al Waleed. So take charge as is your duty.

Send not the Muslims to their destruction for the sake of plunder; and place not the Muslims in a camp without reconnoitring it and knowing what is there.

Send not expeditions except in properly organised units. And beware of taking any steps which may lead to the annihilation of the Muslims.

Allah has tried me with you and tried you with me. Guard against the temptations of this world lest they destroy you as they have destroyed others before you; and you have seen how they felt. [2]

The letter was given to a messenger with instructions to proceed to Syria and hand it personally to Abu Ubaidah.

The next day Umar led the congregational prayer in the mosque of the Prophet. When the prayer was over, he addressed the congregation-the first public address of his caliphate. He started by praising Allah and invoking His blessings on the Prophet; then he continued: *"Lo! The Arab is like a camel which follows its master and waits for him wherever it is made to sit. And by the Lord of the Kabah, I shall carry you on the right path."*

In the rest of his sermon he emphasised various virtues and duties enjoined upon Muslims, and pledged to do his best to further the interests of Islam. Coming to the end of his sermon, he informed the congregation that he had removed Khalid from the command of the army in Syria and appointed Abu Ubaidah in his place.

This announcement was received by the Muslims in hushed silence. Everyone knew that in the heart of Umar there was little love for Khalid, but none had expected Umar to act so harshly against the Sword of Allah, and in such haste, especially after the great victories which Khalid had won for Islam during the last three years. However, Umar was a much feared, albeit respected man, and few would dare to cross him. Moreover, as Caliph he had the authority to appoint and dismiss commanders as he chose, and his decision had to be accepted and obeyed. All remained silent, with a silence more eloquent than words.

1. Ibn Kathir, Al-Bidayah wan-Nihayah, Dar Abi Hayyan, Cairo, 1st ed. 1416/1996, Vol. 7 P. 19.
2. Tabari: Vol. 2, p. 622.

But one youth who was present could not contain himself and leapt to his feet. *"Do you dismiss a man"*, he shouted at Umar, *"in whose hand Allah has placed a victorious sword and with whom. Allah has strengthened His religion! Allah will never forgive you, nor will the Muslims, for sheathing the Sword and dismissing a commander whom Allah has appointed to command."*

Umar knew this youngster, he was from the Bani MakhDhulm-the clan of Khalid. He could also sense the mood of the congregation and knew that its reaction to his announcement was anything but favourable. He decided not to say any more on the subject for the moment., He merely retorted: *"The young man is angry on account of the son of his uncle."* 1 and walked away from the mosque.

Over the day Umar reflected a great deal on the matter of Khalid's dismissal. He came to the conclusion that he would have to explain his action to the Muslims in order to convince them of its justice. Such a dazzling light as Khalid could not be extinguished without offering adequate justification. The following day he again addressed the Muslims:

"I am not averse to Khalid being in command. But he is wasteful and squanders his wealth on poets and warriors, giving them more than they deserve, which wealth could be better spent in helping the poor and the needy among the Muslims. Let none say that I have dismissed a strong man, and appointed a mild man to command, for Allah is with him (i.e. Abu Ubaidah) and will help him." 2

This time no one said anything.

The messenger carrying the fateful letter arrived at Damascus while the siege was in progress and the action against the Roman relief column was still a few days away. He knew the contents of the letter, and being an intelligent man guessed that its effect on the embattled Muslims would be far from healthy. So he told everyone whom he met that all was well and that reinforcements were on their way. Arriving at the tent of Abu Ubaidah, where no one else was present, he handed over the letter.

Abu Ubaidah read the letter and was astounded. He would not have wished this to happen to Khalid. He knew that Khalid was the idol of the army and his presence as commander-in-chief was a factor of the highest importance in making the Muslims so confident of victory against all odds. The impact of the change of command would be most adverse, especially whilst the Muslims were engaged in a stubborn siege which showed no sign of turning in their favour. It would be difficult to convince them of the justice of Khalid's dismissal or the wisdom of its timing. Moreover, Abu Ubaidah did not feel inclined to take over command in the middle of the operation when Khalid had everything so well organised. He therefore decided to say nothing about the death of Abu Bakr or the change of command until after the siege had been successfully concluded. The messenger, on being questioned, assured him that he had not divulged the contents of the Caliph's letter to anyone; and Abu Ubaidah cautioned him to keep the matter to himself.

The Muslims at Damascus remained ignorant of the change of command during the rest of the siege. Even on the day of conquest Abu Ubaidah made no reference to it in his altercation with Khalid, for doing so would have amounted to hitting below the belt and would have belittled Khalid in the presence of friend and foe. Thus it was Khalid who signed the pact with the Damascenes and not Abu Ubaidah. In fact it was not until a few hours after Khalid's return from the raid at the Meadow of Brocade that Abu Ubaidah drew him aside, told him of the death of Abu Bakr and the appointment of the new Caliph, and gave him Umar's letter to read.

1. Tabari: Vol. 2, p. 622.
2.

Slowly Khalid read the letter. It was quite clear: he had been sacked! Abu Ubaidah was the new Commander-in-Chief. Perhaps he should have expected that this would happen if Umar became Caliph; but he had not expected it because he had never considered the possibility of Abu Bakr's death or of Umar's becoming Caliph.
From the date on the letter Khalid saw that it was more than a month old and must have reached Abu Ubaidah at least three weeks before now. He looked up at Abu Ubaidah and asked, *"Why did you conceal this from me? May Allah have mercy upon you!"* Abu Ubaidah replied, *"I did not wish to weaken your authority while you were engaged with the enemy."* 1

For a few moments Khalid remained lost in his thoughts-thoughts of Abu Bakr, his friend, guide and benefactor. Abu Ubaidah looked at him, partly in sympathy, partly in embarrassment. Then Khalid remarked: *"May Allah have mercy upon Abu Bakr! Had he lived, I would not have been removed from command."* 2 Slowly, with bowed head, the Sword of Allah walked away to his tent.

That night Khalid wept for Abu Bakr. 3

The following morning, October 2, 634 (3rd Shaban, 13 Hijri), the army was assembled and informed of the two changes-in the Caliphate and in the command in Syria. On this day the Muslims in Damascus took the oath of allegiance to the new Caliph.

If any resentment or bitterness existed in Khalid's heart-and some must undoubtedly have existed-he showed no sign of it. He remarked casually to his friends, *"If Abu Bakr is dead and Umar is Caliph, then we hear and obey."* 4 There was nothing that Khalid could do to air his grievance without causing serious harm to the Muslim army and the Muslim cause in Syria, for any anti-Umar action would probably have split the army, and this was the last thing that the true soldier and true Muslim would wish.

Once a commander-in-chief is dismissed from his command, he normally does not serve, if he serves at all, in the same theatre where he has been in command. He retires. Or he asks to be transferred or is transferred anyway in consideration for his feelings. Sometimes he is "kicked upstairs." But it was Khalid's destiny to fight and to conquer, and nature had gifted him with all the military virtues needed to fulfil that destiny. Thus we see here the remarkable phenomenon of the greatest general of the time (indeed the greatest general of the first millennium of the Christian Era) being prepared to serve in a lower capacity, even as a common soldier, with the same drive and zeal which he had shown as an army commander. This willingness to serve also reflects the Muslim spirit of the time. And all this became evident a fortnight later in the crisis of Abul Quds.

A week after Abu Ubaidah assumed command of the army, a Christian Arab, seeking the favour of the Muslims, came to the new Commander-in-Chief and informed him that in a few days a great fair would be held at Abul Quds. At this fair visitors and merchants from all the lands in the Asian zone of the Byzantine Empire would come with costly wares to buy and sell. Should the Muslims wish to acquire more spoils, they only had to send a raiding column to pick up all the wealth they wanted (Abul Quds is now known as Abla and lies at the eastern foothills of the Lebanon Range, near Zahle, about 40 miles from Damascus on the road to Baalbeck.) 5 The informer could not say if there would be any Roman soldiers guarding the fair, but there was a strong garrison at Tripolis, on the Mediterranean coast.

Abu Ubaidah spoke to the warriors who sat around him, and asked if anyone would volunteer to take command of a column and raid Abul Quds. He was hoping that Khalid would offer his services for the task, but Khalid remained silent. Then a youth, on whose face the beard had only just begun to grow, volunteered himself with bubbling enthusiasm. This boy was Abdullah, son of Jafar, the Prophet's cousin who had been martyred at Mutah. This young nephew of the Prophet had only just arrived from Madinah and was anxious to win glory in the field. Abu Ubaidah accepted the youth's offer and appointed him commander over a body of 500 horsemen.

1. Balazuri: p.122
2. Yaqubi: *Tareekh*, Vol. 2, p. 140.
3. Waqidi: p. 62.
4. Waqidi: p. 62.
5. Gibbon (Vol. 5, p. 321) calls this place Abyla. It may have been so named in his time, but it is now called Abla.

On October 14, 634 (the 15th of Shaban, 13 Hijri), the column marched by the light of a bright full moon. With young Abdullah rode a dedicated and saintly soldier by the name of Abu Dharr Al Ghifari. The following morning the impetuous boy launched his small group against a Roman force of 5,000 men which was guarding the fair. Since Abdullah sought glory and Abu Dharr sought martyrdom, there was no one to restrain the Muslims; and the result was disastrous. After some heroic fighting, the Muslims were surrounded by the Romans, and it became evident that none would escape. But when the Muslim turned at bay he was a deadly fighter. The veteran soldiers knew how to defend themselves and quickly formed a tight ring to keep the Romans out; and thus surrounded, they continued to fight, their desperate courage imposing caution on the Romans. But their annihilation was only a matter of time.

One Muslim, however, had escaped the Roman encirclement, and realising the gravity of the situation, he galloped off to Damascus for help. Abu Ubaidah was sitting with his generals when this man arrived to report the disaster and ask for immediate help, without which not a single Muslim would return from Abul Quds. Abu Ubaidah was aghast. His thoughts flew to the words of Umar: *"Send not the Muslims to their destruction for the sake of plunder."* Moreover, this was his first military decision as Commander-in-Chief and if Abdullah and his men were not saved, the effect on the army would be devastating. And who could do the job but the Sword of Allah!

Abu Ubaidah turned to Khalid: *"O Father of Sulaiman, I ask you in the name of Allah to go and rescue Abdullah bin Jafar. You are the only one who can do so."*

"I shall certainly do so, Allah willing", replied Khalid. *"I waited only for your command."*

"I felt hesitant to ask you", remarked Abu Ubaidah, alluding to the embarrassment which he felt over the recent change of command.

Khalid continued: *"By Allah, if you were to appoint a small child over me, I would obey him. How could I not obey you when you are far above me in Islam and have been named the Trusted One by the Prophet? I could never attain your status. I declare here and now that I have dedicated my life to the way of Allah, Most High."*

In a voice choking with emotion, Abu Ubaidah said, *"May Allah have mercy upon you, O Father of Sulaiman. Go and save your brothers."* 1

Within half an hour the Mobile Guard was galloping in the direction of Abul Quds with Khalid and Dhiraar in the lead. Of course Khalid saved the trapped Muslims, though many of them had been killed by the Romans. And not only that; he also raided the market of Abul Quds and brought back an enviable amount of booty! He also brought back many wounds on his person, but getting wounded was now such an everyday affair in Khalid's life that he took little notice of them.

The result of the action at Abul Quds left no doubt (if there ever was any) about Khalid's reaction to his dismissal. Abu Ubaidah wrote an account of this action to Umar, giving generous praise to Khalid for the part that he had played in it. But the windows through which the light of such praise could shine at Madinah were closed. They would never open again.

This dual change of personalities-the Caliph at Madinah and the Commander-in-Chief in Syria-was to have its effect on the conduct and pace of military operations. Umar's methods were very different from his predecessor's. While Abu Bakr would give his commanders their mission and area of operations and leave to them the conduct of the campaign, Umar would order specific objectives for each battle. Later in his caliphate he would even lay down such details as who should command the left wing, who should command the right wing, and so on. He also started a system of spies to watch his own generals. These spies were placed in all armies and corps, and everything that any officer said or did was promptly reported to the Caliph. 2

1. Waqidi: p. 66.
2. Tabari: Vol. 2, p. 658.

Umar confirmed the various corps commanders in the roles allotted to them by Abu Bakr. Amr bin Al Aas would command in Palestine, Yazeed in Damascus, Shurahbil in Jordan and Abu Ubaidah in Emessa-after it was taken. These roles included not only the military command of the various corps, but also political control over the provinces. Thus, for instance, Shurahbil was not only the corps commander for operations in Jordan but also the governor of the District of Jordan. And yet Abu Ubaidah remained the Commander-in-Chief of the army as a whole, although he would command the army only when the corps fought together against the Romans. For Khalid there was no role. By the order of Umar he would operate under Abu Ubaidah, and the latter confirmed him as the commander of the corps of Iraq which included the Mobile Guard. In military status Khalid was equal to the other corps commanders; but politically he was now a nobody.

There was inevitably a slowdown in the pace of operations. Abu Ubaidah was a great man and personally a fearless and skilful fighter. Over the next few years he would also become a good general as a result of Khalid's coaching. He would rely heavily on the advice of Khalid, whom he kept beside him as much as possible, but he never possessed the strategic vision or the tactical perception of Khalid. More often than not, he would hold councils of war or write to Madinah to seek the Caliph's decision regarding his next objective. Whereas Khalid would rush like a tornado from battle to battle, using surprise, audacity and violence to win his battles, Abu Ubaidah would move slowly and steadily. Yet, he too would win his battles.

With this new arrangement, with the mutual respect and affection between Abu Ubaidah and Khalid unimpaired, and with Khalid throwing the great weight of his genius behind the new Commander-in-Chief, the conquest of Syria continued.

1. Waqidi: p. 66.
2. Tabari: Vol. 2, p. 658.

Chapter 32

(The Battle of Fahl)

"The plotting of Evil surrounds only its own plotters." [Quran 35:43]

More will be said in a later chapter about the character and abilities of Heraclius and the strategy he used for his attempt to crush the Muslim invaders of his Empire. Here it may just be noted that as an enemy, Heraclius was a man to be reckoned with-not one to give up the struggle while the least hope remained. His next move after the affair of Abul Quds was to put another army in the field, consisting of fresh contingents from Northern Syria, the Jazeera and Europe. This army included the survivors of the Meadow of Brocade. Part of the army gathered at Antioch, while part landed by sea at the Mediterranean ports in Syria and Palestine.

The concentration of this army at Baisan, west of the Jordan River, began in late December 634 (early Dhul Qad, 13 Hijri). From here the army would strike eastwards and cut Muslim communications with Arabia. According to this plan-which was typical of Heraclius-he would avoid a head-on clash with the Muslims at Damascus, put them in a position of strategical disadvantage, and force them to evacuate Damascus. Fahl, just east of the Jordan River, was already occupied by a Roman garrison of moderate size which was engaged by a Muslim cavalry detachment under Abul A'war.

The Muslims received intelligence of the movement of Roman contingents from local agents; and before the concentration of the Romans at Baisan was complete, they knew that the strength of this new army would be about 80,000 men, and that its commanders was Saqalar, son of Mikhraq. It was evident that this force would move eastwards and place itself astride the Muslim lines of communication. A council of war was held by Abu Ubaidah, and it was decided that the Muslims should move and crush this new Roman army, leaving behind a strong garrison to hold Damascus against any threat from the north and west. By now the Muslims had fully rested after their heroic labours. Soon after Abul Quds, more reinforcements had been received from Arabia, while a large number of those who had been wounded in earlier battles had rejoined the Muslim ranks as fit soldiers. This raised the strength of the army to something like 30,000 men, organised in five corps of varying strength.

Now the command arrangement made by Abu Bakr and confirmed by Umar came into effect in a rather unusual way. Yazeed was the commander and governor of the Damascus region, and was consequently left in Damascus with his corps. Shurahbil was the commander appointed for the district of Jordan in which lay Baisan and Fahl. Hence Abu Ubaidah, carrying out the Caliph's instructions to the letter-farther than was probably intended-handed over the command of the army to Shurahbil for the forthcoming operation. In about the second week of January 635, the Muslim army, leaving behind the corps of Yazeed, marched from Damascus under the command of Shurahbil, with Khalid and the corps of Iraq forming the advance guard. In the middle of January the Muslims arrived at Fahl to find the Roman garrison gone, Abul A'war in occupation of the town, and what looked like a marsh stretching on both sides of the Jordan River. [1]

As soon as the Roman garrison of Fahl had heard of the advance of the Muslim army from Damascus, it had left the place in haste, and withdrawing across the river, joined the main body of the Roman army at Baisan. Immediately after, the Romans, not wishing to be disturbed at Baisan before their preparations were complete, dammed the river a few miles south of the Baisan-Fahl line and flooded the low-lying belt which stretched along both banks of the river. The flooded area was determined by the contour line and in places was up to a mile from the river. There were some routes across this inundated area, but they were known only to the Romans. The Muslims knew the desert; they had come to know the hills; but this belt of water and mud which stretched along their front was a new experience and left them nonplussed. However, they decided to attempt a crossing.

1. Fahl is below sea level, and from the town the hillside slopes even further down to the bed of the Jordan Valley. In this area the Jordan River is about 900 feet below sea level.

Shurahbil deployed the army at the foot of the slope below Fahl, facing north-west, with Abu Ubaidah and Amr bin Al Aas commanding the wings. Dhiraar was appointed commander of the Muslim cavalry, while Khalid with his corps was placed in front to lead the advance to Baisan. In this formation the Muslims advanced. But they had not gone far when the Advance Guard got stuck in the mud and had considerable difficulty in extricating itself. Cursing the Romans for this stratagem, the Muslims returned to Fahl and waited. Thus a whole week passed.

Now Saqalar, the Roman commander, decided that the time had come to strike. His preparations were complete and he hoped to catch the Muslims off guard since the marsh would give them, he hoped, a false sense of security. His guides would lead the army through the marsh which the Muslims regarded as impassable. Soon after sunset on January 23, 635 (the 27th of Dhul Qad, 13 Hijri), the Roman army formed up west of the river and began its advance towards Fahl, intending to surprise the Muslims in their camps at night.

But the Muslims had not relaxed their guard. Shurahbil was a watchful general and had deployed the Muslim camp to correspond to the battle positions of the corps, and kept a large portion of each corps in its battle positions during the night. He had also placed a screen of scouts along the marsh to watch and report any movement by the Romans towards Fahl. Thus, as the Romans neared Fahl, they found an army, not resting in its camp, but formed up in battle array. Immediately on contact the battle began.

The two armies fought all night and the whole of the next day-January 24, 635. The Muslim army remained on the defensive and beat off all attempts by the Romans to break through, during one of which Saqalar was killed. By the time darkness had set in again, the Romans decided that they had had enough. They had suffered heavily at the hands of the Muslims, who had stood like a wall of steel in their path; and this wall had not been breached at a single place. Under cover of darkness the Romans disengaged and began to withdraw across the marsh towards Baisan.

This was the moment that Shurahbil was waiting for. He had fought the Romans until they were exhausted, and suffering from the adverse psychological impact of repeated repulses, had started to withdraw. Now was the time to launch the counterstroke. Shurahbil ordered the advance; and in the darkness, the desert-dwellers leapt upon the backs of the Romans!

This time the Roman 'traffic control plan' failed. Thousands of them were lost in the marsh, and as the screaming masses of the Muslims came after them, they gave way to panic and lost all order and cohesion. The Muslims set to with gusto to finish this army and played havoc with their terrified enemy. About 10,000 Romans perished in the Battle of Fahl, which is also known in Muslim history as the Battle of Mud. [1] Some of the Romans arrived safely at Baisan while others, fleeing for their lives in total disorder, dispersed in all directions.

With the defeat of this Roman army, the Muslim army also broke up. Abu Ubaidah and Khalid remained at Fahl, whence they would shortly set out for Damascus and Northern Syria. Shurahbil, with Amr bin Al Aas under command, crossed the marsh and the river, routes through which had now been found, and laid siege to Baisan. After a few days the Romans in the fort made a sally but were slaughtered by Shurahbil. Soon after this sally Baisan surrendered and agreed to pay the Jizya and certain taxes. Shurahbil then went on to Tabariya, which also surrendered on similar terms. This last action was over before the end of February 635 (Dhul Haj, 13 Hijri). There was now no opposition left in the inland part of the District of Jordan.

With the beginning of the fourteenth year of the Hijra, Amr bin Al Aas and Shurahbil turned their attention to Palestine. Here again a change of command took place. Palestine was the province of Amr, and consequently he assumed command of the army, while Shurahbil served under him as a corps commander. But it was some time before this small army of two corps entered Palestine.

1. Most early historians have said that the bulk of the Roman army was destroyed in this battle. Balazuri, however, has placed Roman losses at 10,000 (p. 122); and this is here accepted as the most conservative estimate.

While still in Jordan, Amr had written to the Caliph and given him the latest intelligence about Roman dispositions and strengths in Palestine. The strongest Roman force was at Ajnadein. Umar gave detailed instructions to Amr about the objectives which he was to take, and also wrote to Yazeed to capture the Mediterranean coast. In pursuance of these instructions the Muslim army, excluding the corps of Abu Ubaidah and Khalid, operated against the Romans in Palestine and on the coast as far north as Beirut. The corps of Amr and Shurahbil marched to Ajnadein, and with Amr as army commander, fought and defeated a Roman army in the second Battle of Ajnadein. Thereafter the corps separated. Amr went on to capture Nablus, Amawas, Gaza and Yubna, thus occupying all Palestine, while Shurahbil thrust against the coastal towns of Acre and Tyre, which capitulated to him. Yazeed, with his brother Muawiyah playing an important role under him, advanced from Damascus and captured the ports of Sidon, Arqa, Jabail and Beirut.

The place which took the longest to capture was Caesarea. Umar had given this as an objective to Yazeed; and he and Muawiyah laid siege to it, but Caesarea, reinforced and supplied by the Romans by sea, could not be captured in spite of their best efforts. The siege was raised when the Muslims had to regroup for the Battle of Yarmuk, but was resumed after that battle and continued until the port fell in 640 (19 Hijri).

By the end of 14 Hijri (roughly 635 A.D.), Palestine, Jordan and Southern Syria, with the exception of Jerusalem and Caesarea, were in Muslims hands.

Chapter 33

(The Conquest of Emessa)

"It is said that the Companions said a takbir (Allahu Akbar - Allah is the Greatest!) one day during the siege of Homs (Emessa), by which the town shook, such that some of its walls split asunder. Then they said another takbir, upon which some houses collapsed. Hence the public went to their leaders and said, "Do you not see what has befallen us, the situation in which we are? Will you not make peace with them for us?" So they made peace with them upon terms similar to those of Damascus ..."[1]

In early March 635 (early Muharram, 14 Hijri), Abu Ubaidah and Khalid set off from Fahl to carry the war to the north. They had waited at Fahl while Shurahbil was dealing with Baisan and Tabariya, in case a large scale battle should develop necessitating their participation. Once Tabariya was taken, the possibility of such a battle in Jordan vanished and they were free to depart.

A few miles west and south west of Damascus stretched a grassy plain known in Muslim history as Marj-ur-Rum, i.e. the Meadow of Rome, and towards this plain Abu Ubaidah and Khalid moved with the intention of bypassing Damascus and continuing the advance to Emessa. Yazeed was still in peaceful occupation of Damascus and would remain there a few months yet, before receiving orders from Umar to operate against the Mediterranean coast. At Marj-ur-Rum, Abu Ubaidah again made contact with sizable Roman forces.

On hearing of the Muslim operations at Baisan and Tabariya, Heraclius surmised that the Muslims had chosen Jordan and Palestine as their next strategic objectives and were not interested in Northern Syria. He also heard that only a weak corps of the Muslim army remained at Damascus, and this corps was showing no sign of aggressive intent. He therefore determined to retake Damascus rapidly. With this object in view he sent a Roman force under a general named Theodorus to fight and defeat the Muslim garrison in Damascus and re-occupy the city. This force set off from Antioch, and moving via Beirut, approached Damascus from the west. This movement, however, had hardly begun when Heraclius was informed that Abu Ubaidah and Khalid had left Fahl and were moving north again. They would arrive at Damascus at about the same time as Theodorus, and the Romans would then not have a chance to retake the city. To strengthen the Roman force, Heraclius ordered the detachment of a part of the large garrison of Emessa to reinforce Theodorus. This detachment, under the command of Shans, marched from Emessa on the direct route to Damascus.

The Muslims arrived at Marj-ur-Rum to find Theodorus waiting for them. On the same day Shans also arrived from Emessa and the two armies deployed in battle formation facing each other. In this deployment Abu Ubaidah stood opposite Shans while Khalid stood opposite Theodorus. The strength of the Roman forces here is not known, but it may be assumed that it amounted roughly to two strong corps. It could not have been much less otherwise it is doubtful if the Romans would have accepted battle with the two Muslim corps facing them. For the rest of the day the two armies remained in their battle positions, each waiting for the other to make the first move.

As night fell, Theodorus decided to carry out a skilful strategical manoeuvre. Leaving Shans to face the Muslims, he pulled back his corps under cover of darkness, moved it round the flank of Khalid and by dawn on the next day arrived Damascus. His intention was to keep the main Muslim army busy at Marj-ur-Rum with the corps of Shans, while with his own corps he quickly destroyed the Muslim garrison of Damascus. It was a very clever plan, and the movement was carried out with such perfect organization that it was not until the latter part of the night that the Muslims came to know that half the Roman army facing them was no longer there.

At Damascus, Yazeed's scouts brought word at dawn of the coming of the Romans. On receiving this news, Yazeed immediately deployed his small corps outside the fort facing south-west. Feeling more at home in the open and unused to being besieged in a fort, the Muslims preferred to fight in the plain rather than in the city. Just after sunrise began the battle between Theodorus and Yazeed and soon the Muslims found themselves hard pressed, for the Roman force vastly outnumbered them. But they held their own till about mid-morning. Then, just as the situation had become desperate for Yazeed, the Romans were struck in the rear by a furious mass of Muslim horsemen. This was the corps of Iraq, spearheaded by the Mobile Guard. In a very short time Khalid and his fearless veterans, attacking from the rear, had chopped the Roman corps to pieces. Few Romans escaped the slaughter, and Khalid killed Theodorus in a duel. A large amount of booty, mainly weapons and armour, fell into Muslim hands and was shared by the warriors of Khalid and Yazeed, except for the usual one-fifth reserved for Madinah.

1. Ibn Kathir, Al-Bidayah wan-Nihayah, Dar Abi Hayyan, Cairo, 1st ed. 1416/1996, Vol. 7 P. 65.

Late in the preceding night, when he discovered that half the Roman army had left Marj-ur-Rum, Khalid had correctly guessed that it had gone to Damascus to fight Yazeed. Fearing that Yazeed might not be able to hold out for long, he proposed to Abu Ubaidah that he take his corps to Damascus to help Yazeed while Abu Ubaidah dealt with the remaining Romans under Shans. Abu Ubaidah agreed and early in the morning, Khalid left Marj-ur-Rum to save Damascus, as has just been described. While Khalid was liquidating the corps of Theodorus, Abu Ubaidah attacked the Romans on the Meadow of Rome. Abu Ubaidah killed Shans in a duel, and the plain was littered with Roman dead, but the bulk of the Roman corps got away and withdrew in haste to Emessa.

This action was fought some time in March 635 (Muharram 14 Hijri), and is known as the Battle of Marj-ur-Rum.

Some time was spent at Marj-ur-Rum and Damascus, dealing with the captives and spoils of war and making arrangements for the wounded Muslims. Once these matters had been attended to, Abu Ubaidah sent Khalid with his corps on the direct route to Emessa, while he himself advanced to Baalbeck. The garrison of Baalbeck surrendered peacefully, and Abu Ubaidah proceeded to Emessa to join Khalid, who had laid siege to the fort. [1]

Within a few days of the commencement of the siege a truce was agreed upon. Emessa would pay 10,000 dinars and deliver 100 robes of brocade, and in return the Muslims would not attack Emessa for one year. If, however, any Roman reinforcements arrived to strengthen Emessa, the truce would become invalid. The gates of Emessa were opened as soon as the truce was signed, and thereafter there was free movement of Muslims in and out of the market of Emessa, the inhabitants of which were pleasantly surprised to find that the Muslims paid for whatever they took!

The people of Qinassareen (the ancient Calchis) now heard of the peaceful way in which the citizens of Emessa had avoided battle with the Muslims, and decided to do the same. A truce was not as dishonourable as a surrender and was a convenient way of postponing a difficult decision. Consequently an envoy was sent to Emessa by the governor of Qinassareen, who made a similar truce with Abu Ubaidah for one year. But both governors, of Emessa and Qinassareen, made the truce for reasons of expediency. Both hoped that their garrisons would before long be reinforced by Heraclius, and as soon as that happened they would resume hostilities against the Muslims. The common man in the region, however, was completely won over by the kindness and fair dealing of the Muslims and the absence in them of the arrogance and cruelty which had characterised Roman rule over Syria.

Having temporarily solved the problems of Emessa and Qinassareen, Abu Ubaidah despatched the bulk of his army, in groups, to raid Northern Syria. Muslim columns travelled as far north as Aleppo, and leaving the District of Qinassareen unmolested, raided any locality through which they passed and brought in captives and booty to the Muslim camp near Emessa. Thousands of these captives, however, begged for their freedom and all who agreed to pay the Jizya and pledge loyalty to the Muslims were freed, with their families and goods, and allowed to return to their homes with a guarantee of safety from Muslim raiding columns.

This went on for some months and most of the summer was spent in this manner. Meanwhile Umar was getting impatient at Madinah. The campaign was progressing satisfactorily in Palestine, but in Northern Syria, *i.e.* in Abu Ubaidah's sector, there seemed to be a lull. Consequently, some time in the autumn of 635, Umar wrote a letter to Abu Ubaidah in which he hinted that the general should get on with the conquest of Syria. On receipt of this letter Abu Ubaidah held a council of war, at which it was agreed that the Muslim army should proceed north and conquer more territory. Emessa and Qinassareen could not be touched as they were secure under the terms of the truce; but for other places there was no such truce, and they could be attacked and taken.

1. There are other versions of how Baalbeck was taken, including Waqidi's, according to which a great battle was fought by Abu Ubaidah before Baalbeck surrendered to the Muslims. Other historians, however, have said that Baalbeck surrendered peacefully, and I too feel that this is, what happened.

About early November 635 (middle of Ramazan, 14 Hijri), the Muslim army marched from Emessa to Hama, where the citizens came out of their city to welcome the Muslims. The city surrendered willingly, and the army marched on. One by one the cities of Shaizar, Afamiya (known today as Qalatul-Muzeeq) and Ma'arra Hims (now Ma'arrat-un-Numan) surrendered in peace to the Muslims and agreed to pay the Jizya. At some places the Muslims were received by musicians playing instruments as a sign of welcome. In these areas now, for the first time in Syria, large-scale conversions took place among the local inhabitants. The personality of the gentle, benevolent Abu Ubaidah played an important part in these conversions to Islam.

It was while the Muslims were at Shaizer that they heard of reinforcements moving to Qinassareen and Emessa. The truce was thus violated by the Romans. The arrival of these reinforcements put fresh courage in the hearts of the Romans at Emessa and Qinassareen, and the arrival of winter gave them a further assurance of success. In their forts they would be better protected from the cold than the Muslim Arabs, who were not used to intense cold, and with only their tents to give them shelter would suffer severely from the Syrian winter. In fact Heraclius wrote to Harbees, the military governor of Emessa: *"The food of these people is the flesh of the camel and their drink its milk. They cannot stand the cold. Fight them on every cold day so that none of them is left till the spring."* [1]

Abu Ubaidah decided to take Emessa first, and thus clear his rear of the enemy before undertaking more serious operations in Northern Syria. Consequently the Muslims marched to Emessa with Khalid and the corps of Iraq in the lead. On arrival at the city Khalid found a strong Roman force deployed across his path, but with a quick, violent attack his corps drove it back into the fort. These Romans had followed Heraclius' instructions to *"fight them on every cold day"*, but after their experience in this first clash with Khalid, they decided to let winter do the job! As the Romans withdrew into the fort and closed the gates, Abu Ubaidah arrived with the rest of the army and deployed it in four groups opposite the four gates of Emessa.

Emessa was a circular fortified city with a diameter of rather less than a mile, and it was surrounded by a moat. There was also a citadel atop a hillock inside the fort. Outside the city stretched a fertile plain, broken only on the west by the River Orontes (now Asi).

Abu Ubaidah himself, together with Khalid and his Mobile Guard, camped on the north side, a short distance from the Rastan Gate. [2] The Muslim strength at Emessa was about 15,000 men against which the Roman garrison consisted of something like 8,000 soldiers. Abu Ubaidah left the conduct of the siege in the hands of Khalid, who thus acted as the virtual commander of the Muslims for this operation. It was now late November or early

December (about the middle of Shawal), and the winter descended like a heavy blanket over Emessa.

For more than two months the siege continued with unbroken monotony. Every day there would be an exchange of archery, but no major action took place which could lead to a decision either way. The Romans gloated over the exposed situation of the Muslims, and felt confident that the cold itself would be sufficient to destroy the desert-dwellers or drive them away to warmer climes. The Muslims undoubtedly suffered from the cold but not as severely as the Romans imagined. There was no slackening in their guard and no weakening in their resolve to take Emessa, no matter how long they had to wait.

1. Tabari: Vol. 3, pp. 96-97.
2. The only gate which still exists is the Masdud Gate, to the southwest. The visitor to Emessa today is shown the sites of three other gates: Tadmur (north-east), Duraib (east) and Hud (west); but while the present inhabitants of the city have heard of the Rastan Gate, its location is not known. It was no doubt somewhere in the northern wall, because it faced Rastan, which lies on the road to Hama. Early historians have named the Rastan Gate as one of four, and we do not know which one of the present four gates, as named above, did not then exist. The moat too is still there in many places.

When another few weeks had passed and there was no further retrograde movement by the Muslims, the Romans realised that their opponents had no intention of raising the siege. It was now about the middle of March 636 (the beginning of Safar, 15 Hijri), when the worst of the winter was over. The Roman hope of the cold driving the Muslims away vanished. Supplies were running low, and with the coming of spring and better weather the Muslims would receive further reinforcements and would then be in an even stronger position. Something had to be done quickly. The local inhabitants were all for peace, but Harbees was a loyal son of the Empire and sought glory in battle. He decided to make a surprise sally and defeat the Muslims in battle outside the fort; and with this decision of Harbees matters came to a head. The end was now in sight, though not the kind of end which Harbees had in mind.

Early one morning the Rastan Gate was flung open and Harbees led 5,000 men into a quick attack on the unsuspecting Muslims facing that gate. The speed and violence of the attack took the Muslims by surprise, and although this was the largest of the four groups positioned at the four gates, it was driven back from the position where it had hastily formed up for battle. A short distance back the Muslims reformed their front and held the attack of the Romans, but the pressure became increasingly heavy and the danger of a break-through became clearly evident.

Abu Ubaidah now asked Khalid to restore the situation. Khalid moved forward with the Mobile Guard, took the hard pressed Muslims under his command and redisposed the Muslim army for battle. The surprise of the morning had had a depressing effect on the Muslims, who had already been distressed by the discomfort of the cold; and they took some time to recover from it, but with Khalid present in their midst, they soon regained their spirits and began to give as well as they took. This situation continued till midday. Then Khalid took the offensive and steadily pushed the Romans back, though it was not till near sunset that the Romans were finally driven back into the fort. The sally had proved unsuccessful, but it had the effect of making the Muslims feel a special respect for Harbees and the Roman warriors of Emessa.

The following morning Abu Ubaidah held a council of war. The Muslim officers were in a restrained mood, and did not show their usual enthusiasm. Abu Ubaidah expressed his dissatisfaction with the manner in which the Muslims had given way before the Roman attack, whereupon Khalid remarked that these Romans were the bravest he had ever met. *"Then what do you advise, O Father of Sulaiman?"* asked Abu Ubaidah. *"May Allah have mercy upon you!"*

"O Commander", replied Khalid, *"tomorrow morning let us move away from the fort and. . . ."* [1]

Early the following morning, the Romans saw hectic activity in the Muslim camps around Emessa. Tents were being struck and bundles packed to be loaded onto the camels. Before their eyes the main body of the Muslims began to march away to the south, leaving behind small parties to see to the movement of the families, the baggage and the flocks. Here was deliverance! The Muslims were raising the siege and withdrawing to the south. The winter had got them after all! The Roman soldiers rejoiced at this sight, but Harbees was not a man to be content with a drawn battle. His trained eye could see a military opportunity when it appeared; and such an opportunity had clearly presented itself. He immediately collected 5,000 Roman warriors and led them out of the fort to chase the Muslims. As the Romans approached the main Muslim camp, the few Muslim warriors who were there looked at them with horror and with cries of fear fled southwards, leaving behind the families and the flocks and the baggage!

Harbees decided to leave the camp alone for the moment. The camp could wait. He launched his mounted force into a fast pursuit to catch up with the retreating enemy and strike him down as he fled. He caught up with the Muslims a few miles from Emessa. His leading elements were about to pounce upon the 'retreating enemy' when the Muslims suddenly turned and struck at the Romans with such ferocity that they were taken aback and hard put to defend themselves. As the Muslims turned on the Romans, Khalid shouted a command at which two mounted groups detached themselves from the Muslim army, galloped round the flanks of the surprised Romans and met between them. The plan proposed by Khalid and universally accepted the day before at the council of war had worked; the Romans were now trapped in a ring of steel! Ruefully Harbees thought of the words of a local priest who had tried to warn him as he was leaving Emessa to pursue the Muslims. The priest had said, *"By the Messiah, this is a trick of the Arabs. The Arab never leaves his camels and his family behind!"* [2] But it was now too late.

1. Waqidi: p. 103.
2. Waqidi p.104.

Steadily and systematically the Muslims closed in from all sides, striking with spears and swords. Heaps of Roman bodies began to accumulate on the bloody earth. At first the Romans fought with the courage of wild animals at bay, but as more and more of them fell, their mood turned to dismay and hopelessness. Khalid, striking left and right with his sword, got through with a small group to the centre of the Roman army; and here he saw Harbees still fighting, still refusing to give up. Khalid made for Harbees, but was intercepted by a huge Roman general. The Romans did not know that even if they escaped from this trap they would have nowhere to go. At the time when the Muslims started their attack on the encircled Romans, a group of 500 horsemen under Muadh bin Jabal had galloped back to Emessa to see to it that no escaping Roman got into the fort.

As these horsemen neared Emessa, the terrified inhabitants and the remnants of the Roman garrison which had not joined the pursuit hastily withdrew into the fort and closed the gates. Muadh deployed his men in front of the gates to prevent the Romans in Emessa from coming out and the Romans outside Emessa from getting in. The Muslim camp was now safe.

Khalid and the Roman general squared off. This general has been described by eye-witnesses as a man 'roaring like a lion'. [1] Khalid was the first to strike, and brought down his sword with all his strength on the heavily-armoured head of the Roman; but instead of piercing the helmet, the sword broke and Khalid was left with the hilt in his hand. Before the Roman could strike, Khalid closed in and grappled with him. The two giants held each other in a pitiless embrace; and then Khalid did something that he had never done before: he began to crush the chest of the Roman in his arms. The Roman turned red in the face and was unable to breathe as Khalid's grip tightened. Gasping for breath, the Roman struggled frantically to break the steel-like grip of the Muslim, but the terrible grip only grew tighter. Then the Roman's ribs splintered and the jagged ends plunged into his own flesh. When all movement had ceased in the body of the Roman, Khalid relaxed his grip, and what fell to the ground was a lifeless corps. Khalid had literally crushed his adversary to death in his arms! [2] He now took the Roman general's sword and again his battle cry rang out over the battlefield.

When offering his plan for this feigned withdrawal, Khalid had promised Abu Ubaidah that the Muslims would "tear the Romans apart and break their backs". In this they were eminently successful. It is recorded that only about a hundred Romans got away. [3] The Muslims, on the other hand, lost only 235 dead in the entire operation against Emessa, from the beginning of the siege to the end of this last action.

As soon as this action was over the Muslims returned to Emessa and resumed the siege, but those who were in Emessa had now no stomach for fighting. The local inhabitants offered to surrender on terms, and Abu Ubaidah accepted the offer. This happened around the middle of March, 636 (beginning of Safar, 15 Hijri). The inhabitants paid the Jizya at the rate of one dinar per man, and peace returned to Emessa. No damage was done to the city and nothing was taken by the Muslims as plunder.

Soon after the surrender of Emessa, the Muslims set out once again for the north, intending to take the whole of Northern Syria this time, including Aleppo and Antioch. They went past Hama and arrived at Shaizer. Here a Roman convoy taking provisions to Qinassareen and escorted by a small body of soldiers was intercepted and captured by Khalid. The prisoners were interrogated, and the information they provided stopped the Muslims in their tracks!

The Muslims had fought and defeated every force that Heraclius had thrown against them-all the armies, all the relief columns, all the fortress garrisons. All had bowed before the superior military quality of the Muslim army. But what Heraclius now evidently planned was to unleash a veritable tornado against them, which, if they were not careful, would hurl them in pieces into the Arabian desert.

1. Waqidi: p. 102.
2. *Ibid*.
3. Waqidi: p. 104.

Chapter 34

(The Eve of Yarmuk)

Mahan: *"We know that it is hardship and hunger that have brought you out of your lands. We will give every one of your men ten dinars, clothing and food if you return to your lands, and next year we will send you a similar amount."*

Khalid: *"Actually, what brought us out of our lands is that we are a people who drink blood, and it has reached us that there is no blood tastier than Roman blood."*[1]

The Syrian theatre of operations was like an arena entered by the contestants from opposite sides. Beyond each entrance stretched a sea which was the home ground of the contestant entering from that side. On the west of Syria and Palestine lay the blue expanse of the Mediterranean which was a 'Roman Lake'. On the east and south stretched the desert in whose wastes the Arab was master. The Romans could move with freedom over the Mediterranean in fleets of ships without interference by the Muslims, while the Muslims could move in the desert on fleets of camels with a similar freedom from interference by the Romans. Neither could the Muslims venture into the sea of water nor the Romans into the sea of sand. Within the total arena both sides could manoeuvre with ease.

Thus, for the purpose of fighting a battle in this arena, the ideal location for each side was its home bank where it could deploy with its back to its sea and withdraw in safety in case of a reverse, while at the same time, if victorious, it could pursue and destroy its opponent before he could escape to his refuge. But this advantage favoured the Muslims more than the Romans, for the former could give up the theatre of operations and withdraw to the edge of the desert without loss of face or wealth or territory. The Romans could not give up the theatre of operations as it was their Empire and had to be defended. And this strategical advantage which the Muslims enjoyed, of being able to fight on their home ground, was very much in the mind of Heraclius when he planned the next and greatest operation of this campaign.

Heraclius had come to the throne in 610 when the affairs of the Eastern Roman Empire were at their lowest ebb and the Empire consisted of little more than the area around Constantinople and parts of Greece and Africa. At first he had had to swallow many bitter pills, but then fortune smiled on him, and over a period of almost two decades he re-established the Empire in all its former greatness. He defeated the barbarians of the north, the Turks of the Caucasus and the highly civilised Persians of the Empire of Chosroes; and he did this not only with hard fighting, but also-and this was more important-by masterly strategy and superb organization. Heraclius was a strategist to the fingertips, and it was only his extraordinary organizational ability which made it possible for the Romans to create and put into the field a vast but closely knit imperial army consisting of more than a dozen nations from the Franks of Western Europe to the Armenians of the Southern Caucasus.

Now Heraclius was again being made to swallow bitter pills, and what made the pills still more bitter was the fact that they had been thrust down his throat by a race which the Romans had detested and scorned and regarded as too backward and too wretched to constitute any kind of military threat to the Empire. All the manoeuvres against the

Muslims, though strategically flawless, had ended in defeat. The first concentration of the Roman army at Ajnadein, whence it was to have struck in the rear of the Muslims, was destroyed by Khalid in the first Battle of Ajnadein. Heraclius' attempt to limit Muslim success by a stout defence of Damascus had failed in spite of his best efforts to strengthen the beleaguered garrison. His next offensive manoeuvre, the concentration of a fresh Roman army at Baisan, whence it was again intended to strike in the rear of the Muslims, had also failed, his army being trounced by Shurahbil. Thereafter not only had his attempt to retake Damascus been defeated by Abu Ubaidah and Khalid, but his other defences also crumbled as the Muslims went from victory to victory and took almost all of Palestine and Syria as far north as Emessa.

1. Ibn Kathir, Al-Bidayah wan-Nihayah, Dar Abi Hayyan, Cairo, 1st ed. 1416/1996, Vol. 7 P. 14.

Heraclius decided to organize a massive and overwhelming retaliation. He would raise such an army as had never been seen in Syria, and with this army he would bring the Muslims to battle in such a way that few, if any, would escape his clutches. This was to turn defeat into a glorious triumph.

In late 635, while Emessa was under siege, Heraclius began preparations for this great manoeuvre. Entire corps were gathered from all parts of the Empire and these were joined by princes and nobles of the realm and dignitaries of the church. By May 636, an army of a 150,000 men had been put under arms and concentrated in the area of Antioch and in parts of Northern Syria. This powerful military force consisted of contingents of Russians, Slavs, Franks, Romans, Greeks, Georgians, Armenians and Christian Arabs. [1] No people of the Cross living in the Byzantine Empire failed to send warriors to the new army to fight the invaders in the spirit of a Christian crusade. This force was organised into five armies, each of about 30,000 soldiers. The commanders of these armies were: Mahan, King of Armenia; Qanateer, a Russian prince; Gregory; Dairjan; and Jabla bin Al Eiham, King of the Ghassan Arabs. Mahan[2] commanded a purely Armenian army; Jabla had an exclusively Christian Arab force under him; and Qanateer commanded all the Russians and Slavs. The remaining contingents (all European) were placed under Gregory and Dairjan. [3] Mahan was appointed Commander-in-Chief of the entire imperial army.

At this time the Muslims were split in four groups: Amr bin Al Aas in Palestine, Shurahbil in Jordan, Yazeed at Caesarea, and Abu Ubaidah and Khalid at Emessa and to the north. In this dispersed situation the Muslims were so vulnerable that each of their corps could be attacked in turn without the least chance of fighting a successful battle. And this situation was fully exploited by Heraclius in the plan which he put into execution.

Caesarea was reinforced by sea and built up to a strength of 40,000 men. This force was to tie down Yazeed and his besieging corps so that he would be unable to move to join his comrades. The rest of the imperial army would operate on the following plan:

a. Qanateer would move along the coastal route up to Beirut, then approach Damascus from the west and cut off Abu Ubaidah.
b. Jabla would march from Aleppo on the direct route to Emessa via Hama, and hold the Muslims frontally in the Emessa region. The Christian Arabs would be the first to contact

the Muslim Arabs, and this was probably in the fitness of things. As Heraclius said to Jabla: *"Everything is destroyed by its own kind, and nothing cuts steel but steel."* 4

c. Dairjan would move between the coast and the Aleppo road and approach Emessa from the west, thus striking the Muslims in their flank while they were held frontally by Jabla.

d. Gregory would advance on Emessa from the north-east and attack the Muslims in their right flanks at the same time as they were struck by Dairjan. 5

e. The army of Mahan would advance behind the Christian Arabs and act as a reserve.

Thus the Muslim army would be swallowed up at Emessa by a force perhaps 10 times its size, attacking from all directions, with its escape routes severed. This would be more than even Khalid could handle! After the annihilation of the Muslims at Emessa, the imperial army would advance south while the garrison of Caesarea would advance from the coast and in several battles the Roman armies would attack and destroy each Muslim corps in turn, concentrating against each corps in overwhelming strength.

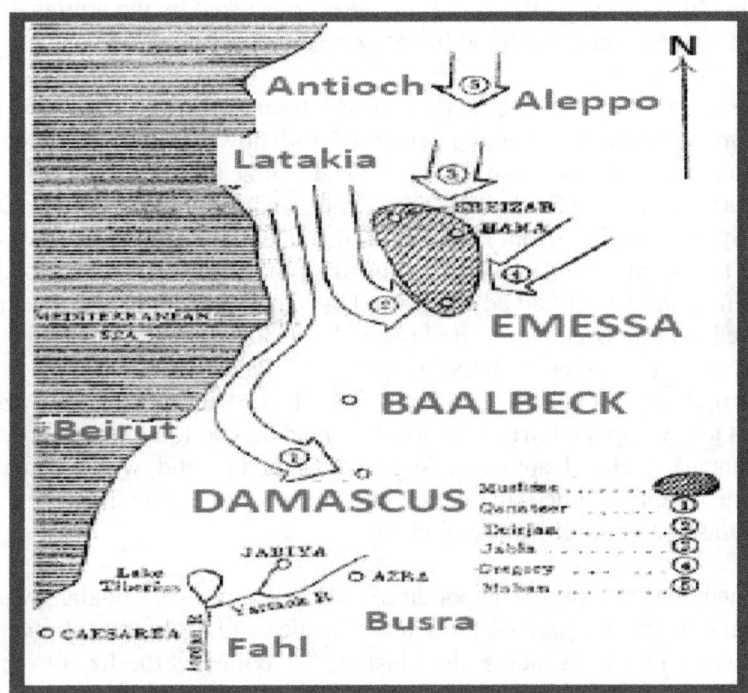

Special services were held all over the Empire for the victory of the imperial army. Generals and bishops exhorted the men to fight in defence of their faith and save their land and its people from the alien invaders. And on this masterly design the imperial army was launched from Antioch and Northern Syria some time in the middle of June 636.

1. Waqidi: p. 100.
2. This monarch's name has also been given as Bahan.

3. Waqidi: p. 106.
4. *Ibid.*
5. Waqidi (p. 107) gives the route of Gregory as "from Iraq". Since most of Western Iraq was now in Muslim hands, this could only mean such an approach as I have suggested.

When the leading elements of Jabla's army arrived at Emessa they found no Muslims. The army of Qanateer hit Damascus from the west in joyful anticipation of the destruction of the Muslims thus trapped in Damascus and the north. But there was not a single Muslim soldier in Damascus and the north. The birds had flown!

It was at Shaizar, through Roman prisoners, that the Muslims first came to know of the preparations being made by Heraclius. The Muslims had established an excellent intelligence system in the land, and no major movement or concentration of enemy forces remained concealed from them. In fact they had agents within the Roman army. As the days lengthened into weeks, the pieces of intelligence brought in by agents were put together like a jigsaw puzzle, and the movement of the Roman armies had hardly got under way when the Muslims knew of it and of the directions taken by the armies. Even the reinforcement of Caesarea and its strength were known.

The Muslims were staggered by the reports, each of which seemed worse than its predecessor. The horizon became darker and darker. Khalid, however, with his unerring sense of strategy at once saw the design of Heraclius and realized how terribly vulnerable the Muslim army was at Emessa and Shaizar. The soundest course was to pull back from North and Central Syria, as well as from Palestine, and concentrate the whole army so that strong, united opposition could be put up against the Roman juggernaut, preferably not far from the friendly desert. Khalid advised Abu Ubaidah accordingly and the Army Commander accepted the proposal. He ordered the withdrawal of the army to Jabiya, which was the junction of the routes from Syria, Jordan and Palestine. Moreover, exercising his authority as Commander-in-Chief in Syria, he ordered Shurahbil, Yazeed and Amr bin Al Aas to give up the territory in their occupation and join him at Jabiya. Thus, before the Romans reached Damascus, Abu Ubaidah and Khalid, with elements of Yazeed's corps, were at Jabiya while the other corps were moving to join them. They had safely extricated themselves from the jaws of death.

The remarkably generous treatment of the populace of Emessa by Abu Ubaidah, when the Muslims left that city, throws light on the sense of justice and truth of this brave and noble general. On the conquest of Emessa, the Muslims had collected the Jizya from the local inhabitants. This tax, as has been explained before, was taken from non-Muslims in return for their exemption from military service and their protection against their enemies. But since the Muslims were now leaving the city and were no longer in a position to protect them, Abu Ubaidah called a meeting of the people and returned all the money taken as Jizya. *"We are not able to help and defend you"*, said Abu Ubaidah. *"You are now on your own."* To this the people replied, *"Your rule and justice are dearer to us than the oppression and cruelty in which we existed before."* [1] The Jews of Emessa proved the most loyal in their friendship, and swore that the officers of Heraclius would not enter the city except by force. Moreover, not content with doing total justice in the matter of the Jizya in his own province, Abu Ubaidah also wrote to the other corps commanders in Syria to return the Jizya to the people who had paid it, and this was done by every Muslim commander before he marched away to join Abu Ubaidah at Jabiya. [2] Such an extraordinary and voluntary return by an all-conquering army of what it has taken according to mutually arranged terms, had never happened before. It would never happen again.

In the middle of July 636, the forward elements of the imperial army, consisting of Christian Arabs, made contact with Muslim screens between Damascus and Jabiya. Abu Ubaidah was now deeply worried. A battle was certain, and one that would decide the fate of the Muslims in Syria. The enemy strength, believed by the Muslims to be 200,000, seemed like a horrible nightmare. Abu Ubaidah worried not for himself but for the Muslim army and the Muslim cause. He called a council of war to brief the officers about the enemy situation and get ideas.

The officers sat in silence, weighed down by the forbidding prospect which faced them. One spoke in favour of a withdrawal into Arabia where the army could wait until this Roman storm has passed and then re-enter Syria, but this proposal was rejected as being tantamount to abandoning all the Muslim conquests in Syria and exchanging the good life of this land for the hardship and hunger of the desert. Others spoke in favour of fighting "here and now", trusting to Allah for victory, and most of the assembled officers favoured this proposal. The mood of the council, however, was not of happy enthusiasm but of grim determination to fight, and if necessary, go down fighting.

1. Balazuri: p. 143.
2. Abu Yusuf: p. 139.

Khalid remained silent while this discussion was in progress. Then Abu Ubaidah turned to him and said, *"O Father of Sulaiman! You are a man of courage and resolve and judgment. What do you think of all this?"*

"What they say is good", replied Khalid. *"I have different views, but shall not oppose the Muslims."*

"If you have other views, speak", said Abu Ubaidah, *"and we shall do as you say."*

Khalid then gave his plan: *"Know, O Commander, that if you stay at this place, you will be helping the enemy against you. In Caesarea, which is not far from Jabiya, there are 40,000 Romans under Constantine, son of Heraclius.* **1** *I advise you to move from here and place Azra behind you and be on the Yarmuk. Thus it would be easier for the Caliph to send reinforcements, and ahead of you there would be a large plain, suitable for the charge of cavalry."* **2**

Khalid did not specifically say so, but the inference was that Constantine, advancing from Caesarea, could attack the Muslims in the rear at Jabiya while they faced the imperial army from the north. The plan was accepted, unanimously and the move put into effect. Khalid, with the Mobile Guard of 4,000 horsemen, was left behind as a rear guard; and instead of staying at Jabiya, he moved forward and clashed with the leading elements of the Roman army. He struck at the head of the Roman column and drove it back towards Damascus. This imposed caution on the Romans, who thereafter made no effort to interfere with the retrograde move of the Muslims. A few days later Khalid rejoined the main body of the Muslim army.

The Muslims, having moved a few miles south-east, established a line of camps in the eastern part of what for want of a better name, we shall call the Plain of Yarmuk. The location of these camps is not known but they were probably south of the present Nawa-Sheikh Miskeen line with a north-west-facing front, so that the Muslims could deploy to receive a Roman attack from the north (Jabiya axis) as well as the north-west (direction of Qunaitra). Here Abu Ubaidah was joined by the corps of Shurahbil, Amr bin Al Aas and Yazeed. Some distance to the east of the Muslims sprawled the lava hills which stretch from north to east of Azra, and the mountains of Jabal-ud-Druz, north and east of Busra.

A few days later the Roman army, preceded by the lightly armed Christian Arabs of Jabla, moved up and made contact with Muslim outposts on the Plain of Yarmuk. The route of the main body of the Roman army is not recorded, but it was almost certainly from the north-west, because the Romans established their camps just north of the Wadi-ur-Raqqad. (Khalid's clash with the Romans on the Jabiya axis may have caused them to switch their axis.) The Roman camp was 18 miles long, and between it and the Muslim camp lay the central and west-central parts of the Plain of Yarmuk. 3 With the arrival of the Romans and the establishment of their camps, the direction of the Roman attack became obvious and Abu Ubaidah adjusted the Muslim camps to correspond to a battlefront running from the Yarmuk to the Jabiya Road. This is what Khalid had advised: the rear towards Azra and a flank on the Yarmuk.

Now the two armies settled down in their respective camps and began to make preparations for battle: reconnaissances, plans, orders, checking of equipment etc. To the Muslims the Romans looked like 'a swarm of locusts'. 4 Hardly had the Romans settled down in camp when a messenger arrived from Heraclius with instructions to the Commander-in-Chief, Mahan the Armenian, not to start hostilities until all avenues of peaceful negotiation had been explored. Mahan was to offer generous terms to the Muslims if they would agree to retire to Arabia and not come back again. Consequently Mahan sent one of his army commanders, Gregory, to hold talks with the Muslims. Gregory rode out to the Muslim camp, in front of which he held a discussion with Abu Ubaidah. The Roman offered to let the Muslims go in peace, taking with them everything which they had acquired in Syria, as long as they would give up all intention of invading Syria again. Abu Ubaidah's answer was in the negative, and the Roman returned empty-handed.

1. According to Gibbon (Vol. 5, p. 333) Constantine, commanding at Caesarea, was the eldest son of Heraclius.
2. Waqidi: p. 109.
3. According to Waqidi (p. 109), the Roman camp was near Jaulan (which is the area between the Wadi-ur-Raqqad and Lake Tiberius and the area to the north), and the distance between the opposing camps was approximately 11 miles (three *farsakh*. A *farsakh* equals 6000 meters.).
4. Waqidi: p. 118.

Mahan next sent Jabla, hoping that as an Arab he would have more success in talking the Muslims into leaving Syria in peace. Jabla tried his best to persuade the Muslims, but like Gregory, returned unsuccessful.

Mahan now realized that a battle was inevitable and nothing could be done to avoid it. Consequently he sent Jabla forward with the bulk of his Arab army to put in a probing attack on the Muslims. This was not so much an offensive as a reconnaissance in force to test the strength of the Muslim front. For such an action the mobile Christian Arab was better suited than his more heavily equipped comrades of the imperial army. This happened some time in late July 636 (middle of Jamadi-ul-Akhir, 15 Hijri).

Jabla moved up with his Arabs and found the Muslims arrayed in battle order. Cautiously the Christian inched his way forward wanting to get as close as possible before ordering a general attack; but before he could give such an order, he found himself assailed by powerful groups of Muslim cavalry operating under the Sword of Allah. After a certain amount of half-hearted resistance the Christian Arabs withdrew confirming Mahan's fear that battle with these Muslims would not be an easy matter.

Thereafter, for almost a month, there was no major action on the Plain of Yarmuk. The cause of this inactivity is not known. We can only guess that the Muslims were not strong enough to take the initial offensive, and the Romans did not feel brave enough to do so. The respite, however, proved beneficial to the Muslims, as during this period a fresh contingent of six thousand Muslims arrived to join them, the majority of whom were from the Yemen. The Muslims now had an army of 40,000 warriors, including 1,000 Companions of the Prophet, and these in turn included 100 veterans of the Battle of Badr-the first battle of Islam. The army included citizens of the highest rank, such as Dhulbair (the Prophet's cousin and one of the Blessed Ten), Abu Sufyan and his wife, Hind.

When a month had passed after the repulse of Jabla, Mahan felt strong enough to take the offensive, but decided to make one more attempt at peace. This time he would hold talks himself. He asked for a Muslim envoy to be sent to his headquarters, and in response to his request, Abu Ubaidah sent Khalid with a few men. Khalid and Mahan met in the Roman camp, but nothing came of these talks as the positions taken by the two sides were too rigid to allow for adjustment. Mahan threatened Khalid with his great army and offered a vast sum of money to all the Muslims, including the Caliph at Madinah; but this made no impression on Khalid, who offered the three alternatives: Islam, the Jizya or the sword. The Armenian chose the last. It appears, however, that as a result of this discussion, both commanders were favourably impressed by each other and the Muslims began to regard Mahan as a fine man except that, to quote Abu Ubaidah: *"Satan has got hold of his reason!"* [1]

As the two leaders parted, they knew that henceforth there would be no parleys. The point of no return had been reached, and the following day the battle would begin.

The rest of the day was spent in feverish activity. Both sides prepared for battle. Plans were finalized and orders issued. Corps and regiments were placed in position so that everyone would know his place in the forthcoming battle. Officers and men checked their armour and weapons.

Both sides offered fervent prayers for victory, beseeching Allah for His help to 'the true faith', and of course they prayed to the same Allah! On the Roman side the priests brandished crosses and exhorted the soldiers to die for Jesus. Tens of thousands of Christians took the oath of death, swearing that they would die fighting and not flee from the enemy. Many of them would remain true to their oath.

The battlefield which stretched between the two camps consisted of the Plain of Yarmuk which was enclosed on its western and southern sides by deep ravines. On the west yawned the Wadi-ur-Raqqad which joined the Yarmuk River near Yaqusa. This stream ran north-east to south-west for 11 miles through a deep ravine with very steep banks, though less so at its upper end. The ravine was crossable at a few places but there was only one main crossing, at a ford, where the village of Kafir-ul-Ma stands today. South of the battlefield ran the canyon of the Yarmuk River, starting at Jalleen and twisting and turning for 15 miles, as the crow flies, down to its junction with the Wadi-ur-Raqqad, beyond which it continued on its way to join the Jordan River south of Lake Tiberius (Sea of Galilee). At Jalleen a stream called Harir, running from the north-east, flowed into, and became the Yarmuk River. On the north the plain continued beyond the battlefield, while to the east it stretched for a distance of about 30 miles from the Wadi-ur-Raqqad to the foot of the Azra hills. The western and central part of this plain was the battlefield.

1. Waqidi: p. 128.

The most significant feature of the battlefield was the existence of the two ravines-the Wadi-ur-Raqqad and the Yarmuk River. Both had banks 1,000 feet high, and while the steepness of the banks was sufficient to make the ravines serious obstacles to movement, they were made even more frightening by the precipices which lined the banks along most of their length. These precipices were sometimes at the bottom, sometimes at the top and sometimes half-way up the bank and created sheer, vertical drops 100 to 200 feet in height. Near the junction of the two ravines, the banks became steeper and the precipices higher-a fearful prospect for anyone who had to cross in haste.

The only dominating tactical feature on the plain of Yarmuk was one named on maps as the Hill of Samain, 3 miles southwest of the present village of Nawa. There was also the Hill of Jabiya, north-west of Nawa, but it lay outside the battlefield and was to play no part in the battle. The Hill of Samein, 300 feet high, so dominated the area around it, and gave such excellent observation over the entire plain, that no general would fail to occupy it should he be the first to deploy his forces on this part of the plain. As a result of this battle the hill was named the *Hill of Jamu'a* (gathering), because part of the Muslim army was concentrated on it. There was no other dominating ground on the plain of Yarmuk.

The plain itself was generally flat, sloping gently from north to south with a certain amount of undulation. One stream which formed an important tactical feature was Allan, running southwards across the plain to join the Yarmuk, and in the last 5 miles of its journey this stream also formed a ravine with steep sides though it was not such a serious obstacle as the bigger ravines. The battlefield was ideal for the manoeuvre of infantry and cavalry and, except for the southern portion of Allan, offered no impediment to movement.

Mahan deployed the imperial army forward of Allan. He used his four regular armies to form the line of battle which was 12 miles long, extending from the Yarmuk to south of the Hill of Jabiya. [1] On his right he placed the army of Gregory and on his left the army of Qanateer. The centre was formed by the army of Dairjan and the Armenian army of Mahan-both under the command of Dairjan. The Roman regular cavalry was distributed equally among the four armies, and each army deployed with its infantry holding the front and its cavalry held as a reserve in the rear. Ahead of the front line, across the entire

12-mile front, Mahan deployed the Christian Arab army of Jabla, which was all mounted-horse and camel. This army acted as a screen and skirmish line, and was not concerned with serious fighting except as its groups joined the army in front of which they were positioned.

The army of Gregory, which formed the right wing, used chains to link its 30,000 foot soldiers. [2] These chains were in 10-men lengths, and were used as a proof of unshakeable courage on the part of the men who thus displayed their willingness to die where they stood. The chains also acted as an insurance against a break-through by enemy cavalry, as has been explained in the chapter on *The Battle of Chains*. All these 30,000 foot soldiers had taken the oath of death.

Although the imperial army established a front of about the same length as the Muslim front, it had the advantage of having four times as many troops and Mahan exploited this numerical superiority by establishing a whole army (Jabla's) as a forward screen and achieving much greater depth in the solid, orderly formations. The Roman ranks stood 30 deep.

1. In terms of present-day geography, the Roman line started from about two miles west of Nawa, and went south-south-west to just west of Seel, then over Sahm-ul-Jaulan to the Yarmuk bank forward of Heet. Of course, these villages probably did not exist then as there is no mention of them in the narrative of this battle.
2. There is also talk of a deep ditch here, but I cannot place it or see its significance, as the Romans are said to have deployed forward of it rather than behind it. It may have been an anti-retreat measure.

Thus the magnificent army of Caesar was arrayed for battle.

When Khalid returned from his talks with Mahan, he informed Abu Ubaidah and the other generals that there would be no more talks, that the issue would be decided by the sword, that the battle would begin the next day. Abu Ubaidah took the news with his usual stoical acceptance of the will of Allah. As Commander-in-Chief he would organise the army for battle and conduct the operation according to his tactical judgement. His military skill was not, however, very great, and he knew it. Khalid knew it, and most of the officers of the army knew it. Abu Ubaidah would fight the battle in a sensible manner, and would react to changing tactical situations like the good, steady general that he was. But with the enemy four times superior in strength, soundness and common sense were not enough. A much finer quality of generalship was required for this battle, and Khalid decided to offer his services to act as the real commander in battle.

"O Commander", said Khalid to Abu Ubaidah, *"send for all the commanders of regiments and tell them to listen to what I have to say."* [1]

Abu Ubaidah got the point. He himself could wish for nothing better. He at once sent an officer to call the regimental and corps commanders to his headquarters; and the officer rode to all the commanders, conveying the message: *"Abu Ubaidah commands that you listen to whatever Khalid says and obey his orders."* [2] The officers understood the meaning of the message and gathered at the headquarters to receive the orders of Khalid. On this tactful note the command of the army was taken over by Khalid, and everyone was satisfied with the arrangement.

Abu Ubaidah remained the nominal commander and somewhat more than that. He continued to deal with matters of administration, led the prayers and saw to various other details of command. He also gave certain orders when his ideas did not clash with the plans and orders of Khalid. But for the purpose of battle, Khalid was now the commander of the Muslim army in Syria, and would remain so until this battle was over.

Khalid immediately set about the reorganization of the army into infantry and cavalry regiments within each corps. The army consisted of 40,000 men, of which about 10,000 was cavalry. This force was now organised by Khalid into 36 infantry regiments of 800 to 900 men each, three cavalry regiments of 2,000 horses each and the Mobile Guard of 4,000 horsemen. The commanders of the cavalry regiments were Qais bin Hubaira, Maisara bin Masruq and Amir bin Tufail. Each of the four corps had nine infantry regiments, which were all reformed on a tribal and clan basis, so that every man would fight next to well-known comrades. Much of Khalid's corps of Iraq was absorbed in the other four corps, while the best of it remained with him as the Mobile Guard.

The army was deployed on a front of 11 miles corresponding roughly to the front of the Roman army. The army's left rested on the Yarmuk River, a mile forward of where the ravine began, while it's right lay on the Jabiya road. 3 On the left stood the corps of Yazeed and on the right the corps of Amr bin Al Aas, and each of these flanking corps commanders was given a cavalry regiment under command. The centre was formed by the corps of Abu Ubaidah (left) and Shurahbil (right). Among the regimental commanders of Abu Ubaidah were Ikrimah bin Abi Jahl and Abdur-Rahman bin Khalid. Behind the centre stood the Mobile Guard and one cavalry regiment as a central reserve for employment on the orders of Khalid. At any time when Khalid was busy with the conduct of the battle as a whole, Dhiraar would command the Mobile Guard. Each corps pushed out a line of scouts to keep the Romans under observation. (For the dispositions of the two armies.

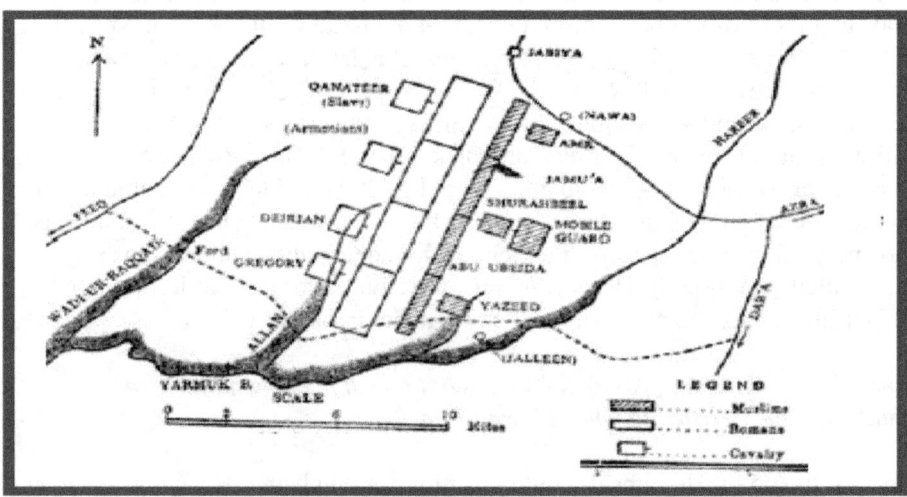

Compared with the Romans, the Muslim army formed a thin line, only three ranks deep, but there were no gaps in the ranks which stretched in unbroken lines from edge to edge. All the spears available in the army were issued to the front rank, and in battle the men would stand with the long spears at the ready, making it impossible for an assailant to get to grips without braving the frightening points of the spears. The archers, most of whom were Yemenis, stood interspersed in the front rank. On the first approach of the enemy the archers would open up and bring down as many of the Romans as possible. As the assailants clashed with the Muslims, they would be killed with spears, and thereafter the men would draw their swords.

1. Waqidi: p. 129.
2. *Ibid*.
3. In terms of present-day geography, the Muslim line started from about a mile west of Nawa and went south-south-west to over the Hill of Jamu'a, then between Seel and Adwan, then between Sahm-ul-Jaulan and Jalleen, to just short of the Yarmuk.

The flanking corps would use their own cavalry regiments as corps reserves to re-establish their positions in case they were pushed back by the Romans. Khalid with his Mobile Guard and one cavalry regiment would provide the local reserve for the two central corps and also be available as an army reserve to intervene in the battle of the flanking corps as required.

The situation of the two armies with regard to flanks was similar. Each had its southern flank on the Yarmuk and this flank could not be turned. The northern flank of both armies was exposed, and on this side outflanking movements were possible. The difference in the situation of the two armies lay in their respective rears. Behind the Muslims stretched the eastern extension of the Plain of Yarmuk, beyond which rose the broken Azra hills and the Jabal-ud-Druz; and into this region the Muslims could withdraw in safety and be invulnerable in case of a reverse. Behind part of the Roman position, however, lay the forbidding ravine of the Wadi-ur-Raqqad-deep and precipitous. As a discouragement to retreat this was fine and would probably make the Romans fight more desperately; but in case the Romans were worsted in battle and cut off from the northern escape route, the ravine would prove an abyss of death. Against it they would be caught like mice in a trap. However, the Romans had no intention of losing this battle.

This topographical situation was uppermost in Khalid's mind when he formulated his plan of battle. Initially the Muslims would stand on the defensive and receive and hold the Roman attack until it had lost its impetus and the enemy was worn out. Then the Muslims would go on to the offensive and drive the Romans towards the Wadi-ur-Raqqad. The terrible ravine would be the anvil on which the Muslim hammer would fall, crushing the Roman army to powder! At least, so Khalid planned!

The women and children were placed in camps stretching in a line in the rear of the army. Behind the men of each regiment stood their women and children. 1 Abu Ubaidah went round the camps and addressed the women: *"Take tent poles in your hands and gather heaps of stones. If we win all is well. But if you see a Muslim running away from battle, strike him in the face with a tent pole, pelt him with stones, hold his children up before him and tell him to fight for his wife and children and for Islam."* 2 The women prepared accordingly.

As the army formed up in its battle position, Khalid, Abu Ubaidah and other generals rode round the regiments and spoke to the officers and men. Khalid gave a set speech before each regiment: *"O men of Islam! The time has come for steadfastness. Weakness*

and cowardice lead to disgrace; and he who is steadfast is more deserving of Allah's help. He who stands bravely before the blade of the sword will be honoured, and his labours rewarded, when he goes before Allah. Lo! Allah loves the steadfast!" [3]

While Khalid was going past one of the regiments, a young man remarked, *"How numerous are the Romans and how few are we!"* Khalid turned to him and said, *"How few are the Romans and how numerous are we! An army's strength lies not in numbers of men but in Allah's help, and its weakness lies in being forsaken by Allah"* [4]

Other commanders and elders, while exhorting the men to fight, recited verses from the Quran, the most popular one being: **"How many a small group has overpowered a large group by Allah's help, and Allah is with the steadfast." [Quran: 2:249.]** They spoke of the fire of hell and the joys of paradise, and quoted the example set by the Holy Prophet in his battles. For good measure they also reminded the soldiers of the hunger of the desert and the good life of Syria!

The night that followed was hot and sultry. It was the third week of August 636 (second week of Rajab, 15 Hijri.).[5] The Muslims spent the night in prayer and recitation of the Quran, and reminded each other of the two blessings which awaited them: either victory and life or martyrdom and paradise. The Holy Prophet had established a tradition after Badr of reciting the chapter of Al Anfal from the Quran before battle, and all night the verses of this chapter could be heard wherever Muslims sat, singly and in groups.

The fires in the two camps burned merrily the whole night and could be seen for miles like twinkling stars descended to earth. But there was no merriment in the hearts of those who sat in the light of these fires. The thought of the ordeal that awaited them had driven all joy from their minds. They were brave men, these soldiers who awaited the morrow, these Romans and Arabs, these Europeans and Asians, these Christians and Muslims. They were lions and eagles and wolves. But they were also human beings and thought of their wives and children to whom they would bid farewell in a few hours-perhaps for the last time.

This was the eve of Yarmuk …. the greatest battle of the Century…. one of the decisive battles of history …. and perhaps the most titanic battle ever fought between the Crescent and the Cross.

1. According to some reports, the families were put on a hill well to the rear. This, as we shall see from the course of battle, could not have been so.
2. Waqidi: pp. 129-30.
3. *Ibid*: p. 137.
4. Tabari: Vol. 2, p. 594.
5. The only thing recorded in the early accounts about the date of this battle is the month- Rajab, 15 Hijri. My statement recording the week in which the battle began is the result of calculations made from the timing of earlier events narrated in this chapter.

Chapter 35

(Al Yarmuk)

> *"Did you not see us victorious upon the Yarmuk,*
> *The way we prevailed in the campaigns of 'Iraq ?*
> *The virgin cities we conquered, as well as*
> *The Yellow Meadow, on our galloping steeds.*
> *We conquered before that Busra, which was*
> *Impenetrable even to the flying crows.*
> *We killed those who stood against us*
> *With flashing swords, and we have their spoils.*
> *We killed the Romans until they were reduced*
> *Upon the Yarmuk, to emaciated leaves.*
> *We smashed their army as they rushed headlong*
> *To the Neck-Breaker, with our sharp steel.*
> *By morning they tumbled into it, reaching*
> *The mysterious matter that defies the senses."*
> [Al-Qa'qa' bin Amr, commander in Khalid's army]1

At dawn the Muslim corps lined up for prayers under their respective commanders. As soon as the prayers were over, every man rushed to his assigned place. By sunrise both armies stood in battle order, facing each other across the centre of the Plain of Yarmuk, a little less than a mile apart.

There was no movement and little noise in the two armies. The soldiers knew that this was a fight to the finish, that one of the two armies would lie shattered on the battlefield before the fight was over. The Muslims gazed in wonder at the splendid formations of the Roman legions with banners flying and crosses raised above the heads of the soldiery. The Romans looked with something less than awe at the Muslim army deployed to their front. Their confidence rested on their great numbers, but during the past two years the performance of the Muslims in Syria had instilled a good deal of respect in the hearts of the Romans. There was a look of caution in Roman eyes. Thus an hour passed during which no one stirred and the soldiers awaited the start of a battle which, according to the chroniclers, "began with sparks of fire and ended with a raging conflagration", and of which "each day was more violent than the day before." 2

Then a Roman general by the name of George emerged from the Roman centre and rode towards the Muslims. Halting a short distance from the Muslim centre, he raised his voice and asked for Khalid. From the Muslim side Khalid rode out, delighted at the thought that the battle would begin with himself fighting a duel. He would set the pace for the rest of the battle.

As Khalid drew near, the Roman made no move to draw his sword, but continued to look intently at Khalid. The Muslim advanced until the necks of the horses crossed, and still George did not draw his sword. Then he spoke, in Arabic: *"O Khalid, tell me the truth and do not deceive me, for the free do not lie and the noble do not deceive. Is it true that Allah sent a sword from heaven to your Prophet ? ... and that he gave it to you ? ... and that never have you drawn it but your enemies have been defeated?"*

"No!" replied Khalid.

"Then why are you known as the Sword of Allah?"

Here Khalid told George the story of how he received the title of Sword of Allah from the Holy Prophet. George pondered this a while, then with a pensive look in his eyes, asked,

"Tell me, to what do you call me?"

"To bear witness", Khalid replied, *"that there is no Allah but Allah and Muhammad is His Slave and Messenger; and to believe in what he has brought from Allah."*

"If I do not agree?"

"Then the Jizya, and you shall be under our protection."

"If I still do not agree?"

"Then the sword!"

George considered the words of Khalid for a few moments, then asked, *"What is the position of one who enters your faith today?"*

"In our faith there is only one position. All are equal."

"Then I accept your faith!" [3]

To the astonishment of the two armies, which knew nothing of what had passed between the two generals, Khalid turned his horse and Muslim and Roman rode slowly to the Muslim army. On arrival at the Muslim centre George repeated after Khalid: *"There is no Allah but Allah; Muhammad is the Apostle of Allah!"* (A few hours later the newly-converted George would fight heroically for the faith which he had just embraced and would die in battle.) On the auspicious note of this conversion began the Battle of Yarmuk.

Now came the phase of duels between champions and this suited both sides, for it acted as a kind of warming up. Scores of officers rode out of the Muslim army, some on instructions from Khalid and others on their own, and throwing their individual challenges, engaged the Roman champions who emerged to fight them. Practically all these Romans were killed in combat, the honours of the day going to Abdur-Rahman bin Abi Bakr, who killed five Roman officers, one after the other.

1. Ibn Kathir, Al-Bidayah wan-Nihayah, Dar Abi Hayyan, Cairo, 1st ed. 1416/1996, Vol. 7 P. 20.
2. Waqidi: p. 133
3. Tabari: Vol. 2, p. 595

This duelling went on till midday. Then the Roman Commander-in-Chief, Mahan, decided that he had had enough of this and that if it went on very much longer, not only would he lose a large number of officers, but also the moral effect on his army would be quite bad. He would have a better chance of success in a general battle in which sheer weight of numbers would favour his army. But he was rightly cautious, for a false step at the beginning of battle could have far-reaching effects on its course. He would attempt a limited offensive on a broad front to test the strength of the Muslim army, and if possible, achieve a breakthrough wherever the Muslim front was weak.

At midday the 10 forward ranks of the Roman army, i.e. one-third of the infantry of each of the four armies, advanced to battle. This human wave moved slowly forward, and as it came within range of the Muslim archers, was subjected to intense archery, which caused some casualties. The wave continued to advance and before long struck the Muslim front rank. Soon the Muslims had dropped their bloody spears and drawn their swords, and both sides were locked in combat.

But the Roman assault was not a determined one, and the soldiers, many of whom were unused to battle, did not press the attack, while the fury with which the hardened Muslim veterans struck at them imposed caution. On some parts of the front the fighting was more violent than on others, but on the whole the action of this day could be described as steady and moderately hard. The Muslims held their own. The Romans did not reinforce their forward infantry, and at sunset the action ended with the two armies separating and returning to their respective camps. Casualties were light on this day, though higher among the Romans than the Muslims.

The night was spent in peace. The Muslim women greeted their men with pride, and wiped the sweat and blood from their faces and arms with their head coverings. The wives said to their husbands: *"Rejoice in tidings to paradise, O Friend of Allah!"* [1] The Muslims now felt more confident for they had inflicted worse punishment on the enemy than they had taken themselves, and prayers and recitation of the Quran continued for most of the night. During the night, however, a few Roman parties came forward into the no-man's-land to pick up their dead and this led to some patrol clashes, but otherwise there was no engagement to disturb the peace of the night.

Mahan had got nowhere. He called a council of war at which plans for the next day were discussed. He would have to do something different if success were to be achieved and Mahan decided to launch his next attack at the first light of dawn, after forming up during the hours of darkness, in the hope of catching the Muslims off their guard, before they were prepared for battle. Moreover, he would attack in greater strength. The two central armies would put in holding attacks to tie down the Muslim centre, while the two flanking armies would launch the major thrusts and either drive the flanking corps off the battlefield or push them into the centre. To have a grand stand view of the battle, Mahan had a large pavilion placed on a hillock behind the Roman right, from where the entire plain could be seen. Here Mahan positioned himself with his court and a bodyguard of 2,000 Armenians, while the rest of the army prepared for the surprise dawn attack.

Soon after dawn the Muslims were at prayer when they heard the beating of drums. Messengers came galloping from the outposts to inform the commanders that the Romans were attacking. The Muslims were certainly caught unawares, but Khalid had ordered the placing of a strong outpost line in front during the night, and these outposts caused sufficient delay in the Roman advance to enable the Muslims to don their armour and weapons and get into battle position before the flood hit them. Moreover, the speed with which the Muslims got into position was faster than the Romans had anticipated. The sun was not yet up on this second day of battle when the two armies clashed.

The battle of the central corps continued steadily for most of the day with no break in the Muslim line. Here, in any case, the Romans were not pressing hard as this was meant to be a limited attack to hold these Muslim corps in their position. Thus the centre remained stable. But on the flanking corps fell the heaviest blows of the Roman army, and these corps bore the brunt of the fighting.

On the Muslim right the army of Qanateer, consisting mainly of Slavs, attacked the corps of Amr bin Al Aas. The Muslims held on bravely and the attack was repulsed. Qanateer attacked for the second time with fresh troops, and again the Muslims repulsed him. But when Qanateer attacked for the third time, again using fresh regiments, the resistance of the now tired Muslims broke, and the bulk of the corps fell back to the camp, while part of it retired to the centre, *i.e.* towards the corps of Sharhabeel.

1. Waqidi: p. 133.

As the corps fell back in some disorder, Amr ordered his cavalry regiment of 2,000 horse to counter-attack and throw back the Romans. The cavalry went into battle with great dash and for some time checked the Roman advance, but was unable to hold it for long. It was repulsed by the Romans and turned away from battle, also making for the Muslim camp. As the horsemen reached the camp along with the foot soldiers, they found a line of women waiting for them with tent poles and stones in their hands. The women screamed: *"May Allah curse those who run from the enemy!"* And to their husbands they shouted: *"You are not our husbands if you cannot save us from these infidels."* Other women began to beat drums and sang an improvised song:

> *O you who run from a constant woman*
> *Who has both beauty and virtue;*
> *And leave her to the infidel,*
> *The hated and evil infidel,*
> *To possess, disgrace and ruin!* [2]

What these Muslims received from their women was not just stinging rebukes; they were actually assaulted! First came a shower of stones, then the women rushed at the men, striking horse and rider with tent poles; and this was more than the proud warriors could take. Indignant at what had happened, they turned back from the camp and advanced in blazing anger towards the army of Qanateer. Amr now launched his second counter attack with the bulk of his corps.

The situation on the Muslim left was only a little less serious. Here too the initial Roman attack was repulsed, but in a second attack the Romans broke through the corps of Yazeed. This was the army of Gregory, with chains, more slow-moving than the others but also more solid. Yazeed too used his cavalry regiment to counter attack and it too was repulsed; and after a period of stiff resistance the warriors of Yazeed fell back to their camp, where the women awaited them, led by Hind and Khaulah. The first Muslim horseman from the left wing to arrive at the camp was Abu Sufyan, and the first woman to meet him was none other than Hind! She struck at the head of his horse with a tent pole and shouted: *"Where to, O Son of Harb? Return to battle and show your courage so that you may be forgiven your sins against the Messenger of Allah."* [3]

Abu Sufyan had experienced his wife's violent temper before and hastily turned back. Other warriors received the same treatment from these women as the soldiers of Amr had received from theirs, and soon the corps of Yazeed returned to battle. A few women ran forward alongside the horses and one of them actually brought down a Roman with her sword. As the warriors of Yazeed turned again to grapple with the army of Gregory, Hind took up her song of Uhud:

We are the daughters of the night;
We move among the cushions
With a gentle feline grace
And our bracelets on our elbows.
If you advance we shall embrace you;
And if you retreat we shall forsake you
With a loveless separation. 4

One may question the propriety of Hind singing such a provocative song, but she felt that she was young enough to do so. After all, she was not a day over 50!

1. Waqidi: p. 140.
2. *Ibid*.
3. *Ibid*: p. 141.
4. *Ibid*: p. 140.

It was now about midday. While the Muslim flanking corps were fighting their battle, Khalid was watching these actions from his position in the centre. So far he had done nothing to help these corps, and had refused to be drawn into battle with his central reserve before it was absolutely necessary. But as the corps returned to battle from the camps to which they had retreated, Khalid decided to launch his cavalry reserve to assist them and quicken the re-establishment of the Muslim positions.

He first turned to the right wing and with his Mobile Guard and one cavalry regiment struck at the flank of the army of Qanateer at the same time as Amr counter-attacked again from the front. Very soon the Romans, attacked from two sides, turned and beat a hasty retreat to their original position. Amr regained all the ground that he had lost and reorganized his corps for the next round.

As soon as this position was restored, Khalid turned to the left wing. By now Yazeed had begun a major counter attack from the front to push the Romans back. Khalid detached one regiment under Dhiraar and ordered him to attack the front of the army of Deirjan in order to create a diversion and threaten the withdrawal of the Roman right wing from its advanced position. With the rest of the army reserve he attacked the flank of Gregory.Here again the Romans withdrew under the counter-attacks from front and flank, but more slowly because with their chains the men could not move fast.

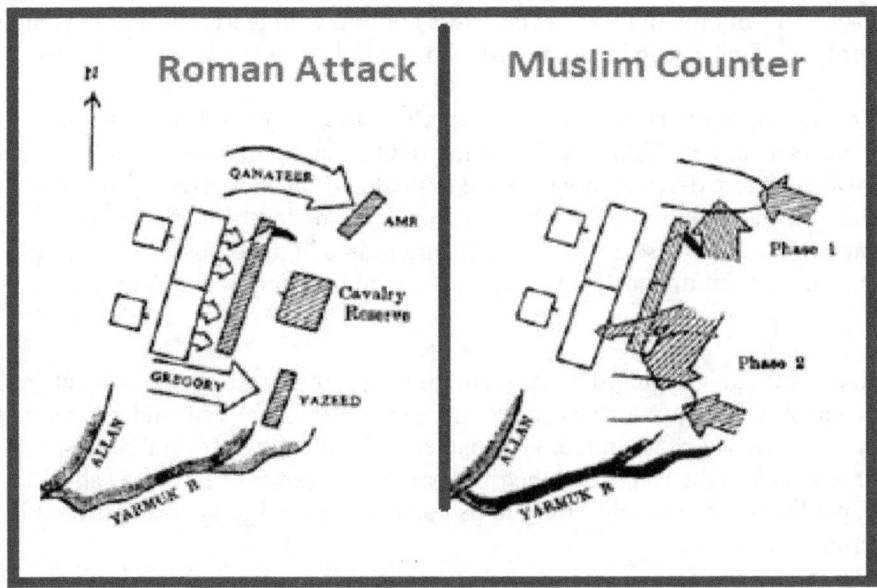

While the Roman right was falling back, Dhiraar broke through the army of Deirjan and got to its commander who stood well forward with his body-guard. Here Dhiraar killed Deirjan. But soon after, the pressure against him became so heavy that he was forced to retire to the Muslim line.

Before sunset the two flanking armies of the Romans had been pushed back. At sunset the central armies also broke contact and withdrew to their original positions and both fronts were restored along the lines occupied in the morning. The Muslims had faced a critical situation but had regained their lost ground. The right wing of the Muslims suffered more severely than the other corps, as the most vicious fighting had taken place in the sector of Amr. However, the day's fighting ended with the Muslims winning this bout on points.

The night that followed was again a quiet one. The Muslim women got busy dressing wounds, preparing food, carrying water and so on. On the whole, Muslim spirits were high as they had been attacked by the bulk of the Roman army and had thrown the attackers back from their positions. The Muslims had remained on the defensive, the counter attacks being no more than part of the general defensive posture.

In the Roman camp, however, the mood hardened. Thousands of Romans had been slain on this day, and the Muslims had not only repulsed the flanking armies which had penetrated their positions but had actually attacked the Roman centre (Dhiraar's charge) and broken through, killing the army commander. This was a great loss, for Deirjan was a distinguished and highly esteemed general. Mahan appointed another general, one by the name of Qureen, to command Deirjan's army, and transferred the command of the Armenians to Qanateer, the commander of the Roman left. This was necessary, for in the next day's battle the major Roman effort would be made against the Muslim right and right centre.

The battle had got beyond the stage of 'sparks of fire'. It had not yet reached the stage of 'raging conflagration', but the fire was nevertheless burning with fearful heat as the battle entered its third day. This was to be, for the Muslims, a right-handed action.

The army of chains made no move on this day as it had suffered more heavily on the previous day than the army of Qanateer. The army of Qureen made a limited effort on the front of Abu Ubaidah as a diversionary measure to tie down Muslim reserves. But the Armenians and the left wing of the Roman army, both now under the command of Qanateer, struck with extreme severity at the Muslim right and the corps of Sharhabeel, selecting as the main point of attack the junction between Sharhabeel and Amr bin Al Aas.

The initial attack was again repulsed by Amr and Sharhabeel, but the Roman advantage of numbers, against which the Muslims could only put up the same tired soldiers, soon began to tell. Thus, shortly before midday, Qanateer broke though in several places. The corps of Amr fell back to the camp, and the right part of Sharhabeel's front was also pushed back, while his left still held firmly to its position. Several gaps now appeared in the Muslim front.

Again the Muslim women came into action with tent poles and stones and sharp tongues; and again the Muslims recoiled from them to face the Romans. One of these Muslims confided to his comrades: *"It is easier to face the Romans than our women!"* [1] The bulk of the two corps re-established a second line and held the Roman efforts to break through. Amr even took the offensive and struck at the Romans with his cavalry and infantry, intending to dislodge them from their forward positions, but had little success.

At the stage a Muslim lady came running to Khalid. She had suddenly got a bright military idea and wanted Khalid to get the benefit of it - just in case he did not know. *"O Son of Al Waleed"* said the lady, *"you are among the noblest of the Arabs. Know that the men only stay with their commanders. If the commanders stand fast the men stand fast. If the commanders are defeated the men are defeated."* [2]

Khalid thanked her politely for the advice and assured her that in this army the commanders would not be defeated!

Now Khalid launched his cavalry reserve against the flank of Qanateer. At the same time Amr's cavalry regiment manoeuvred from the right and struck Qanateer in his left flank, while the infantry of Amr and Sharhabeel counter-attacked frontally. This time the Roman opposition to the Muslim counter-attack proved much more stubborn and hundreds of Muslims fell in combat, but by dusk the Romans were pushed back to their own position and the situation restored as at the beginning of the battle.

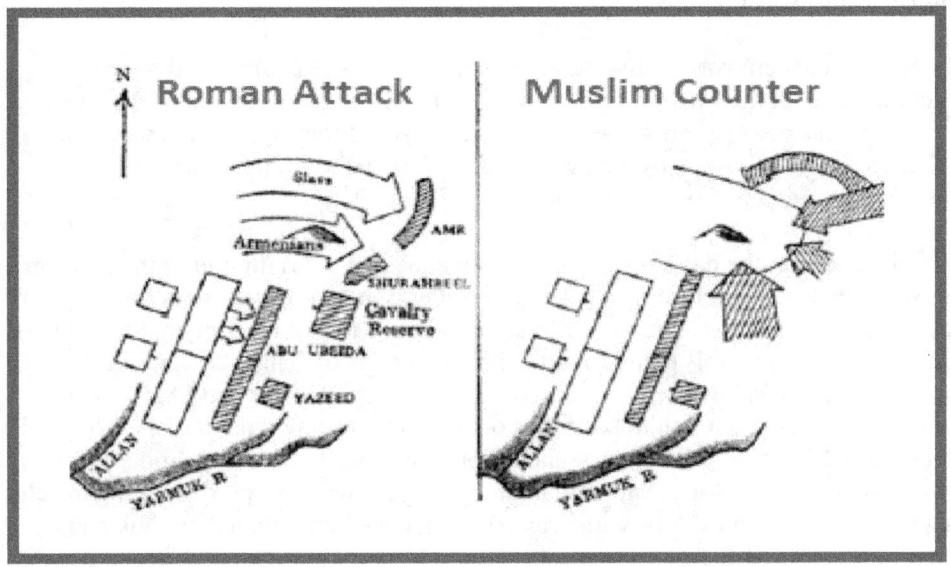

1. Waqidi: p. 142.
2. *Ibid.*

This had proved a harder day than the day before. However, the losses of the Romans far outnumbered those of the Muslims, and at the end of the day's fighting Muslim spirits were even higher, while Roman morale had suffered a serious blow. The Romans were now getting desperate. All their attacks had failed, in spite of a heavy toll in human lives, and they were in no better position than at the start of the battle. Mahan upbraided his generals, who promised to do better the next day. The next day would in fact be the most critical day of the battle.

Khalid and Abu Ubaidah spent the night walking about the Muslim camp, offering encouragement to the tired Muslims and speaking to the wounded. Being wounded in this battle did not mean getting evacuated to the rear. A Muslim had to be badly wounded indeed before he could expect to rest from fighting. A moderate wound meant a few hours' rest, and then back to the front!

The fourth day of battle dawned with an atmosphere tense with expectation. The Romans knew that this day would prove decisive, for now they were going to make their greatest effort to shatter the Muslim army which had so far withstood all assaults. If even this attack failed, then all prospects of further offensive action would disappear. It was now or never.

Khalid also knew that the battle had reached a critical stage, and that this day's operations would give a final indication of success or failure. Thousands of Romans had been killed so far, and if on this day also the Romans were repulsed with bloody losses, they would be unlikely to take the initiative again. Thereafter the counter-offensive could be launched. The Muslim strength was now somewhat depleted. The archers, positioned in the front rank, had suffered the heaviest losses, and now only 2,000 of them remained in fit state for battle. These were re-allocated at the scale of 500 to each corps. The Muslims were also more tired than the Romans because of their fewer numbers, but courage was never higher in the Muslim army.

Khalid's greatest concern was for his right. However, he was reassured by the thought that the commander of the Muslim right was Amr bin Al Aas, who in generalship was second only to Khalid. Amr had so far seen the heaviest fighting of this battle and was destined to continue to do so. Anyway, known as the shrewdest of the Arabs, Amr was more than a match for any Roman general.

Mahan decided to start the day's operation with an attack on the right half of the Muslim front as was done the day before. Once this part of the front was driven back and Muslim reserves committed in this sector, he would strike with the rest of his army at the left half of the Muslim front. With this plan of battle, the two armies of Qanateer were set in motion, and the Slavs and the Armenians sprang at the corps of Amr and Sharhabeel. Amr was pushed back again, but not as far as on the previous day; this time the Muslims were not going to face the ire of their women! Some distance behind its original position, the corps of Amr held the Slavs; and here manoeuvre gave way to a hard slogging match in which the Muslims, led by Amr with drawn sword, gave better than they took and inflicted severe losses on their opponents.

In the sector of Sharhabeel, however, the Armenians broke through and pushed the Muslims back towards their camp. The Armenians were strongly supported by the Christian Arabs of Jabla, and this proved the most serious penetration of the Muslim front. Sharhabeel was able to slow down the advance of the Armenians but could not repulse it. Soon it was clear that the corps would not be able to hold out for very long. It now became necessary for Khalid to enter this sector with his reserve.

What Khalid feared most was an attack in strength on a broad front. In case the enemy broke through at several places, there would be no way of expelling him as the army reserve could not be everywhere at the same time. On the second day of battle Khalid had been able to restore the situation on both flanks by first striking at one and then at the other penetration; but if the Romans got through in strength at many places, this could not be done. Consequently, when he saw the initial success of the enemy against Amr and Sharhabeel, he ordered Abu Ubaidah and Yazeed to attack on their front and thus forestall a Roman attack on the Muslim left in case such an attack was intended. This was to be a spoiling attack. By mid-morning the corps of Abu Ubaidah and Yazeed had engaged the armies of Qureen and Gregory, and at the time when Sharhabeel's position became delicate, both these corps were pressing hard against the right half of the Roman front.

Khalid, feeling more assured about his left, decided to strike against the Armenians. He divided the army reserve into two equal groups of which he gave one to Qais bin Hubeira and kept the other with himself. Leading his own cavalry group, Khalid galloped round behind the corps of Sharhabeel and appeared against the northern flank of the Armenian salient. Now began a three-pronged counter attack against the Armenians and Christian Arabs: Khalid from the right, Qais from the left and Sharhabeel from the front. The fighting became vicious in this part of the battlefield as the enemy resisted stoutly, and for several hours a bitter struggle raged between the Muslims and the Christians; but at last the Armenians broke under the blows of the Muslim cavalry and infantry and fell back to their own position, losing heavily in the process. In this action, which lasted the whole afternoon, the Christian Arabs proved the heaviest losers.

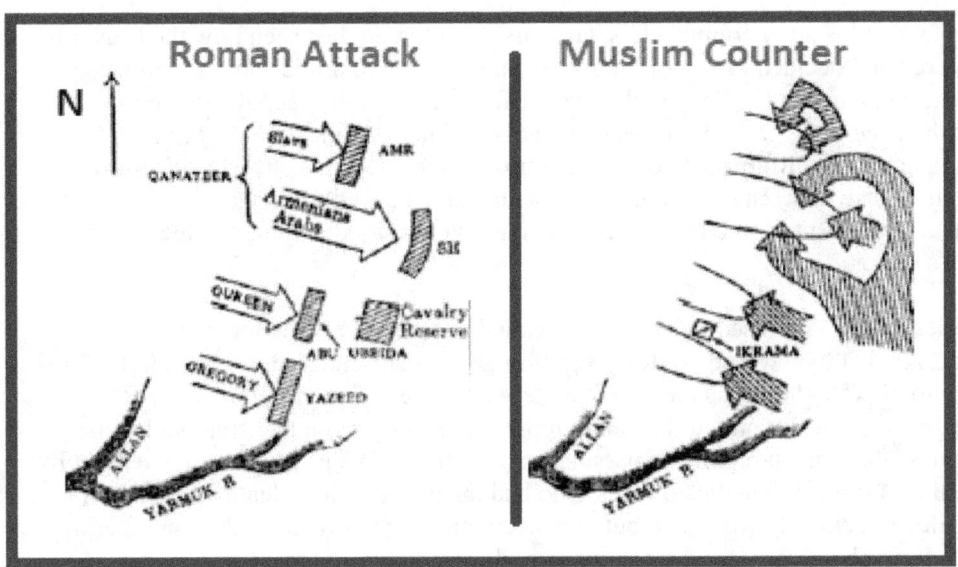

As the Armenians pulled back, Amr bin Al Aas renewed his efforts to dislodge the Slavs from the position which they had taken; and the Slavs, denied the support of the Armenians on their flank, also retired. The positions of Sharhabeel and Amr were now restored. But this action on the Muslim right was not completed till the evening; and while it was in progress an equally critical and more fierce battle was being fought on the left side of the Muslim front. What made the latter action so dangerous was the fact that the army reserve was heavily committed on the right and could do nothing to help Abu Ubaidah and Yazeed, who had to rely entirely on their own resources.

On the orders of Khalid, the two left corps had advanced to attack the Romans on their front and were in contact when Khalid moved the Mobile Guard to deal with the Armenians. Initially these corps enjoyed some success and the Romans were pushed back, but this action had not proceeded far when the Muslims found themselves subjected to a merciless barrage of archery. Thousands of Roman archers opened up on the Muslims, and so rapid and intense was the flight of arrows that according to some accounts, *"arrows fell like hailstones and blocked the light of the sun!"* [1] Many a Muslim was wounded by these arrows, the wounds varying from light to severe, and each of 700 Muslims lost an eye. From the sectors of Abu Ubaidah and Yazeed rose the lament: *"O my eye! O my sight!"* [2] Abu Sufyan also is believed to have lost an eye in this action. [3] As a result of this calamity, this fourth day of battle became known as the Day of Lost Eyes, [4] a tribute to Roman marksmanship. And this was undoubtedly the worst day of battle for the Muslim army.

The Muslims of the left now fell back. Their own bows were ineffective against the Roman archers because of their shorter range and fewer numbers; and the only way to avoid further casualties was to withdraw out of range of the Roman archers, which Abu

Ubaidah and Yazeed promptly did. As the two sides disengaged, both fronts stood still and the Muslims wisely refrained from advancing again. There was in fact a certain amount of consternation among the Muslims as a result of the arrow wounds and lost eyes.

But Mahan and his army commanders, Gregory and Qureen, had seen how the Muslims had suffered and decided to exploit their advantage. The two armies now advanced to assault the Muslims before they could recover from their repulse and the two bodies of men clashed again. As a result of the Roman assault the Muslims fell back to their own position and here the Romans, knowing that this was the decisive day of battle, attacked with even greater fury. The corps of Abu Ubaidah and Yazeed were again pushed back a short distance, except for the regiment of Ikrimah which stood at the left edge of Abu Ubaidah's sector.

The fearless Ikrimah refused to retreat, and called to his men to take the oath of death with him, *i.e.* that they would go down fighting and not surrender their position. In response to his call 400 of his men immediately took the oath and fell upon the Romans like hungry wolves. Not only did Ikrimah repulse the Romans on his front but he also lashed out at the Roman regiments passing on his flanks. This position was never lost by the Muslims. Of the 400 dedicated men who had taken the oath of death, everyone was either killed or seriously wounded, but they accounted for many times their number of Romans. Ikrimah and his son, Amr, were mortally wounded.

1. Waqidi: pp. 146, 148.
2. *Ibid*: p. 149.
3. We have already noted the loss of Abu Sufyan's eye at Taif. However, some sources indicate that this happened at Yarmuk and not at Taif.
4. Waqidi: p. 148.

The corps of Abu Ubaidah and Yazeed did not this time reach the camp. They did not have to, for the women themselves, many of them carrying swords, rushed forward and joined their men. Even the women understood that on this phase hung the fate of the battle. They came with swords and tent poles for the Romans and water for the Muslim wounded and thirsty. Among them were Khaulah and the wife of Zubair and Umm Hakeem, who shouted to the women: *"Strike the uncircumcised ones in the arm!"* [1] The women rushed through the Muslim corps to the front rank, determined to fight ahead of their men this time; and this proved the turning point in this sector.

The sight of their women fighting alongside, and some even ahead of them, turned the Muslims into raging demons. In blind fury they struck at the Romans in an action in which there was now no manoeuvre and no generalship - only individual soldiers giving of their superhuman best. Striking with sword and dagger, the valiant men of Abu Ubaidah and Yazeed hurled the Romans back from their positions, and the Romans retreated fast before the terrible blows of the infuriated Muslims.

The battle of this day reached its climax along the entire front in the late afternoon. At this time all the generals were engaged in combat like their men, and every corps commander proved his right to be the leader of brave men. Several Romans bit the dust under the blows of Muslim women. Khaulah took on a Roman warrior, but her adversary proved a better swordsman and struck her on the head with his sword, as a result of which she collapsed in a heap with blood dying her hair red. When the Romans were pushed back, and the other women saw her motionless body, they wailed in sorrow and searched frantically for Dhiraar, to inform him that his beloved sister was dead. But Dhiraar could not be found till the evening. When he did arrive where his sister lay, Khaulah sat up, smiling. She was all right, really!

By dusk the days' action was over. Both armies stood once again on their original lines. It had been a terrible day - one that the veterans of Yarmuk would never forget and on which the Romans came very near victory. But many of them paid with their lives for a success which they were not destined to gain. The most crippling losses had been suffered by the chained men, the Armenians and the Christian Arabs. The Muslims had suffered more than on previous days, and those who were not wounded were fewer in number than those who were, but a glow of pride and satisfaction warmed their hearts, especially Khalid's who knew that the crisis was over. The tide had turned.

1. Waqidi: p 149. According to Balazuri (p. 141) these words were uttered by Hind.

One incident remains to be narrated before we come to the end of this account of the Day of Lost Eyes. During a pause in the fighting in Sharhabeel's sector, Khalid suddenly appeared deeply worried, and this surprised his men who had never seen him so. But they understood when he ordered the men to look for his red cap which he had dropped on the battlefield. A search was at once carried out and the cap found, for which Khalid was profuse in his thanks. There were some men who did not know about this cap and asked Khalid what was so wonderful about it. Thereupon Khalid told the story of the red cap:

When the Messenger of Allah had his head shaved on the last pilgrimage, I picked up some of the hair of his head. He asked me, *"What will you do with this, O Khalid?"* I replied, *"I shall gain strength from it while fighting our enemies, O Messenger of Allah."* Then he said, *"You will remain victorious as long as this is with you."*

I had the hair woven into my cap, and I have never met an enemy but he has been defeated by the blessing of the Messenger of Allah, on whom be the blessings of Allah and peace. 1

This is the story of Khalid's red cap - the one possession with which he would not part.

Darkness had fallen when Khalid sat on the blood-spattered earth at the left edge of Abu Ubaidah's sector. On one knee rested the head of Ikrimah, his nephew and dear, dear friend. On the other knee lay the head of Amr, son of Ikrimah. Life was ebbing fast from the bodies of father and son. Khalid would now and then dip his fingers into a bowl of water and let the water drip into the half-open mouths; and he would say: *"Does the Son of Hantamah think we do not get martyred?"* 2 Thus died Ikrimah and his son, in the dearly loved arms of the Sword of Allah. The man who for years had been the most blood-thirsty enemy of Islam earned final redemption in martyrdom. The greatest glory on the Day of Lost Eyes, a day such as the Muslims would never again see in Syria, went to Ikrimah bin Abi Jahl.

The night was spent in peace, if there could be peace for exhausted, wounded men who had driven their bodies to perform feats of strength and endurance which the human body was never intended to perform. Normally Abu Ubaidah would nominate a general as duty officer for the night, whose task it would be to go round the guards and the outposts and check the vigilance of the sentries. But on this night the generals themselves were so tired that Abu Ubaidah, kind and considerate as ever, did not have the heart to ask any of them to carry out this onerous task. Although his own sword dripped with the blood of several Romans and his need for rest was no less than that of the others, Abu Ubaidah decided to act as duty officer himself. Along with a few selected Companions of the Prophet he began his round. But he need not have worried. Everywhere that he went he found the generals up and mounted, going about and talking to the sentries and the wounded. Zubair was doing the rounds accompanied by his wife, also on horseback!

1. Waqidi: p. 151.
2. Tabari: Vol.2, p. 597. The Son of Hantamah was Umar, and by 'we' Khalid meant the Bani Makhzum.

Early on the fifth day of battle the two armies again formed up on their lines - the same lines which they had adopted before the start of battle. But on this day the soldiers did not stand so erect, nor look so imposing. Next to each unwounded man stood a wounded one. Some could hardly stand, but stand they did. Khalid looked intently at the Roman front for any sign of movement and wondered if the Romans would perhaps attack once again. But there was no movement, not for an hour or two. Then one man emerged from the Roman centre. This was an emissary of Mahan who brought a proposal for a truce for the next few days so that fresh negotiations could be held. Abu Ubaidah nearly accepted the proposal but was restrained by Khalid. On Khalid's insistence he sent the envoy back with a negative reply, adding: *"We are in a hurry to finish this business!"* 1

Now Khalid knew. He had guessed right. The Romans were no longer eager for battle. The rest of the day passed uneventfully while Khalid remained busy giving orders for the counter-offensive and carrying out some reorganisation. All the cavalry regiments were grouped together into one powerful mounted force with the Mobile Guard acting as its hard core. The total strength of this cavalry group was now about 8,000 horse.

The next day the sword of vengeance would flash over the Plain of Yarmuk.

The sixth day of battle dawned bright and clear. It was the fourth week of August 636 (third week of Rajab, 15 Hijri). The stillness of the morning gave no indication of the holocaust that was to follow. The Muslims were now feeling more refreshed, and knowing of their commander's offensive intentions and something of his plans, were eager for battle. The hopes of this day drowned the grim memories of the Day of Lost Eyes. To their front stretched the anxious ranks of the Roman army - less hopeful but still with plenty of fight in them.

As the sun rose over the dim skyline of the Jabal-ud-Druz, Gregory, the commander of the army of chains, rode forward, but from the centre of the imperial army. He had come with the mission of killing the Muslim army commander in the hope that this would have a demoralising effect on the Muslim rank and file. As he drew near the Muslim centre, he shouted a challenge and asked for *"none but the commander of the Arabs"*. 2

Abu Ubaidah at once prepared to go forth. Khalid and the others tried to dissuade him, for Gregory had the reputation of being a powerful fighter, and looked it too. All felt that it would be better if Khalid went out in response to the challenge, but Abu Ubaidah was adamant. He gave the army standard to Khalid, and with the words, *"If I do not return you shall command the army, until the Caliph decides the matter,"* 3 set out to meet his challenger.

The two generals met on horseback, drew their swords and began to duel. Both were splendid swordsmen and treated the spectators to a thrilling display of swordsmanship with cut, parry and thrust. Romans and Muslims held their breath. Then, after a few minutes of combat, Gregory drew back from his adversary, turned his horse and began to canter away. Shouts of joy rose from the Muslim ranks at what appeared to be the defeat of the Roman, but there was no such reaction from Abu Ubaidah. With his eyes fixed intently on the retreating Roman, he urged his horse forward and followed him.

1. Waqidi: p. 153.
2. *Ibid*: p. 153.
3. *Ibid*.

Gregory had hardly gone a few hundred paces when Abu Ubaidah caught up with him. Now Gregory, who had deliberately controlled the pace of his horse to let the Muslim overtake him, turned swiftly and raised his sword to strike at Abu Ubaidah. His apparent flight had been a trick to throw his opponent off guard. But Abu Ubaidah was no novice; he knew more about sword play than Gregory would ever learn. The Roman raised his sword, but that is as far as he got. He was struck at the base of his neck by Abu Ubaidah, and the sword fell from his hand as he crashed to the ground. For a few moments Abu Ubaidah sat still on his horse, marvelling at the enormous size of the Roman general. Then, leaving behind the bejewelled and gold-encrusted armour and weapons of the Roman, which he ignored with his habitual disregard for worldly possessions, the saintly soldier turned and rode back to the Muslim front.

On the return of Abu Ubaidah, Khalid galloped off to join the cavalry which had been positioned behind the corps of Amr bin Al Aas. As he arrived at his place he gave the signal for the general attack and the entire Muslim front surged forward. The Muslim centre and left engaged the Roman armies on their front but did not press the attack. On the right the cavalry galloped round to the flank of the Roman left. From here Khalid despatched a regiment to engage and hold the Roman cavalry of the left, and with the rest of the Muslim cavalry struck at the flank of the Roman left wing (the Slavs) at the same time as Amr assaulted their front with extreme violence. The Slavs were stout fighters, and for some time defended themselves courageously, but getting no support from their cavalry and assailed from front and flank, they at last gave way. Recoiling from the blows of Khalid and Amr, they fell back into the centre -the Armenians.

As the Roman left wing crumbled, Amr moved his corps forward, swung it to the left, and came up against the left and now exposed flank of the Armenians, in whose ranks there was considerable disorder as a result of the disorganised arrival of the broken Slavs. Meanwhile Khalid wheeled his cavalry and engaged the Roman cavalry of the left, which had been held in check by the regiment he had detached a little earlier. The second phase of the Muslim offensive began with Sharhabeel attacking the front of the Armenians while Amr assailed their flank. Then Khalid struck at the Roman cavalry of the left and drove it back from its position. This cavalry group, having got a severe mauling from Khalid, galloped away to the north and to safety. It had had enough of battle.

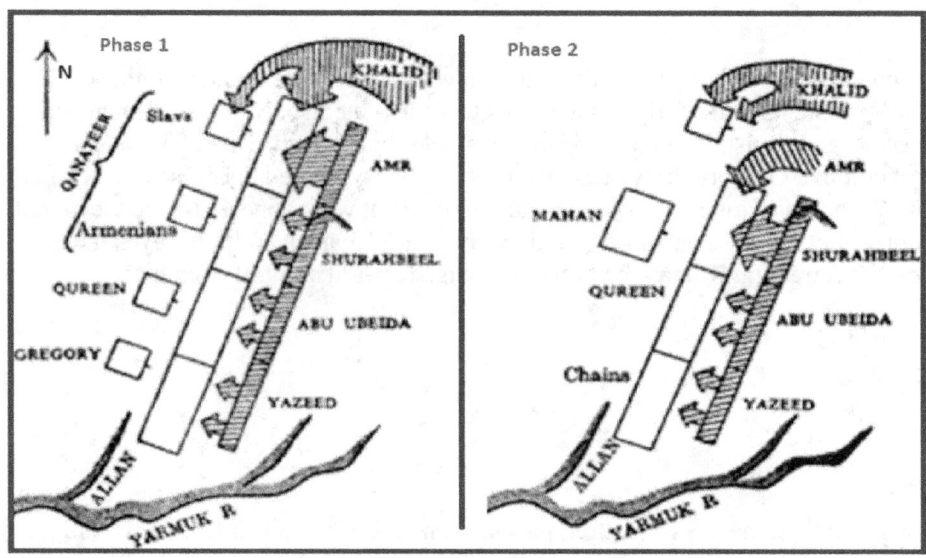

I shall not attempt to explain Khalid's plan as it will become evident to the reader as we proceed with the course of the battle. But one point that needs especial mention is Khalid's intention with regard to the enemy cavalry. He had determined to drive the Roman cavalry off the battlefield so that the infantry, which formed the bulk of the Roman army, would be left without cavalry support and thus be helpless when attacked from flank and rear. In fast-moving operations the cavalry was the dominant partner, and without it the infantry would be at a great disadvantage, unable to move fast or to save itself by a rapid change of position.

At about the time when the Roman cavalry of the left was being driven away by Khalid, Mahan had concentrated the remainder of his cavalry into one powerful, mobile army behind the Roman centre to counter attack and regain lost positions. But before the massed Roman cavalry could start any manoeuvre, it was assailed in front and flank by the Muslim cavalry. For some time, urged on by the intrepid Mahan, the Romans fought gallantly; but in this type of fluid situation the regular, heavy cavalry was no match for the light, fast-moving horsemen of Khalid who could strike, disengage, manoeuvre and strike again. At last the Roman cavalry, seeing no other way of survival, broke contact and fled to the north, taking with it the protesting Mahan. In this manner the Roman cavalry abandoned the infantry to its fate. With Mahan, altogether 40,000 mounted troops got away, consisting partly of regular Roman cavalry and partly of the mobile Christian Arabs of Jabla bin El Eiham.

In the cavalry actions of this morning there was no sign of Dhiraar. The Muslims missed the familiar sight of the half-naked warrior in the kind of battle in which he would have revelled. They did not know where he was; and Khalid would not tell!

Meanwhile the Armenians were stoutly resisting Amr and Sharhabeel's attempts to crush them. The two Muslim corps had made some headway but not much; and this is understandable, for the Armenians were very brave fighters indeed. [1] Abu Ubaidah and

Yazeed were also attacking the Romans on their front (though their role was as yet secondary- a holding operation), but were held by the army of Qureen and the army of chains. It was at this stage that Khalid, having driven the Roman cavalry from the battlefield, turned on the Armenians and charged them in the rear. In the face of the three-pronged attack the Armenians disintegrated. Abandoning their position, they fled to the South-West-the only direction open to them, and were much relieved and surprised that the Muslim cavalry made no effort to interfere with their movement as it could easily have done. They travelled in the direction in which they saw safety. Unknown to them, this was also the direction which Khalid wanted them to take.

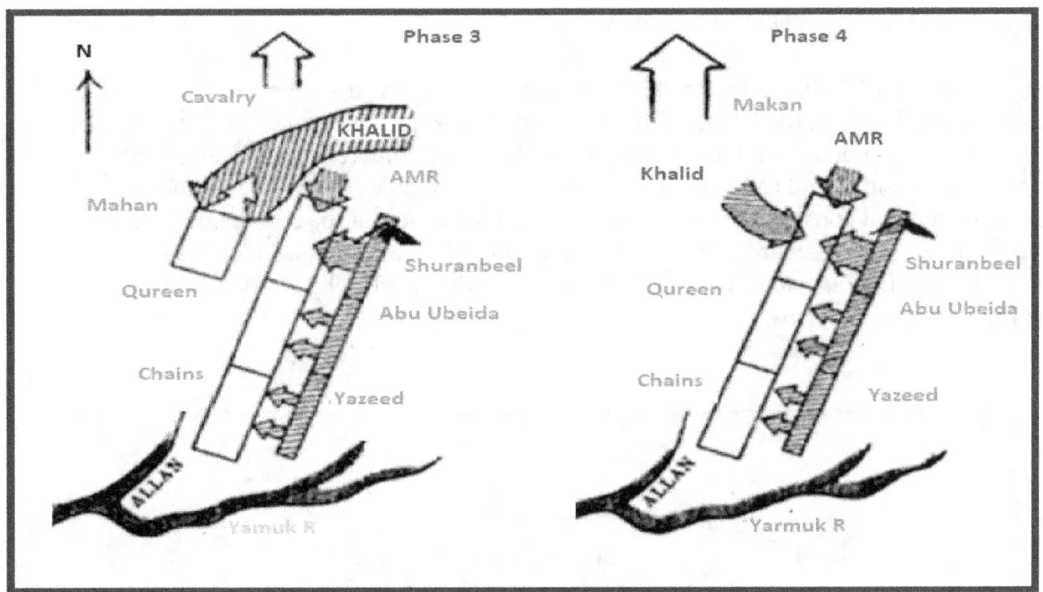

As the Armenian army collapsed, and mingling in a confused mass with the survivors of the Slav army of Qanateer fled towards the Wadi-ur-Raqqad, the remaining Roman armies realised the hopelessness of their position. Their flank and rear were completely exposed. Consequently they also began to withdraw, and with discipline and good order made their way westwards. Here again the Roman movement was not intercepted by Khalid.

1. Gibbon, in his Decline and Fall of the Roman Empire, describes the Armenians as "the most warlike subjects of Rome"

The sun had not yet reached its zenith when the Roman infantry was in full retreat- part of it fleeing in panic and part withdrawing in good order. It made for the Wadi-ur-Raqqad. After the retreating Romans came the Muslim corps, now reformed into orderly lines with shorter fronts. The Muslim cavalry moved to the north of the Roman army so that none may escape in that direction, though before this escape route could be fully sealed, thousands of Slavs and Armenians did manage to get away. In this manner the Muslims closed in on the already defeated Army of Caesar. [1]

As the Romans fled the field of battle, their only desire was to put as much distance as possible between themselves and the Muslims. They knew that the northern escape route was closed by the Muslim cavalry; but another channel of escape was available where the Raqqad was crossed, at a ford, by a good road. Towards this ford the officers guided their men. As the leading regiment arrived at the ford, it rushed down the eastern slope of the ravine and began to cross the stream. The eastern slope was not so bad here as in other parts of the ravine; but the western slope was much steeper, and near the top it became precipitous on either side of the road, creating a bottleneck where a few brave men could hold up an army.

Overjoyed at their escape from the Plain of Yarmuk, the men in the lead laboured up the road on the western bank of the ravine. It was only when they got near the top that they noticed a group of Muslims standing above them with drawn swords. At their head stood a lean, young warrior, naked above the waist!

During the night Khalid had sent Dhiraar with 500 horsemen from the Mobile Guard to make a wide detour of the Roman left, get behind the Wadi-ur-Raqqad, and occupy a blocking position on the far bank of the ravine. Dhiraar, guided by a Christian Arab named Abu Jueid, 2 had carried out the move with admirable efficiency. Unknown to the Romans-who had considered the crossing of the Raqqad too far back to be of tactical significance - he had secured the western bank of the ravine and concealed his men near the ford. Now Dhiraar stood with his men on top of the western bank, looking down at the tired, panting Romans.

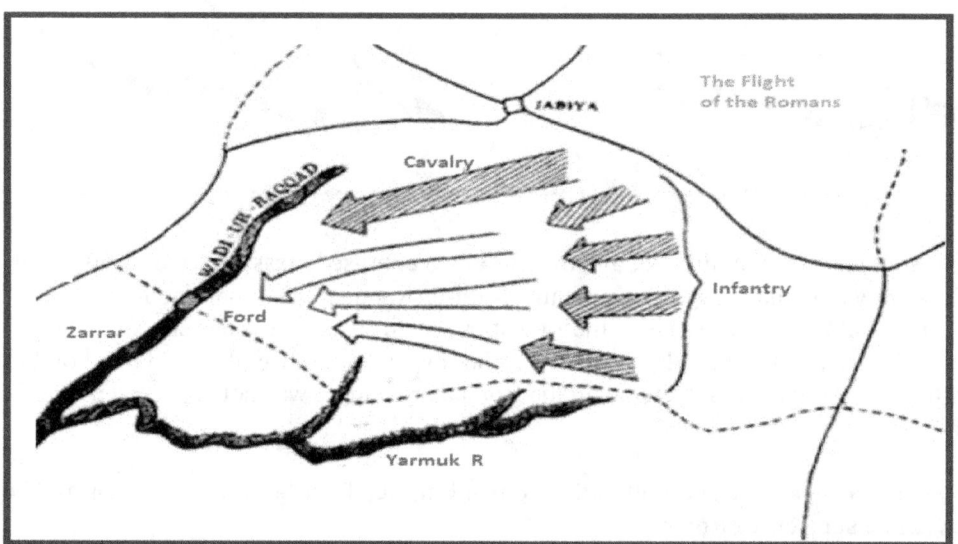

Soon a volley of stones hit the Romans. A few of them managed to get to the top, but were cut down instantly, Finding themselves under a hail of stones, the leading elements fell back on those behind them, these on those behind them, and these again on those behind them. As Dhiraar charged at the Romans, they went sliding down-a screaming, twisting, rolling avalanche-to the bottom of the ravine.

The Romans still on the eastern bank stopped when they saw the horror that had befallen the leading regiment. It was clear that this escape route was also closed. Nothing could be done to dislodge Dhiraar because of the narrowness of the crossing which allowed no room for manoeuvre; so the Roman army turned to defend itself against the impending attack from the east. The generals who still remained with the army hastily deployed the regiments for defence with their backs to the Wadi-ur-Raqqad and their right flank resting on the Yarmuk River. They were caught between two calamities-the ravine and the Muslims-and could not decide which was worse!

1. The statement made by some later Western writers that the Roman defeat was due to Khalid's exploitation of a violent sand storm which blew in the faces of the Romans is utterly incorrect. No Muslim historian has mentioned such a storm. Gibbon (Vol. 5, p. 327) states that there was "a cloud of dust and adverse wind", but only a child would imagine that the Muslim army, which still numbered about 30,000 fit soldiers, deployed on an 11 mile front, could be thrown into action so quickly, in such a superbly conceived manoeuvre, merely to exploit a dust storm. And this in the days when communication was by horse-rider! This is nothing but a proud Western historian's attempt by to find an excuse for the Roman defeat.
2. Waqidi: p. 152.

In the late afternoon of this sixth day of battle, began the last phase of the Muslim attack. Only a third of the Roman army remained in this crowded corner of the Plain of Yarmuk; against it the Muslims were arranged in a neat semi-circle, with the infantry on the east and the cavalry on the north. The Muslim strength here was less than 30,000 men. The time for generalship and manoeuvre was over. The skill of the general had placed the troops in the ideal situation for combat, and it was up to the soldiers to fight and win. The generals drew their swords and became warriors like the rest, as the lions of the desert moved in for the final kill.

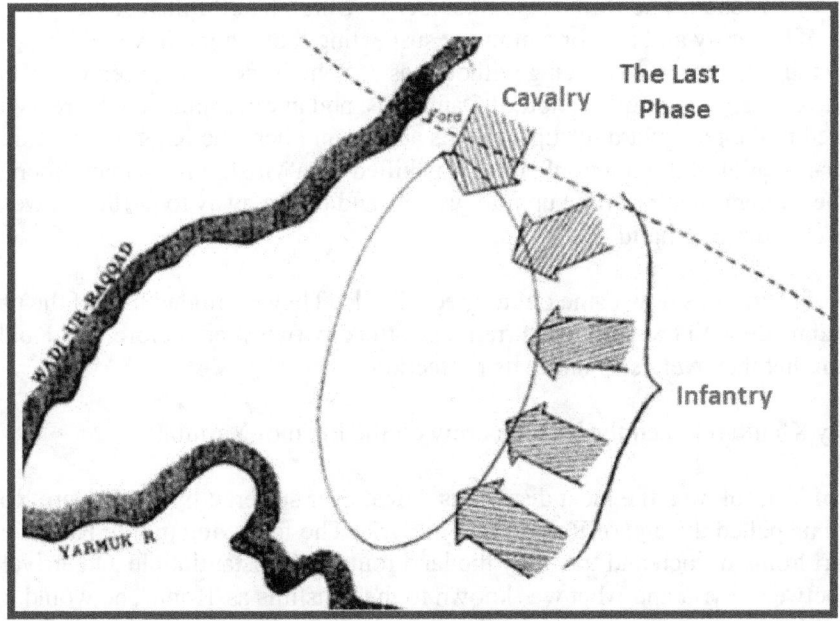

The attackers struck with sword and spear at the confused, seething mass in front of them. At places the Romans were too closely packed for elbow-room to use their weapons; but their front rank fought with heroic, if futile, courage to stem the tide. Soon it was struck down, and the next rank and the next, as the Muslims advanced-cutting, slashing, stabbing, thrusting. In the dust and confusion the Romans ran into each other, and those not agile enough fell and suffered a painful death under the trampling feet of their own comrades.

The Muslim cavalry, rejoined by Dhiraar's detachment, pressed the Romans farther into the corner where they lost all freedom of action. Khalid's horsemen now began to use the knees and hooves of their horses to knock down the exhausted defenders. The screams of the Romans mingled with the shouts of the Muslims as the last resistance collapsed, and the battle turned into a butchery and a nightmare of horrors. For the last time the Romans broke and fled in disorder. Those who still retained a desire to fight were carried away by their panic-stricken comrades, especially in the army of chains in which groups of 10 fought, moved and fell together.

Moving like stampeding cattle, the Roman rabble reached the edge of the ravine. The view to the bottom was terrifying, but so was the last wild charge of the Muslims. Those coming in the rear pressed blindly against those on the edge of the ravine, and rank after rank, the Roman army began to fall down the precipice. The blood-curdling screams of some continued until they hit the bottom, while the screams of others were cut short as their bodies crashed against jutting rocks and then continued their descent as shapeless, bloody lumps.

It was almost dark when the last of the Romans ceased to move. The day of 'the raging conflagration' had ended. Khalid's greatest battle was over. [1]

Early next morning, while the rest of the army gathered the spoils of war and buried the martyrs, Khalid set off with the Muslim cavalry on the road to Damascus in the hope of catching up with Mahan. The Roman Commander-in-Chief, heartbroken at the annihilation of his army and not for a moment suspecting that a pursuit would be launched by the Muslims, was moving without haste. Some time in the afternoon Khalid overtook the Romans a few miles short of Damascus, and at once attacked the rear-guard. Mahan rushed to the rear-guard to supervise its action, and here the King of Armenia, the Commander-in-Chief of the imperial army, was killed by a Muslim horseman. Soon after his death, the Roman cavalry broke up into groups, and riding away to north and west, escaped the clutches of Khalid.

The people of Damascus now came out to greet Khalid. They reminded him of the pact which he had made with them on the surrender of the city two years before, and Khalid assured them that they were still under its protection.

The next day Khalid rejoined the Muslim army on the Plain of Yarmuk.

The Battle of Yarmuk was the most disastrous defeat ever suffered by the Eastern Roman Empire, and it spelled the end of Roman rule in Syria. The following month Heraclius would depart from Antioch and travel by the land route to Constantinople. On arrival at the border between Syria and what was known to the Muslims as 'Rome', he would look back towards Syria and, with a sorrowing heart, lament: *"Salutations to thee, O Syria! And farewell from one who departs. Never again shall the Roman return to thee except in fear. Oh, what a fine land I leave to the enemy!"* [2]

As an example of a military operation, the Battle of Yarmuk combined many tactical forms: the frontal clash, the frontal penetration, counter-attack and repulse, the flank-attack, the rear-attack and the outflanking manoeuvre. Khalid's plan of remaining on the defensive until he had worn down the Romans had worked admirably. During the defensive phase, lasting four days, every offensive blow by Khalid had been a limited tactical manoeuvre to restore his defensive balance. Only when it was certain that the Romans were badly hurt and no longer capable of fighting offensively, did he launch his counter-offensive, on the last day of battle. On this day he had rolled up the Roman position from a flank, but only after he had separated the cavalry from the infantry and rendered the latter helpless. Then he had driven the Roman infantry into the corner formed by the Wadi-ur-Raqqad and the Yarmuk River, having already positioned Dhiraar at the crossing of the ravine so that none might escape, and launched his last, all-destroying assault. Against the anvil of the Wadi-ur-Raqqad the Muslim hammer had crushed the Roman army to powder.

1. There is a disagreement about the two basic points in this battle: the strength of the opposing forces and the exact location of the battlefield. For an explanation see Notes 12 and 13 in Appendix B.
2. Tabari: Vol. 3, p. 100; Balazuri: p. 142.

It is known that the Muslims lost 4,000 men in this battle, and those who did not carry wounds on their persons were few indeed; but the Roman casualty figures vary. Waqidi's estimate is exaggerated to an unacceptable degree. Tabari, in one place, [1] gives the Roman dead as 120,000 but elsewhere quotes Ibn Ishaq's estimate of 70,000. [2] Balazuri also gives the Roman dead as 70,000. [3] This last figure appears to be reasonable-about 45 per cent of the Roman army. Of these 70,000 about half fell on the plain and half fell into the ravine. Some 80,000 men got away, most of them horse and camel-mounted, including those who escaped before the Muslim ring was closed. Many may even have succeeded in crossing the Wadi-ur-Raqqad at places where it was not so precipitous.

The Battle of Yarmuk was a glorious victory for Islam; and the Plain of Yarmuk and the Wadi-ur-Raqqad provided ample, if gruesome, evidence of it. Tens of thousands of Roman bodies lay scattered, singly and in heaps, on the plain and at the bottom of the ravine. The worst signs of carnage were visible at the corner of the plain and in the ravine itself. Broken, maimed and mutilated bodies could be seen everywhere, lying in grotesque shapes and postures. Blood-covered bodies without limbs lay on the blood-spattered earth, staring with sightless eyes at the eternity of death. Thousands of Romans sprawled with broken swords in their hands, true to the oath of death which they had taken on the eve of battle. And mingled with the soldiers, lay countless priests, still clutching their crosses. The nauseating stench of decaying flesh rose and poisoned the air over the Plain of Yarmuk.

A vast and heroic battle had been fought; a great and terrible victory had been won.

1. Tabari: Vol. 2, p. 596.
2. *Ibid*: Vol. 3, p. 75.
3. Balazuri: p. 141.

Chapter 36

(The Completion of the Conquest)

> *"There are many virtues of al-Sham and its people established by the Book, the Sunnah and the traditions of the people of knowledge, and this is one of the things I relied upon in my encouraging the Muslims to fight the Tatars, my order to them to remain in Damascus, my forbidding them from fleeing to Egypt, and my inviting the Egyptian military to Syria and consolidating the Syrian military there ..."*
> [Ibn Taymiyyah, speaking of the events of 700-702AH, when Damascus was successfully defended against the ravaging Mongol army]1

After Yarmuk the remnants of the Roman army withdrew in haste to Northern Syria and the northern part of the Mediterranean coast. The vanquished soldiers of Rome, those who survived the horror of Yarmuk, were in no fit state for battle. The victorious soldiers of Islam were in no fit state for battle either. Abu Ubaidah sent a detachment to occupy Damascus, and remained with the rest of his army in the region of Jabiya for a whole month. During this period the men rested; spoils were collected, checked and distributed; the wounded were given time to recover. There was much to be done in matters of administration, and this kept the generals occupied.

In early October 636 (late Shaban, 15 Hijri), Abu Ubaidah held a council of war to discuss future plans. Opinions of objectives varied between Caesarea and Jerusalem. Abu Ubaidah could see the importance of both these cities, which had so far resisted all Muslim attempts at capture, and unable to decide the matter, wrote to Umar for instructions. In his reply the Caliph ordered the Muslims to capture Jerusalem. Abu Ubaidah therefore marched towards Jerusalem with the army from Jabiya, Khalid and his Mobile Guard leading the advance. The Muslims arrived at Jerusalem around early November, and the Roman garrison withdrew into the fortified city.

For four months the siege continued without a break. Then the Patriarch of Jerusalem, a man by the name of Sophronius, offered to surrender the city and pay the Jizya, but only on condition that the Caliph himself would come and sign the pact with him and receive the surrender. When the Patriarch's terms became known to the Muslims, Sharhabeel suggested that instead of waiting for Umar to come all the way from Madinah, Khalid should be sent forward as the Caliph. Umar and Khalid were very similar in appearance; 2 and since the people of Jerusalem would only know Umar by reports, they could perhaps be taken in by a substitute. The Muslims would say that actually the Caliph was already there-and lo, he comes!

On the following morning the Patriarch was informed of the Caliph's presence, and Khalid, dressed in simple clothes of the poorest material, as was Umar's custom, rode up to the fort for talks with the Patriarch. But it did not work. Khalid was too well known, and there may have been Christian Arabs in Jerusalem who had visited Madinah and seen both Umar and Khalid, noting the differences. Moreover, the Patriarch must have wondered how the great Caliph happened to be there just when he was needed! Anyhow, the trick was soon discovered, and the Patriarch refused to talk. When Khalid reported the failure of this mission, Abu Ubaidah wrote to Umar about the situation, and invited him to come to Jerusalem and accept the surrender of the city. In response the Caliph rode out with a handful of Companions on what was to be the first of his four journeys to Syria.

Umar first came to Jabiya, where he was met by Abu Ubaidah, Khalid and Yazeed, who had travelled thither with an escort to receive him. Amr bin Al Aas was left as commander of the Muslim army besieging Jerusalem. Khalid and Yazeed were magnificently attired in silk and brocade and rode gaily caparisoned horses-and the sight of them infuriated Umar. Dismounting from his horse, he picked up a handful of pebbles from the ground and threw them at the two offending generals, *"Shame on you"*, shouted the Caliph, *"that you greet me in this fashion ! It is only in the last two years that you have eaten your fill. Shame on what abundance of food has brought you to! By Allah, if you were to do this after 200 years of prosperity, I should still dismiss you and appoint others in your place."* 3

1. Introduction to *Manaqib al-Sham wa Ahlih* (The Virtues of al-Sham and its People) by Ibn Taymiyyah.
2. Waqidi: p. 162, Isfahani: Vol. 15, pp. 12, 56.
3. Tabari: Vol. 3, p. 103.

Umar was dressed in simple, patched garments as he was wont to wear in the time of the Holy Prophet. Becoming Caliph had made no difference to his austere and unspoiled way of life, and he continued to abhor luxury and ostentation.

Recovering from their discomfiture, Khalid and Yazeed hastily opened their robes and showed the armour and weapons which they wore underneath. *"O Commander of the Faithful!"* they cried. *"These are only garments. We still carry our weapons"*. 1 Umar was sufficiently mollified by this reply. Now Abu Ubaidah walked up, dressed as simply and unaffectedly as always, and the Caliph and the Commander-in-Chief shook hands and embraced each other.

From Jabiya, Umar proceeded to Jerusalem, accompanied by his generals and the escort. His arrival at Jerusalem was a great moment for the Muslim soldiers, and they rejoiced at the sight of their ruler.

Next day, at about noon, Umar sat with a large group of Companions, talking of this and that. Soon it would be time for the early afternoon prayer. Bilal the Negro was also present. Bilal, who has been mentioned in the second chapter of this book, had suffered many tortures in the early days of Islam at the hands of the unbelieving Qureish, but had remained steadfast in his faith. When the institution of the Adhan (the Muslim call to prayer) was adopted in 2 Hijri, the Prophet appointed Bilal as the Muazzin; and thereafter, five times a day, the powerful and melodious voice of Bilal could be heard at Madinah, calling the Faithful to prayer. Over the years Bilal had risen in stature as a saintly Muslim, and had become one of the closest and most venerated Companions of the Prophet. But on the death of the Holy Prophet, Bilal had fallen silent; he would not call the Adhan any more.

It now occurred to some of the Companions that perhaps the conquest of the holy city of Jerusalem was an important enough occasion for Bilal to break his silence. They asked Umar to urge him to call the Adhan, just this one time! Umar turned to Bilal: *"O Bilal! The Companions of the Messenger of Allah implore you to call the Adhan and remind them of the time of their Prophet, on whom be the blessings of Allah and peace."* 2 For a few moments Bilal remained lost in thought. Then he looked at the eager faces of the Companions and at the thousands of Muslim soldiers who were gathering for the congregational prayer. Then he stood up. Bilal would call the Adhan again!

The glorious voice of the illustrious Muazzin beat upon the vast multitude. As he called the opening words, *Allah is Great*, the minds of the Faithful turned to memories of the dearly loved Muhammad and tears welled up in their eyes. When Bilal came to the words, *Muhammad is the Apostle of Allah*, his audience broke down and sobbed.

1. Tabari: Vol. 3, p. 103.
2. Waqidi: p. 165.

On the following day the pact was drawn up. [1] It was signed on behalf of the Muslims by Caliph Umar and witnessed by Khalid, Amr bin Al Aas, Abdur-Rahman bin Auf and Muawiyah. Jerusalem surrendered to the Caliph, and peace returned to the holy city. This happened in April 637 (Rabi-ul-Awwal, 16 Hijri). After staying 10 days at Jerusalem, the Caliph returned to Madinah.

Following the Caliph's instructions, Yazeed proceeded to Caesarea and once again laid siege to the port city. Amr and Sharhabeel marched to reoccupy Palestine and Jordan, which task was completed by the end of this year. Caesarea, however, could not be taken till 640 (19 Hijri), when at last the garrison laid down its arms before Muawiyah. Abu Ubaidah and Khalid, with an army of 17,000 men, set off from Jerusalem to conquer all of Northern Syria.

Abu Ubaidah marched to Damascus, which was already in Muslim hands, and then to Emessa, which welcomed his return. His next objective was Qinassareen, and towards this the army advanced with Khalid and the Mobile Guard in the lead. After a few days the Mobile Guard reached Hazir, 3 miles east of Qinassareen, and here it was attacked in strength by the Romans. [2]

The Roman commander at Qinassareen was a general named Meenas-a distinguished soldier who was loved by his men. Meenas knew that if he stayed in Qinassareen, he would be besieged by the Muslims and would eventually have to surrender, as at present no help could be expected from the Emperor. He therefore decided to take the offensive and attack the leading elements of the Muslim army well forward of the city and defeat them before they could be joined by the main body. With this plan in mind, Meenas attacked the Mobile Guard at Hazif with a force whose strength is not recorded; He either did not know that Khalid was present with the leading elements of the Muslim army or did not believe all that he had heard about Khalid.

For Khalid to throw his cavalry into fighting formation for battle was a matter of minutes, and soon a fierce action was raging at Hazir. The battle was still in its early stages when Meenas was killed; and as the news of his death spread among his men, the Romans went wild with fury and attacked savagely to avenge their beloved leader's death. But they were up against the finest body of men of the time. Their very desire for vengeance proved their undoing, for not a single Roman survived the Battle of Hazir. [3] The Mobile Guard took this encounter in its stride as one of its many victories.

As soon as the battle was over, the people of Hazir came out of their town to greet Khalid. They pleaded that they were Arabs and had no intention to fight him. Khalid accepted their surrender, and advanced to Qinassareen.

When Umar received reports of the Battle of Hazir, he made no attempt to conceal his admiration - for the military genius of Khalid. *"Khalid is truly the commander,"* [4] Umar exclaimed. *"May Allah have mercy upon Abu Bakr. He was a better judge of men than I have been."* [5] This was Umar's first admission that perhaps he had not judged Khalid rightly.

1. According to some reports, the pact was actually signed at Jabiya with representatives of the Patriarch, and after signing the pact there, Umar travelled to Jerusalem and received the surrender.
2. Hazir still exists-a large farming village.
3. Tabari: Vol. 3, p. 98.
4. Literally: "Khalid has made himself commander", i.e., that the role comes naturally to him.
5. Tabari: Vol. 3, p. 98.

At Qinassareen the part of the Roman garrison which had not accompanied Meenas to Hazir shut itself up in the fort. As soon as Khalid arrived, he sent a message to the garrison: *"If you were in the clouds, Allah would raise us to you or lower you to us for battle."* [1] Without further delay Qinassareen surrendered to Khalid. The Battle of Hazir and the surrender of Qinassareen took place about June 637 (Jamadi-ul-Awwal, 16 Hijri). [2]

Abu Ubaidah now joined Khalid at Qinassareen, and the army marched to Aleppo, where a strong garrison under a Roman general named Joachim held the fort. This general, following the same line of thought as the commander of Qinassareen, set out to meet the Muslims in the open and clashed with the Mobile Guard 6 miles south of the city. A bloody engagement took place here, in which the Romans were worsted; and Joachim; now wiser, pulled back in haste and regained the safety of the fort.

Aleppo consisted of a large walled city and a smaller but virtually impregnable fort outside the city atop a hill, a little more than a quarter of a mile across, surrounded by a wide moat. The Muslims moved up and laid siege to the fort. Joachim was a very bold commander and launched several sallies to break the siege, but received heavy punishment every time. After a few days of this, the Romans decided to remain in the fort and await such help as Heraclius might be able to send. Heraclius however, could send none; and four months later, around October 637, the Romans surrendered on terms. The soldiers of the garrison were allowed to depart in peace; but Joachim would not go. He became a Muslim and elected to serve under the banner of Islam. In fact, over the next few weeks, he proved a remarkably able and loyal officer, and fought gallantly under various Muslim generals.

Once Aleppo was taken, Abu Ubaidah sent a column under Malik Ashtar to take Azaz on the route to 'Rome'. The region which the Muslims called Rome included the area which is now Southern Turkey east of the, Taurus Mountains. Malik, assisted by Joachim, captured Azaz and signed a pact with the local inhabitants, whereafter he returned to Aleppo. The capture and clearance of Azaz was essential to ensure that no large Roman forces remained north of Aleppo, whence they could strike at the flank and rear of the Muslims as the next major operation was launched. As soon as Malik rejoined the army, Abu Ubaidah marched westwards to capture Antioch.

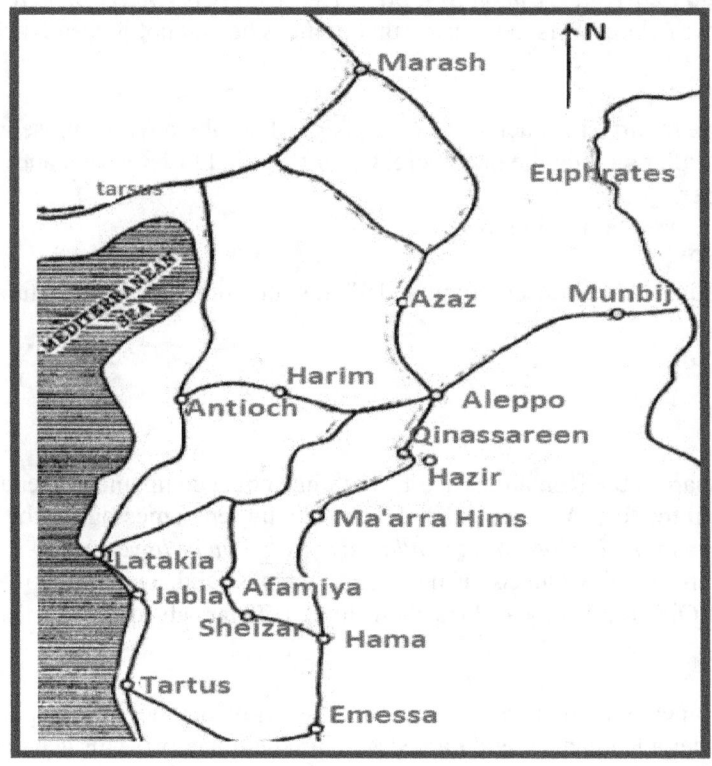

The army moved via Harim and approached Antioch from the east. Some 12 miles from the city, at Mahruba, where a bridge of iron spanned the River Orontes (now known as Nahur-ul-Asi), the Muslims came up against a powerful Roman army-the defenders of Antioch. A major battle was fought here, the details of which are not recorded, and the Romans were soundly thrashed by Abu Ubaidah, Khalid again playing a prominent role with his Mobile Guard. With the exception, of Ajnadein and Yarmuk, the Roman casualties here are believed to have been the highest in the Syrian Campaign, and the remnants of the Roman army went fleeing in disorder to the city. The Muslims moved up and laid siege to Antioch, but not many days had passed before the greatest city of Syria, the capital of the Asian Zone of the Eastern Roman Empire, surrendered to the Muslims. Abu Ubaidah entered Antioch on October 30, 637 (the 5th of Shawwal, 16 Hijri). The defeated Roman soldiers were allowed to depart in peace.

1. Tabari: Vol. 3, p. 98.
2. Qinassareen lay in a South-South-Westerly direction from Aleppo, 20 miles by road and about 18 as the crow flies. It was built on a low ridge which runs astride the present Aleppo-Saraqib road, but most of it was on the Southern slope of the Eastern part of the ridge, *i.e.* on the East side of the road. The ridge is now known as Al Laees, and this is also the name of a small village which stands on that was probably the South-Eastern corner of Qinassareen. The visitor to Qinassareen today imagines that he can see the ruins of the city-ancient ruins such as one sees in many places in Syria. But on closer examination he finds that they are not ruins but immense whitish rocks and caves shaped by nature into semblance of ruins. Actually nothing remains of Qinassareen-not a stone, not a brick.

316

Following the surrender of Antioch, Muslim columns moved south along the Mediterranean coast and captured Latakia, Jabla and Tartus, thus clearing most of North-Western Syria of the enemy. Abu Ubaidah next returned to Aleppo, and during this move his columns subdued what remained of Northern Syria. Khalid took his Mobile Guard on a raid eastwards up to the Euphrates in the vicinity of Munbij, but found little opposition. In early January, 638, he rejoined Abu Ubaidah at Aleppo.

All of Syria was now in Muslim hands. Abu Ubaidah left Khalid as commander and administrator at Qinassareen, and returned with the rest of his army to Emessa, where he assumed his duties as governor of the province of Emessa, of which Qinassareen was then part. From Qinassareen Khalid would keep watch over the northern marches.

By the end of 16 Hijri (corresponding roughly to 637 A.D.) all Syria and Palestine was in Muslim hands, except for Caesarea which continued to hold out. The various Muslim commanders settled down to their duties as governors of provinces: Amr bin Al Aas in Palestine, Sharhabeel in Jordan, Yazeed in Damascus (but currently engaged at Caesarea) and Abu Ubaidah in Emessa. Khalid had a lower appointment as administrator in Qinassareen under Abu Ubaidah. This state of peace continued for a few months until the mid-summer of 638, when clouds again darkened the sky over Northern Syria. This time the Christian Arabs of the Jazeera took to the warpath.

Heraclius was no longer able to attempt a military comeback in Syria. In fact he was now more worried about the rest of his Empire, which, after the destruction of his army at Yarmuk and Antioch, was extremely vulnerable to Muslim invasion. He had few military resources left with which to defend his domains against an army which marched from victory to victory. To gain time for the preparation of his defences it was essential to keep the Muslims occupied in Syria, and he did this by inciting the Arabs of the Jazeera to take the offensive against the Muslims. Bound to him by ties of religion, they submitted to his exhortations; and gathering in tens of thousands, began preparations to cross the Euphrates and invade Northern Syria from the east.

Agents brought Abu Ubaidah information on the preparations being made in the Jazeera. As the hostile Arabs began their move, Abu Ubaidah called a council of war to discuss the situation. Khalid was all for moving out of the cities as one army and fighting the Christian Arabs in the open, but the other generals favoured a defensive battle at Emessa. Abu Ubaidah sided with the majority, and pulled in the Mobile Guard from Qinassareen and other detachments from places which they had occupied in Northern Syria. He concentrated his army as Emessa and at the same time informed Umar of the situation.

Umar had no doubt that Abu Ubaidah and Khalid would hold their own against the irregular army which now threatened them; but he nevertheless decided to assist them, and did so in a most unusual manner. He sent instructions to Sad bin Abi Waqqas, the Muslim Commander-in-Chief in Iraq, to despatch three columns into the Jazeera: one under Suheil bin Adi directed at Raqqa, another under Abdullah bin Utban directed at Nuseibeen and a third under Ayadh bin Ghanam operating between the first two.
At the same time Umar ordered the despatch of 4,000 men under Qaqa bin Amr from Iraq to Emessa, along the Euphrates route, to reinforce Abu Ubaidah.

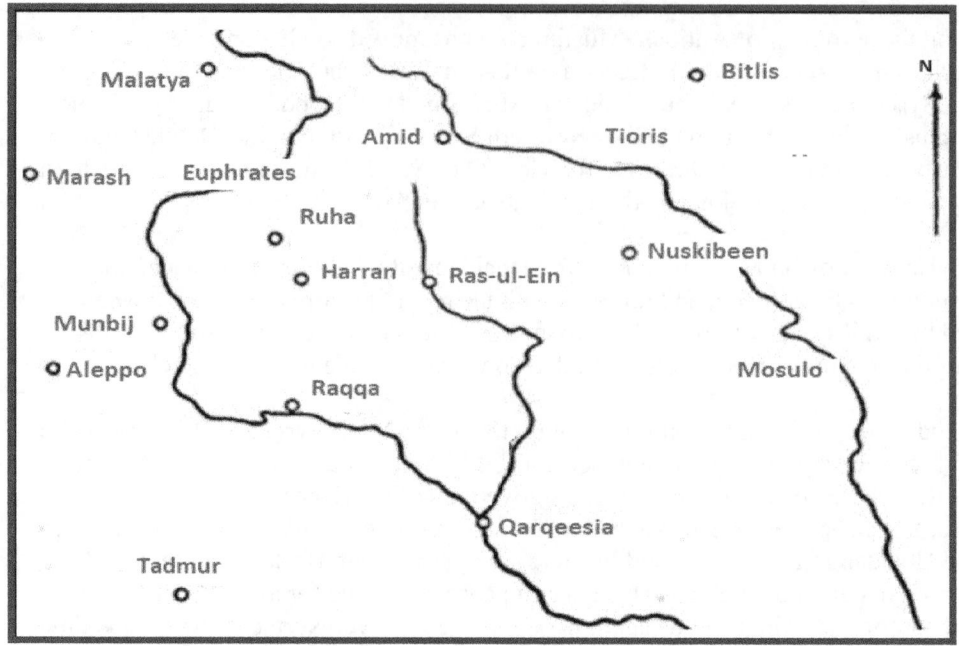

The Christian Arabs arrived at Emessa to find the Muslims safely fortified, and not knowing what else to do, laid siege to the city. But hardly had the siege begun when messengers came galloping from the Jazeera to inform them that three Muslim columns were marching from Iraq towards the Jazeera. The Christian Arabs now realised the absurdity of their situation: while they were fighting the Muslims in Syria, pulling Heraclius' chestnuts out of the fire for him, their own land was about to fall to the Muslims coming from another direction. They abandoned the siege and hastened back to the Jazeera, which was the only sensible thing to do. Qaqa arrived at Emessa three days after the departure of the Christian Arabs.

As soon as the three Muslim columns from Iraq heard of the return of the Christian Arabs, they halted on their route to await further instructions from Sad. Their mission had been accomplished. With this neat, indirect manoeuvre Umar had repulsed the invading army of the Jazeera, without shooting an arrow!

The abortive attempt of the Arabs of the Jazeera to fight the Muslims did no damage to the Muslims in Syria. It did, however, arouse the anger of the Muslims and made them conscious of the fact that they could not regard Syria as being safely in their possession until neighbouring lands were cleared of all hostile elements. These elements existed in the Jazeera and in the region east of the Taurus Mountains; and they would have to be destroyed or subdued in order to create a zone of security beyond the borders of Syria.

Umar decided to deal with the Jazeera first. He ordered Sad to arrange for its capture, and appointed Ayadh bin Ghanam as the commander of this theatre of operations. Sad instructed Ayadh to continue the invasion of the Jazeera with the forces under his command, and the Muslims from Iraq resumed their forward march late in the summer of 638. Ayadh operated with three columns, and over a period of a few weeks overran the region between the Tigris and the Euphrates up to Nuseibeen and Ruha (now Urfa). It was a bloodless operation.

As soon as this part of the Jazeera was occupied, Abu Ubaidah wrote to Umar and asked for Ayadh to be put under his command, so that he could use him for raids across the northern border. Umar agreed to this request, and Ayadh moved to Emessa with part of the Muslim force sent from Iraq to the Jazeera.

In the autumn of 638, Abu Ubaidah launched several columns, including two commanded by Khalid and Ayadh, to raid Roman territory north of Syria up to as far west as Tarsus. Khalid's objective was Marash, and he arrived here and laid siege to the city which contained a Roman garrison. By now the presence of Khalid was sufficient to strike terror in the hearts of the Romans; and a few days later Marash surrendered on condition that the garrison and the populace be spared. As for material wealth, the Muslim could take all they wished. And the Muslims did. Khalid returned to Qinassareen laden with spoils such as had seldom been seen before. Just the spoils of Marash were sufficient to make the soldiers of this expedition rich for life.

Had Khalid acquired the quality of thrift in his youth, he would have been one of the richest men of his time. It was the custom in those days that a warrior who won a duel took all the possessions of his vanquished foe, and this reward was apart from his normal share of the spoils taken in battle. Khalid had fought more duels than anyone else in the Muslim army and won each one of them. Moreover, his adversaries were usually generals, more richly equipped than others, especially the Persian and Roman generals who wore jewels and gold ornaments with their dress. Thus more wealth came into the hands of Khalid than of others; but it slipped through his fingers like sand. He would live well and give generously. Whatever wealth was gained in one battle lasted only till the next. Khalid had acquired a large retinue of slaves. He had married many times and had dozens of children; and the upkeep of his household took a good deal of money. Then there were the soldiers. After every battle Khalid would pick out warriors who had done better than others and give them extra gifts from his own pocket.

1. Some narrators have described a considerable amount of fighting in the Jazeera; but most early historians agree that it was a peaceful occupation.

This was known to the austere and frugal Caliph, who regarded it not as generosity but as extravagance. 1

On Khalid's return from Marash the same thing happened; he gave lavishly to his soldiers. And by now a number of unscrupulous persons had arisen in the Muslim army who would approach successful generals, sing their praises and receive gifts- in true Oriental fashion. One such man was Ashath bin Qais, chief of the Kinda, who has been mentioned in Part II of this book. (He had led the apostate revolt of his tribe in the Yemen, and saved himself at the last minute by betraying his own followers!) Ash'as was a great poet. He came to Khalid at Qinassareen and recited a fine poem in praise of the great conqueror; and in return Khalid gave him a gift of 10,000 dirhams. Within a fortnight the agents of the Caliph had informed him of this episode; and Umar was furious. This, thought Umar, was the limit!

Ash'as did not know that when he recited his eloquent poetry, he was in fact digging the grave of Khalid's military career.

1. Khalid's earnings from his duels and from his share of the spoils of war were not part of his pay, the scale of which for a corps commander was between 7,000 and 9,000 dirhams a year (Abu Yusuf: p. 46).

Chapter 37

(Farewell to Arms)

*"You were better than a million people,
When the faces of men were downcast.
Brave? You were braver than the tiger,
Damr bin Jahm, father of Ashbal.
Generous? You were more generous than
The unstoppable deluge flowing between mountains."*
[Lubabah the Younger, mother of Khalid, eulogising him]

"Have women ever stopped mourning for anyone like Khalid?"
[Umar bin Al-Khattab]

"Women will no longer be able to give birth to the likes of Khalid bin Al-Waleed."
[Abu Bakr]1

Some time before his expedition to Marash, Khalid had a special bath. Just as he did everything well, Khalid also bathed well. He had with him a certain substance prepared with an alcoholic mixture which was supposed to have a soothing effect on the body when applied externally. Khalid rubbed his body with this substance and thoroughly enjoyed his bath, from which he emerged glowing and refreshed.

A few weeks later he received a letter from the Caliph: *"It has come to my notice that you have rubbed your body with alcohol. Lo, Allah had made unlawful the substance of alcohol as well as its form, just as He has made unlawful both the form and substance of sin. He has made unlawful the touch of alcohol in a bath no less than the drinking of it. Let it not touch your body, for it is unclean."* 2

This, pondered Khalid, was carrying the Muslim ban on alcohol a bit too far. Like all Companions, Khalid was thoroughly conversant with the Holy Book and knew that the Quranic verses on alcohol dealt only with the drinking of it, and that the injunction against strong drink was intended to eliminate the evils of drunkenness and alcoholism. The Quran said nothing about the external application of oils and ointments treated with alcohol. Khalid wrote back to Umar and explained the method of preparation of the offending substance with the alcoholic mixture and the cleaning of it by boiling. He added: *"We kill it so that it becomes like bathwater, without alcohol."* 3 In this matter of the interpretation of the Quranic verses on alcohol Umar was not on a strong wicket. So he contented himself with writing to Khalid: *"I fear that the house of Mugheerah 4is full of wrong-doing. May Allah not destroy you on account of it!"* 5 And there the matter rested. We do not know whether Khalid ever again had such a bath; probably not. But it is clear that the goodwill which Khalid had gained in the eyes of Umar as a result of the Battle of Hazir was washed away by Khalid's rejection of Umar's opinions regarding the external application of substances treated with alcohol.

Shortly after Khalid's capture of Marash, in the autumn of 638 (17 Hijri), Umar came to know of Ash'as reciting a poem in praise of Khalid and receiving a gift of 10,000 dirhams. This was more than the Caliph could take. This, thought Umar, was the limit! He immediately wrote a letter to Abu Ubaidah: *"Bring Khalid in front of the congregation, tie his hands with his turban and take off his cap. Ask him from what funds he gave to Ash'as. . . .from his own pocket or from the spoils acquired in the expedition? If he confesses to having given from the spoils, he is guilty of misappropriation. If he claims that he gave from his own pocket, he is guilty of extravagance. In either case dismiss him, and take charge of his duties."* 6

This was no ordinary letter. Though the method described by Umar for arraigning the accused was the normal custom of the Arabs, the accused in this case was no ordinary accused. The instructions of the Caliph would have to be carried by a Companion of high standing, and Umar selected Bilal the Muazzin for the task. He entrusted the letter to Bilal, briefed him about how he was to proceed in the matter of Khalid, and ordered him to journey with all speed to Emessa.

Bilal arrived at Emessa and handed the letter to Abu Ubaidah, who read it and was aghast. He could hardly believe that this was to be done to the Sword of Allah; but the Caliph's orders had to be obeyed, and Abu Ubaidah sent for Khalid.

Khalid left Qinassareen without the least suspicion of what lay in store for him. He imagined that he was being called for another council of war, that perhaps there was to be another expedition to 'Rome' or even a full scale invasion of the Byzantine Empire. He looked forward eagerly to more battles and more glory. Arriving at Emessa, he went to the house of Abu Ubaidah, and here for the first time he came to know the purpose of Abu Ubaidah's call. The Commander-in-Chief briefly explained Umar's charge against him, and asked if he would confess his guilt. Khalid was astounded by Abu Ubaidah's statement, and saw it not as a simple matter of a question or a charge, but as an attempt on the part of his old rival, Umar bin Al Khattab, to bring about his undoing. He asked Abu Ubaidah for a little time to consult his sister, and Abu Ubaidah agreed to wait.

1. Ibn Kathir, Al-Bidayah wan-Nihayah, Dar Abi Hayyan, Cairo, 1st ed. 1416/1996, Vol. 7 P. 141.
2. Tabari: Vol. 3, p. 166.
3. *Ibid*.
4. Mugheerah was the grandfather of Khalid.
5. Tabari: Vol. 3, p. 166.
6. *Ibid*: Vol. 3, p. 167.

Khalid went to the house of his sister, Fatimah, who was then in Emessa, explained the position to her and sought her advice. Fatimah confirmed his suspicions. *"By Allah,"* she said, *"Umar will never be pleased with you. He wants nothing more than that you should confess to some error, so that he can dismiss you."* 1

"You are right", said Khalid. He kissed his sister on the head, and returning to Abu Ubaidah, informed him that he would not confess. Thereafter the two generals walked in

silence to a place where a large number of Muslims were gathered, most of whom rushed to shake Khalid's hand. At one end stood a raised platform, and on this Abu Ubaidah and Khalid sat down, facing the congregation. On one side of Abu Ubaidah sat Bilal the Negro.

For a few minutes there was complete silence. The Muslims had no idea of the purpose of the congregation; nor had Khalid. He did not connect Umar's charge against him with this gathering, for it never occurred to him that he would face a public trial. Bilal looked questioningly at Abu Ubaidah, but Abu Ubaidah turned his face away. He had obeyed the Caliph's instructions as far as he considered necessary. If a man like Khalid, who had rendered military services to the new Muslim State as no other general had done, was to be subjected to public humiliation, he, Abu Ubaidah, would have nothing to do with it. Bilal could do as he wished.

Bilal understood Abu Ubaidah's reluctance. He stood up, faced Khalid, and in a voice which could be heard by the entire congregation, called out: *"O Khalid! Did you give Ash'as 10,000 dirhams from your own pocket or from the spoils?"*

Khalid stared at Bilal in shocked silence. He could hardly believe his ears!

Bilal repeated his question; but Khalid, for once in his life, was left dumbfounded. When another minute had passed with no reply from Khalid, Bilal walked up to him, and with the words, *"The Commander of the Faithful has ordered this"*, took off Khalid's turban and cap and with the turban tied Khalid's hands behind his back. Again the Caliph's messenger spoke: *"What do you say? From your pocket or from the spoils?"*

Only now did Khalid find his speech. *"No!"* he protested. *"From my own pocket."*

When he heard these words, Bilal untied Khalid's hands, replaced Khalid's cap, and with his own hands tied Khalid's turban on his head. He said, *"We hear and obey our rulers. We honour and serve them."* [2] Then he returned to his place and sat down.

For a few minutes pin drop silence reigned in the assembly. Abu Ubaidah and Bilal sat staring at the floor. Then Khalid stood up, still shaken by what had happened. He did not know the result of the trial, whether he was dismissed or still in command of his corps. Not wishing to embarrass the gentle Abu Ubaidah with questions, he walked away from the assembly, mounted his horse and rode to Qinassareen. [3]

Bilal returned to Madinah and gave Umar an account of the proceedings. The Caliph now awaited a letter from Abu Ubaidah confirming that he had dismissed Khalid from command at Qinassareen; but when another week had passed and no such letter arrived, Umar guessed that Abu Ubaidah was reluctant to inform Khalid of his dismissal. The Caliph decided to deal with the matter of Khalid's dismissal himself, and wrote to Khalid to report to him at Madinah.

On receiving Umar's letter, Khalid came to Emessa and questioned Abu Ubaidah about his position. The Commander-in-Chief told him that he was dismissed from office by order of the Caliph. *"May Allah have mercy upon you"* said Khalid. *"Why did you do this to me? You concealed a matter from me which I would have liked to know before today."*

1. Ibid: Vol. 2, p. 624. Yaqubi: *Tareekh*, Vol. 2, p. 140.
2. Tabari: Vol. 3, p. 167.
3. According to one version, which appears to be mistaken, Umar ordered Abu Ubaidah to confiscate half of whatever Khalid possesses; and Abu Ubaidah carried out the Caliph's instructions with such meticulous accuracy that a pair of shoes that Khalid wore, one shoe was taken away and the other left with him! (Tabari: Vol. 2, p. 625; Yaqubi: *Tareekh*, Vol. 2, p. 140).

There was sorrow in the eyes of Abu Ubaidah, and a great deal of affection and commiseration, as he replied, *"By Allah, I knew that this would hurt you. I would never hurt you if I could find a way."* 1

Khalid went back to Qinassareen, got the Mobile Guard together, and addressed the warriors whom he had led to victory and glory in battle after battle-warriors who had followed him with unquestioning loyalty and faith. He informed them that he had been dismissed from command, and that he was now proceeding to Madinah on the instructions of the Caliph. Then he bade farewell to the Mobile Guard-a body of men which under Khalid had not known the meaning of defeat.

From Qinassareen he rode again to Emessa, said his farewells, and then continued his journey to Madinah. He was going to Madinah not as a hero returning home from the wars to receive honours from a grateful government, but as a man under disgrace.

Khalid arrived at Madinah and proceeded towards the house of the Caliph. But he met Umar in the street. As these two strong men drew closer to each other-the greatest ruler of the time and the greatest soldier of the time-there was no fear in the eyes of either. Umar was the first to speak. He extemporised a verse in acknowledgement of Khalid's achievements and recited it:

> *You have done;*
> *And no man has done as you have done.*
> *But it is not people who do;*
> *It is Allah who does.* 2

In reply Khalid said, *"I protest to the Muslims against what you have done. By Allah, you have been unjust to me, O Umar!"*

"Whence comes all this wealth?" countered Umar.

"It is what is left of my share of the spoils. Whatever exceeds 60,000 dirhams is yours." 3

Umar had a check made of all Khalid's possessions, which consisted mainly of military equipment and slaves, and found that it was valued at 80,000 dirhams. He confiscated the surplus of 20,000 dirhams.

When this had been done, Umar said to Khalid, *"O Khalid! By Allah, you are honourable in my eyes, and you are dear to me. You will not have cause to complain of me after this day."* 4 The point was academic, however, for there was not much more that could be done to Khalid!

After a few days, Khalid left Madinah for Qinassareen, never to return to Arabia. Hardly had he left, when the people of Madinah came to Umar and appealed to him to return Khalid's property to him. To this Umar replied, *"I do not trade with what belongs to Allah and the Muslims."* 5 But after this, according to Tabari, Umar's heart was 'cured' of Khalid.

Very soon it became evident to Umar that his treatment of Khalid was being deeply resented by the Muslims. It was openly said that Khalid had suffered because of Umar's personal hostility towards him. This popular disapproval of Umar's action became so widespread that the Caliph found it necessary to write to all his commanders and administrators:

I have not dismissed Khalid because of my anger or because of any dishonesty on his part, but because people glorified him and were misled. I feared that people would rely on him. 6 *I want them to know that it is Allah who does all things; and there should be no mischief in the land.* 7

In this letter Umar, unwittingly paid, Khalid the highest compliment that any general could hope to earn: that his men regarded him as a god! But Khalid returned to Qinassareen an embittered man. The Destroyer of the Apostasy, the Conqueror of Iraq and Syria, came home as a nobody-dismissed and disgraced. As his wife greeted him at the door, he said: *"Umar appointed me over Syria until it turned to wheat and honey; then dismissed me!"* 8

1. Tabari: Vol. 3, p. 167.
2. *Ibid*: Vol. 3, p. 168.
3. Tabari: Vol. 3, p. 167.
4. *Ibid*.
5. Tabari: Vol. 2, p. 625.
6. *i.e.* rather than Allah, for victory.
7. Tabari: Vol. 3, p. 167.
8. Tabari: Vol. 3, p. 99.

Khalid's campaigning days were over. The Sword of Allah-the sword that Allah had drawn against the infidels-which Abu Bakr had refused to sheathe, was at last sheathed by Caliph Umar.

Little remains to be told. After his dismissal Khalid had less than four years to live, and these were not very pleasant years. Financially, though not impoverished, he was severely restricted. In 15 Hijri, Umar had started the institution of allowances to all Muslims, varying in extent according to their position in Islam and the services rendered by them in war. All those who had accepted the Faith after the Truce of Hudebiya and before the Apostasy received an annual allowance of three thousand dirhams, 1 and this category included Khalid. The sum was enough to enable a man and his family to live modestly; but with Khalid, born an aristocrat and accustomed to giving away thousands of dirhams on an impulse, it did not go far. He took his family to Emessa, where he bought a house and settled down to retirement.

His dismissal was a terrible blow to him. But as if this were not enough, Khalid suffered even more grievous losses in the plague which struck soon after his return from Madinah, and which claimed most of those nearest and dearest to him.

The plague started at Amawas in Palestine in Muharram or Safar, 18 Hijri (January or February 639), and spread rapidly across Syria and Palestine, striking down Christians and Muslims in its path. The Caliph was deeply grieved by the sufferings of the Muslims in Syria and concerned especially about Abu Ubaidah, and thought to save the Trusted One of the Nation by asking him to visit Madinah. But Abu Ubaidah saw through Umar's letter and knew that the Caliph would detain him in Madinah until the epidemic had spent itself. The man who had not abandoned his soldiers in the thick of battle was not going to abandon them in the plague. He refused to visit Madinah, and for his loyalty to his men paid with his life.

Thousands of Muslims died in the Plague of Amawas, and these included the noblest and best: Abu Ubaidah, Sharhabeel, Yazeed, Dhiraar-Khalid's dearest friends. And yet this was not the end of his sufferings, for he lost 40 sons in the epidemic! The terrible pestilence thus took away most of those whom Khalid loved, those who could have added comfort and cheer to his years of retirement. We only know of three sons who survived Khalid: Sulaiman, who fell in battle in the latter part of the Egyptian Campaign; Muhajir, who fought and died under Ali at Siffeen; and Abdur-Rahman, who survived to live to a mature age and appeared to be endowed with his father's military prowess. But he too met an untimely death at the hands of a poisoner in 46 Hijri, during the caliphate of Muawiyah. It is recorded that the assassination was engineered by Muawiyah, who was jealous and fearful of the great prestige of the son of the Sword of Allah. [2] The assassin was later killed, as an act of vengeance by Abdur-Rahman's son. We do not know how many daughters Khalid had, but the male line of descent from Khalid is believed to have ended with his grandson, Khalid bin Abdur-Rahman bin Khalid.

After the death of three of the original corps commanders, Amr bin Al Aas took command of the army and immediately dispersed it in the hills of Syria and Palestine. By so doing he was able to save much of the army, but not before 25,000 Muslims had fallen before the foul breath of the plague. The epidemic had not yet ended when Umar appointed Ayadh bin Ghanam as military governor of Northern Syria, and Muawiyah of Damascus and Jordan, while Amr remained in command in Palestine.

When Abu Bakr was planning the Campaign of the Apostasy, he discussed with Amr bin Al Aas the appointment of various generals as corps commanders. The Caliph said, *"O Amr, you are the shrewdest of the Arabs in judgement. What is your opinion of Khalid?"* Amr replied, *"He is a master of war; a friend of death. He has the dash of a lion and the patience of a cat!"* [3]

But the patience of a cat was not enough for a man of Khalid's temperament at this stage of his life. What makes patience possible and bearable in a cat is the prospect of a victim for supper. If there were no victim in sight even a cat could not bear to be patient; and Khalid now had no prospects, nothing to be patient for. He could fight no more battles, kill no more enemies. In enforced obscurity Khalid mourned the loss of his comrades and his sons.

1. Tabari: Vol. 3, p. 109. Balazuri: p. 437.
2. Tabari: Vol. 4, p. 171; Isfahani: Vol. 15, pp. 12-13.
3. Yaqubi: Tareekh, Vol. 2, p. 129.

The conquests of Islam continued. After the plague, in 18 Hijri, Ayadh again invaded the Jazeera; and by the end of the following year had completed its subjugation, after several battles, as far north as Samsat, Amid (now Diyar Bakr) and Bitlis. He even raided successfully as far as Malatya. News from the eastern front was just as thrilling. By the time of Khalid's dismissal, Sad bin Abi Waqqas had conquered most of what is now Iraq and parts of present-day South-Western Persia-Ahwaz, Tustar, Sus. On this front further advances were made, though the last great battles against the still formidable Persians were not fought till after Khalid's death. In 640 (19 Hijri) Caesarea surrendered to the Muslims and Amr bin Al Aas invaded Egypt.

Like all Muslims, Khalid gloried in the conquests of Islam; but each victory also reminded him that he had not taken part in the battle. The news that reached him at Emessa was, to him, bitter-sweet. He was like an ardent lover who sees his beloved before him but is unable to move towards her. Thus lived, for the last few years of his life, the man whom Gibbon, in his *Decline and Fall of the Roman Empire*, has described as "....*the fiercest and most successful of the Arabian warriors.*" [1]

Fortunately, in Khalid's relations with Umar there was a marked change for the better. Umar was no longer the harsh, impetuous, hot-tempered man that he had once been. With the burdens of the caliphate on his shoulders, he had mellowed and grown more patient. He was still stern and puritanical, but he imposed no burden upon others which he did not carry himself. He was strict with the strong, kind to the weak, generous to widows and orphans. He sat with the poor and often spent the night sleeping on the steps of the mosque. At night he would walk the streets of Medina with a whip in his hand, and Umar's whip was feared more than the sword of another man. He lived on salted barley bread, dry dates and olive oil, and allowed no better fare to his family. His clothes were made of the poorest material, patched in many places. Unshakeable in his resolve to do justice, he had his own son, Ubaidullah, whipped for drinking.

Khalid, now having more time for reflection, saw the great virtues and enviable qualities of his old rival. He forgave him. One day he said to a visitor, *"Praise be to Allah who took Abu Bakr away. He was dearer to me than Umar. Praise be to Allah who appointed Umar in authority He was hateful to me, but I grew to like him."* [2] This change in attitude was so great that when he died, Khalid named Umar as his heir, to receive whatever he left. Time, mercifully, healed the wounds.

Khalid spent a good deal of his time thinking of his battles, as old soldiers are wont, to do. He would relieve the battles and duels in which he had challenged the greatest champions of the world and made them bite the dust. He was naturally proud of his victories, but there was no vanity or conceit in Khalid's mind. He attributed his victories to the help of Allah and to his red cap, in which was woven the hair of the Holy Prophet. When not thinking of his battles, his mind would be occupied by memories of his fellow generals-Abu Ubaidah, Sharhabeel, Yazeed, Amr bin Al Aas; and his valiant champions like Abdur-Rahman bin Abi Bakr, Raafe bin Umairah and the incomparable Dhiraar bin Al Azwar, whose feats of skill and daring, like his own, would glow for ever in the pages of history. He did not, however, know his place in history as we do now.

Khalid was the most versatile soldier history has ever known-a true military genius. He had the strategical vision of a Changez Khan and a Napoleon, the tactical brilliance of a Timur and a Frederick the Great, and the individual strength and prowess of the half-legendary Rustam of Persia. In no other case in history do we see such diverse military virtues combined in one man. Khalid was one of only two great generals in history who never suffered a defeat. The other was Changez Khan, but Changez Khan was not a champion fighter like Khalid, even though his conquests covered a far greater region of the earth. Combined with Khalid's strategical and tactical genius was the extreme violence of his methods. To him a battle was not just a neat manoeuvre leading to a military victory, but an action of total violence ending in the total annihilation of the enemy. The manoeuvre was only an instrument for bringing about the enemy's destruction.

---

1. While some sources have stated that Khalid fought under Ayadh in the Jazeera, most early historians have quoted other sources to indicate that after death of Abu Ubaidah, Khalid did not serve under anyone. I accept the latter version as correct.
2. Tabari: Vol. 2, p. 598.

Khalid was the only man who inflicted a tactical defeat on the Holy Prophet-at Uhud. He was the first Muslim commander to leave Arabia and conquer foreign lands; the first Muslim to humble two great empires, one after the other. Almost all his battles are studies in military leadership, especially Uhud, Kazima, Walaja, Muzayyah, Ajnadein and Yarmuk. His finest battle was Walaja, while his greatest was undoubtedly Yarmuk.

Khalid was essentially a soldier. He also administered the territories which he conquered, but this he did as a routine responsibility of a high-ranking general, who had not only to conquer territory but also to rule it as a military governor. His plans and manoeuvres show a superb military intellect; but towards such things as learning and culture he was in no way inclined. Khalid was pure, unadulterated, undiluted, unspoilt soldier. It was his destiny to fight great battles and vanquish mighty foes.... to attack, kill, conquer. This destiny became apparent only when, with the rise of Islam, the prospect of holy war arose in Arab lands. And it was only after he had accepted the new faith and submitted to the Prophet that this destiny came into full play. Wherever Khalid marched, enemies stood up to oppose him, as if some unkind fate had condemned them to death by his sword. Wherever Khalid passed, he left behind a trail of glory. From the Battle of Uhud up to the time of his dismissal, over a period of 15 years, Khalid fought 41 battles (excluding minor engagements), of which 35 were concentrated in the last seven years. And he never lost a single one! Such was Khalid, the irresistible, all-conquering master.

It is interesting to speculate what would have happened if he had remained in command of the Muslim army in Syria and had been launched to conquer the Byzantine Empire, Since Khalid never lost a battle, there is no doubt that he would have taken the whole of Asia Minor and reached the Black Sea and the Bosphorus. But it was not to be. By the end of 17 Hijri Khalid's race was run. Thereafter the stage of history was crowded by other players.

In 641, Ayadh bin Ghanam died. In this year, too, died Bilal the Muazzin and Khalid's defeated foe, Heraclius, Emperor of Rome. The following year it was Khalid's turn to go.

Some time in 642 (21 Hijri), at the age of 58, Khalid was taken ill. We do not know the nature of his illness, but it was a prolonged one and took the strength out of him. As with all vigorous, active men upon whom an inactive retirement is suddenly thrust, Khalid's health and physique had declined rapidly. This last illness proved too much for him; and Khalid's sick bed became his death bed. He lay in bed, impatient and rebellious against a fate which had robbed him of a glorious, violent death in battle. Knowing that he had not long to live, it irked him to await death in bed.

A few days before his end, an old friend called to see him and sat at his bedside. Khalid raised the cover from his right leg and said to his visitor, *"Do you see a space of the span of a hand on my leg which is not covered by some scar of the wound of a sword or an arrow or a lance?"*

The friend examined Khalid's leg and confessed that he did not. Khalid raised the cover from his left leg and repeated his question. Again the friend agreed that between the wounds farthest apart the space was less than a hand's span.

Khalid raised his right arm and then his left, for a similar examination and with a similar result. Next he bared his great chest, now devoid of most of its mighty sinews, and here again the friend was met with a sight which made him wonder how a man wounded in so many places could survive The friend again admitted that he could not see the space of one hand span of unmarked skin.

Khalid had made his point. *"Do you not see?"* he asked impatiently. *"I have sought martyrdom in a hundred battles. Why could I not have died in battle?"*

"You could not die in battle", replied the friend.

"Why not?"

"You must understand, O Khalid," the friend explained, *"that when the Messenger of Allah, on whom be the blessings of Allah and peace, named you Sword of Allah, he predetermined that you would not fall in battle. If you had been killed by an unbeliever it would have meant that Allah's sword had been broken by an enemy of Allah; and that could never be."*

Khalid remained silent, and a few minutes later the friend took his leave. Khalid's head could see the logic of what his visitor had said, but his heart still yearned for a glorious death in combat. Why, oh why could he not have died a martyr in the way of Allah!

On the day of his death, Khalid's possessions consisted of nothing more than his armour and weapons, his horse and one slave-the faithful Hamam. On his last day of life he lay alone in bed with Hamam sitting in patient sorrow beside his illustrious master. As the shadows gathered, Khalid put all the torment of his soul into one last, anguished sentence: *"I die even as a camel dies. I die in bed, in shame. The eyes of cowards do not close even in sleep."* [1]

Thus died Khalid, son of Al Waleed, the Sword of Allah. May Allah be pleased with him!

The news of Khalid's death broke like a storm over Madinah. The women took to the streets, led by the women of the Bani Makhzum, wailing and beating their breasts. Umar had heard the sad news and now heard the sounds of wailing. He was deeply angered. On his very first day as Caliph, he had given orders that here would be no wailing for departed Muslims. And there was logic in Umar's point of view. Why should we weep for those who have gone to paradise? the blissful abode promised by Allah to the Faithful! Umar had enforced the order, at times using his whip. 2

Umar now heard sounds of wailing. He stood up from the floor of his room, took his whip and made for the door. He would not permit disobedience of his orders; the wailing must be stopped at once! He got to the door, but there he paused. For a few silent moments the Caliph stood in the doorway, lost in thought. This was, after all, no ordinary death; this was the passing away of *Khalid bin Al Waleed*. Then he heard the sounds of mourning from the next house-his own daughter, Hafsa, widow of the Holy Prophet, was weeping for the departed warrior. 3

Umar turned back. He hung up his whip and sat down again. In this one case he would make an exception. *"Let the women of the Bani Makhzum say what they will about Abu Sulaiman, for they do not lie"*, said the Caliph. *"Over the likes of Abu Sulaiman weep those who weep."* 4

In Emessa, to the right of the Hama Road, stretches a large, well-tended garden which has lawns studded with ornamental trees and flower beds and is traversed by footpaths. At the top end of the garden stands the Mosque of Khalid bin Al Waleed. It is an imposing mosque, with two tall minarets rising from its north-western and north-eastern corners. The inside of the mosque is spacious, about 50 yards square, its floor covered with carpets and the ceiling upheld by four massive columns. Each of the four corners of the ceiling is formed as a dome, but the highest dome is in the centre, at a considerable height, and from this dome several chandeliers are suspended by long metal chains. In the north-west corner of the mosque stands Khalid's shrine-the last resting place of Abu Sulaiman.

The visitor walks up the garden, crosses the courtyard of the mosque, takes off his shoes and enters the portals. As he enters, he sees to his right the shrine of Khalid. The actual grave is enveloped by an attractive domed marble structure which gives the impression of a little mosque within the larger one. The visitor, if so inclined, says a prayer and then loses himself in contemplation of the only man who ever carried the title of the Sword of Allah.

And if the visitor knows something about Khalid and his military achievements, he lets his imagination wander and pictures of an attack by Khalid flicker through his mind. He sees a long, dark line of horsemen emerge from behind a rise in the ground and charge galloping at a body of Roman troops. The cloaks of the warriors fly behind them and the hooves of their horses pound the earth pitilessly. Some carry lances; others brandish swords; and the Romans standing in the path of the charge tremble at the sight of the oncoming terror, for they are standing in the way of the Mobile Guard, whom none may resist and survive to tell the tale. The line of charging horsemen is not straight, for it is impossible to keep it straight at such a mad, reckless pace. Every man strives to get ahead of his comrades and be the first to clash with the infidel; strives to get ahead of all but the Leader, for no one may, or possibly could, overtake the Leader.

1. Ibn Qutaibah: p. 267.
2. Tabari: Vol. 2, p. 614.
3. Yaqubi: Tareekh: Vol. 2, p. 157.
4. Isfahani: Vol. 19, p. 89.

The Leader gallops ahead of the Muslims. A large, broad-shouldered, powerfully-built man, he is mounted on a magnificent Arab stallion and rides it as if he were part of the horse. The loose end of his turban and his cloak flutter behind him and his large, full beard is pressed against his chest by the wind. His fierce eyes shine with excitement-with the promise of battle and blood and glory- the glory of victory or martyrdom. His coat of mail and the iron tip of his long lance glint in the clear sunlight, and the earth trembles under the thundering hooves of his fiery charger. Perhaps beside him rides a slim young warrior, naked above the waist.

The visitor sees all this with the eyes of his mind. And with the ears of his mind he hears, just before the Mobile Guard hurls itself at the Romans in a shattering clash of steel and sinew, the roar of *Allah-o-Akbar* as it issues from the throats of the Faithful and rends the air. And rising out of this roar, he hears the piercing cry of the Leader:

I am the noble warrior;
I am the Sword of Allah
Khalid bin Al Waleed!

Appendix A : Bibliography

1. Ibn Hisham: *Seerat-un-Nabawi*, Cairo, 1955.
2. Waqidi: *Maghazi Rasulillah*: Cairo, 1948; *Futuh-ush-Sham*, Cairo, 1954.
3. Ibn Sad: *Tabaqat-ul-Kubara*, Cairo, 1939.
4. Ibn Qutaibah: *Al Ma'arif*, Cairo, 1960.
5. Al-Yaqubi: *Tareekh-ul-Yaqubi*, Beirut, 1960;
Al Buldan, *Lieden*, 1892.
6. Al-Baladhuri: *Futuh-ul-Buldan*, Cairo, 1959.
7. Dinawari: *Akhbar-ul-Tiwal*, Cairo, 1960.
8. At-Tabari: *Tareekh-ul-Umam wal Muluk*, Cairo, 1939.
9. Al-Masudi: *Muruj-uz-Dhahab*, Cairo, 1958; *Al Tanbeeh wal Ashraf*, Cairo, 1938.
10. Ibn Rusta: *A'laq-un-Nafeesa*, Lieden, 1892.
11. Isfahani: *Al Aghani*, Cairo, 1905.
12. Yaqut: *Mu'jam-ul-Buldan*, Teheran, 1965.
13. Abu Yusuf: *Kitab-ul-Kharaj*, Cairo, 1962.
14. Edward Gibbon: *Decline and Fall of the Roman Empire*, London, 1954.
15. Alois Musil: *The Middel Euphrates*; New York, 1927.

Ibn Hisham. His abridgement of the last pioneering work, Seerah Rasoolullah, by Muhammad bin Ishaq, is invaluable. Portions of Ibn Ishaq have recently been recovered and published. Muhammad bin Ishaq (who died in 150 or 151AH), is unquestionably the principal authority on the Seerah (Prophetic biography) and Maghazi (battles) literature. Every writing after him has depended on his work, which though lost in its entirety, has been immortalised in the wonderful, extant abridgement of it, by Ibn Hisham. Ibn Ishaq was one of the Tabieen (second generation who saw the Sahabah but not the Prophet SAWS himself) of humble beginnings as a former slave. Ibn Ishaq's work is notable for its excellent, rigorous methodology and its literary style is of the highest standard of elegance and beauty. This is hardly surprising when we recall that Ibn Ishaq was an accomplished scholar not only in Arabic language but also in the science of hadith. For this reason, most of the isnad (chains of narration) that he gives in his Seerah are also to be found in the authentic books of hadith. Ibn Ishaq, like Bukhari and Muslim, travelled very widely in the Muslim world in order to authenticate the isnad of his hadith. It is reported that Ibn Ishaq saw and heard Saeed bin Al-Musayyib, Aban bin Uthman bin Affan, Az-Zuhri, Abu Salamah bin Abdur-Rahman bin Awf and Abdur-Rahman bin Hurmuz Al-Araj. It is also reported that Ibn Ishaq was the teacher of the following outstanding authorities among others:

(a) Yahya bin Saeed Al-Ansari
(b) Sufyan Ath-Thawri
(c) Ibn Jurayh
(d) Shu'bah bin Al-Hajjaj
(e) Sufyan bin Uyainah
(f) Hammad bin Zaid

Al-Waqidi. The second most authoritative book on Seerah is that of Al-Maghazi by Muhammad bin Umar Al-Waqidi Al-Aslami (who lived from 130 to 207AH and is buried in Baghdad). This book was widely read in various parts of the Muslim world. While som

Ibn Sad. The third authoritative work on Seerah is Ibn Sad's *Tabaqat-ul-Kubara* (nine volumes). Ibn Sad was both the student and the scribe/secretary of Al-Waqidi. The quality and scholarly excellence of the *Tabaqat-ul-Kubara* of Ibn Sad say a great deal about the academic competence of his teacher and patron.

Al-Yaqubi. (Ahmad bin Jafar bin Wahb, died 292AH). Al-Yaqubi's work is unique for its examples of the Prophet's (SAWS) sermons, not to be found elsewhere, especially those containing instruction and admonition.

Al-Baladhuri. (Ahmad bin Yahya bin Jabir, died in 279AH). The work of this early historian is valuable for the texts it contains of certain important agreements which the Prophet (SAWS) concluded with some groups and individuals- among others, the texts of his agreements with the Christians of Najran, his agreement with the people of Maqna, his book to Al-Mundhir bin Sawi and to Akaydar Dawmah.

At-Tabari. (Ibn Jareer, died in 310AH) in his monumental world history *Tareekh-ul-Umam wal Muluk.* At-Tabari was not merely a historian, but also an unrivalled authority on the Arabic language and grammar, on hadith and fiqh, and on the tafseer (exegesis) and interpretation of the Quran. Evidence of the excellence of his scholarship, his prodigious and untiring intellectual genius, is provided by his major works which run into many lengthy volumes each.

Al-Masudi. (Abul-Hasan Ali bin Al-Husain bin Ali Al-Masudi, died in 346AH). A well-known Arab historian, descendent of one of the Companions of the Prophet (SAWS), Abdullah bin Masood (RA), author of two books on history including long sections on Seerah, both mentioned above.

Appendix B : Notes

Note 1: The Blessed Ten

These 10 men were known as *Ashrat-ul-Mubashira*-literally: "the ten who have been given the good news". Long before their death these men had been told by the Prophet that they would enter into paradise. I have translated the term idiomatically as the Blessed Ten. For those interested, these 10 men were:

          Ali
          Abu Bakr
          Uthman
          Zubair bin Al Awwam
          Abdur-Rahman bin Auf
          Sad bin Abi Waqqas
          Talha bin Ubaidullah
          Abu Ubaidah bin Al Jarrah
          Saeed bin Zaid
          Umar

The list shows the order of their conversion. There were others who became Muslims before some of these (like Zaid bin Harithah, who was the second male to accept Islam), but they were not among the Blessed Ten.

Note 2: Khalid and the Yemen

Early historians have quoted some sources as saying that Khalid was also sent to the Yemen to convert the province to Islam; that he had no success; that the Prophet then sent Ali, placing Khalid under his command; and that Ali successfully accomplished the conversion of the Yemen. It is true that Ali went and converted the Yemenis, but I do not believe that Khalid was sent before him to the Yemen. The conversion of the Yemen was too big a job for a new convert like Khalid; it needed a Companion of high standing, like Ali. Moreover, if as stated, Khalid had no success in the Yemen, he was bound to have shed blood; and we know that there was no bloodshed in the Yemen at the hands of Khalid. Finally, Khalid returned from Najran in January 632 (Ibn Hisham; Vol 2, p. 594), while Ali left for the Yemen in December 631 (Ibn Sad: p. 687). So the report of Khalid going to the Yemen could not be true.

Perhaps the nearness of Najran to the Yemen led some chroniclers to confuse Khalid's actual mission to the former with an imagined mission to the latter.

Note 3: Dates of the Campaign of Apostasy

We know the dates of events more or less accurately up to and including the organisation of the 11 corps by Abu Bakr at Zhu Qissa. Thereafter the early historians give no dates.

We do know, however, that the entire campaign was completed by the end of 11 Hijri. We know that the Battle of Yamamah was fought in the early part of winter ("The cold has come"-Tabari: Vol. 2, p. 518.) We also, know the chronological sequence of the various battles and other events. And there are certain pointers, like reinforcements for Battle E being sent on conclusion of Battle C, and so on.

From this information, I have worked out the approximate dates of the various events that have been described, with the help of my military experience and judgement: for instance, it would take so long for a force to move from A to B; it would take so long to prepare for battle; it would take so long to complete administrative actions after battle before the next operation could be launched, and so on.

Because such estimates cannot be entirely accurate, I have given all dates after the formation of the 11 corps in terms of weeks rather than days. There may still be some inaccuracy, but probably of no more than a week or so.

Note 4: Plan for Invasion of Iraq

Early historians quote certain sources as narrating another version of the Caliph's plan for the invasion of Iraq. This version is as follows:

a. Abu Bakr instructed two generals to enter Iraq: Khalid from below, via Uballa, and Ayadh bin Ghanam from above, via Muzayyah. Ayadh was then somewhere in Northern Arabia, between Nibbaj and the Hijaz.

b. Hira was given as the common objective to both generals. Whichever of them got to Hira first would become the Commander-in-Chief and the other would serve under him.

c. The Commander-in-Chief would then leave sufficient forces to guard Hira as a base, and proceed with the rest of the army to fight the Persians in the imperial capital, Ctesiphon.

I reject this version for the following reasons:

(i) Abu Bakr, who relied heavily on Khalid to fight the main battles of the apostasy, would not plan the new campaign in a way that might place Khalid under the command of an untried general like Ayadh bin Ghanam, who was in any case not in the same class as Khalid.

(ii) Abu Bakr believed strongly in the principle of concentration of force, as is evident from his conduct of the Campaign of the Apostasy. It is unlikely that he would violate that principle now by splitting the available forces into two and launching them in two widely separated areas whence they would be unable to support each other. A two-pronged invasion, sometimes desirable and sound, would be a grave error in this case, when the enemy was so much stronger and operated on interior lines. The Muslims, split into two forces, would suffer defeat in detail.

(iii) If this version were true and Ctesiphon was indeed the ultimate objective of the invasion, Khalid would certainly have attacked it. Actually, we know from Tabari that towards the end of the campaign, Khalid wanted to attack Ctesiphon but did not do so for fear of the Caliph's disapproval. And there is no mention anywhere of Abu Bakr ever cancelling his supposed order to attack Ctesiphon or, alternatively, reminding Khalid of it.

(iv) As it happened, Ayadh was stuck at Daumat-ul-Jandal and was later helped out by Khalid. If his objective were "Iraq from above" there was no need for him to engage in serious hostilities in a region far from his objective, especially when other and better routes to that objective were available to him. Ayadh bin Ghanam did have a force under his command, but its strength was not very great. His objective was probably none other than Daumat-ul-Jandal. Abu Bakr may, of course, have had the intention of sending him to Iraq later to support Khalid, for after Daumat-ul-Jandal he did proceed to Iraq with Khalid and served under his command.

Note 5: The Battle of the River

The description of this battle is taken exclusively from Tabari, but Tabari places this battle at Mazar and names it after both Mazar and the River (Sinyy). Balazuri (p. 243) also mentions a battle at Mazar, while the accounts of Ibn Ishaq and Waqidi make no mention of action in the Kazima-Uballa-Mazar region.

Mazar, according to Yaqut(Vol. 4, p. 468) was four days' journey from Basra on the road to Wasit (which was founded in 83 Hijri); and it was on the One-Eyed Tigris. Mazar is now believed to have stood on the site of the present Azeir, on the right bank of the Tigris. Azeir is actually about eighty miles from Basra and this places it correctly as the site of Mazar.

But it should be noted that Mazar was about 70 miles beyond the *River* and 25 miles beyond the Euphrates. Thus it could not be reached without an elaborate arrangement of boats for crossing the rivers; and there is no mention of Khalid crossing any river to go to Mazar. The arrangements necessary for the collection of boats for the crossing of the entire Muslim army would surely have attracted the attention of historians, as Khalid's use of boats after Ulleis did, and an event like the crossing of the Euphrates could not have gone unnoticed by the chroniclers.

I believe that Khalid never crossed this river. No Arab commander (of those days, that is) would cross a large river and move far beyond it, thus putting a major obstacle between himself and the desert, while powerful enemy forces were free to manoeuvre against his rear. This would be even more so when the enemy was the formidable imperial Persian. The desert strategy of Khalid, and of other Arab commanders after him, was based on the principle of staying close to the fringes of the desert. Moreover, since Khalid's objective was Hira, his crossing the *River* and the Euphrates to go into Central Iraq, thus deviating from his objective, would not make sense. Hence, in my view, this battle was the Battle of the *River* only, and not the Battle of Mazar also.

Note 6: Location of Khanafis etc.

There is no certainty about the location of Khanafis, Huseid, Muzayyah, Saniyy and Zhumail, and I accept the possibility of error in my geographical reconstruction of this operation. Historical indications about these locations are given below.

Khanafis: According to Yaqut (Vol. 2, p. 473), the place was near Anbar and in the vicinity of Baradan. Musil, in one part of his book (p. 309), suggests that it was west of the Euphrates, but elsewhere places it at the present Kazimein, 5 miles above Baghdad, stating that Baradan was a district 4 leagues from Baghdad.

I place it definitely west of the Euphrates (though its exact location is a guess) on the basis of Tabari and geographical indications:

a. Tabari states (Vol. 2, p. 580) that Khalid left Hira on the route to Khanafis and caught up with Qaqa and Abu Laila at Ain-ut-Tamr. If Khanafis were east of the Euphrates, there could be no question of Ain-ut-Tamr being on the way to it.

b. In 13 Hijri, Muthanna wanted to raid Baghdad and Khanafis from Ulleis (Tabari: Vol. 2, pp. 655-6), and was told by guides that the two were some days journey apart and that Khanafis was quicker to get to. Muthanna went to Khanafis by "the land route" and then continued to Anbar. Thus if Khanafis were at or near Kazimein, it would not be some days away from Baghdad and could not be reached by reached by "the land route".

I place Bardan as the Wadi-ul-Ghadaf, which was also known as Wadi-ul-Burdan, and runs 25 miles north of Ain-ut-Tamr. If there was a place named Bardan or Burdan, it no longer exists, but the *Wadi* suggests its general geographical location. This is supported by Tabari's statement (Vol. 2, p. 581) that on his march from Ain-ut-Tamr to Muzayyah, Khalid went past Bardan. I have placed Khanafis a little north of the *Wadi*.

Huseid: Yaqut (Vol. 2, p. 281) mentions this as a small valley between Kufa and Syria. Hence this was definitely west of the Euphrates, but must have been east of Khanafis because of the direction of Mahbuzan's withdrawal. Had Khanafis been to the east, Mahbuzan would have withdrawn towards Ctesiphon and not to Muzayyah. Musil also (p. 309) places Huseid west of the Euphrates, but its exact location is uncertain.

Muzayyah: Yaqut (Vol. 4, p. 560) calls this place a hill in Najd, on the bank of the Wadi-ul-Jareeb, in the area of the Rabee'a, which was one of the tribes of the Taghlib group in Iraq. (*Najd* means a high tableland, and this was not the Najd in Arabia.) Tabari (Vol. 2, p. 580) places it between Hauran and Qalt. Musil (p. 303) locates it at Ain-ul-Arnab, 270 kilometres from Hira, 115 kilometres north-west of Anbar and 75 kilometres south-south-east of Aqlat Hauran. Ain-ul-Arnab is correctly located now 25 miles west of the present Heet. In the absence of any other indications, I accept this as Muzayyah.

Saniyy and *Zhumail*: Both Yaqut and Tabari (and Musil, who follows Yaqut in this) are well off the mark in their location of these places. Yaqut (Vol. 1, p. 937) places Saniyy east of Risafa and Zhumail near Bashar, east of Risafa and east of the Euphrates. Tabari (Vol. 2, p. 582) also locates them east of Risafa. Risafa is in Syria, north-north-east of Palmyra and 15 miles south of the Euphrates. Musil (p. 312) places the two battlefields in the foothills of the Bisri Range, which runs north-east of Palmyra, and regards Bashar as another name for Bisri.

I reject these locations emphatically. It is unthinkable that Khalid should take his army 250 miles from Muzayyah (about 400 miles from his base at Hira-almost a six weeks turn-round-merely to tackle two Arab concentrations which in no way concerned him. Moreover, the Bisri Range and Risafa were deep in Syria, and there could be just no question of Khalid taking such liberties with the Caliph's instructions, which confined his area of operations to Iraq.

I have found Zhumail. It is marked on certain maps as a ruined town 35 miles north-west of Ain-ut-Tamr, on the bank of the *Wadi Zhumail*. The one-million scale map (Dept. of Survey, War Office and Air Ministry, London, 1962) spells it as *Thumail*, and this is a much more sensible location for the place. Saniyy must have been between this place and Muzayyah.

It is possible of course, that there was another Risafa somewhere in Western Iraq, in which case the reference to it by Yaqut and Tabari would be correct.

Note 7: Dates of Campaign in Iraq

The dates of some battles are given in the early accounts, but for others they are missing. We know that Khalid's march from Yamamah and the Battle of Kazima took place in Muharram; the Battles of the River, Walaja and Ulleis were fought in Safar; and Hira was conquered in Rabi-ul-Awwal. We also know the approximate date of Khalid's arrival at Firaz and the exact date of the Battle of Firaz.

The dates of the remaining battles, from Anbar to Zhumail, are not given. I have assessed these dates according to military judgement, as for the Campaign of the Apostasy. And since these are approximations, I have given these dates in weeks rather than in days.

Note 8: The Camel and Water

Early writers have described how, before this march began, many of the camels were made to drink vast quantities of water which they stored inside their bellies. These animals were then slaughtered on the march, a few every day, and the water in their bellies used for the horses.

This is an old legend and, strangely enough, believed up to this day. Actually the camel carries no water as a reservoir in any part of its body. The fact is that the muscle tissue of the camel holds a higher percentage of water than is the case with other animals, and thus the camel can go without water for longer periods than other animals without suffering dehydration. The legend of the water reservoir inside the camel, which the rider can use for survival as a last resort, is just that-*a legend*!

Note 9: The Route of the Perilous March

Certain present-day writers have given Khalid's route as the southern one, *i.e.* via Daumat-ul-Jandal. According to this version, Khalid marched from Hira to Daumat-ul-Jandal and then to Quraqir (and there is such a place about 70 miles north-west of Daumat-ul-Jandal). Thence he made the perilous march across unchartered desert to reach Eastern Syria.

Indeed some of the early writers have mentioned Daumat-ul-Jandal. Balazuri gives Daumat-ul-Jandal as the stage a next after Arak and just before Tadmur, which is clearly impossible. Tabari mentions one source as saying that from Hira Khalid went to Quraqir via Daumat-ul-Jandal, though in his main account of the march he makes no reference at all to Daumat-ul-Jandal. Waqidi and Yaqubi in their description of this route entirely omit both Daumat-ul-Jandal and Quraqir.

The fact is that some of the early narrators confused this march with Khalid's operations against Daumat-ul-Jandal, which have been described in Part III of this book. They have even stated that it was on this march that he captured Daumat-ul-Jandal, which is incorrect.

The operative condition in Abu Bakr's instructions to Khalid was to march *with speed*. Going via Daumat-ul-Jandal, Khalid could not fulfil this condition as the move would take a long time to complete. And if he did go to Daumat-ul-Jandal, why should he not take the direct caravan route to Syria and join the Muslim forces deployed in the area of Busra and Jabiya? On this route there was no enemy on the way, his fears of which we have mentioned. Why should he risk the annihilation of the entire army by traversing an unchartered, waterless desert only about 20 or 30 miles east of the main caravan route? And why should he bypass the Muslim forces, which it was his task to contact and take under command, go deeper into hostile territory in Eastern Syria, and then fight his way back again to join the Muslim forces? This does not make sense.

Moreover, according to both Tabari (Vol. 2, p. 601) and Balazuri (p. 118), Khalid marched from Hira to Quraqir via Sandauda and Muzayyah. These two places are north-west of Hira, the distance to Muzayyah being about 200 miles. Why should Khalid travel this extra distance north-westwards from Hira and then turn south again to go to Daumat-ul-Jandal? This makes even less sense!

The correct general alignment of the route is given by Waqidi and Yaqubi and also by Tabari, except for his passing reference to one source relating to the route through Daumat-ul-Jandal. The route that I have described, i.e. the one leading directly from Iraq to Syria, was the quickest route; and Khalid's subsequent operations in Eastern Syria, which followed his entry into Syria, only make sense in the context of a march on this route. In fact clear indications of this route are given by:

a. Khalid's fears of being held up by Roman garrisons (Tabari: Vol. 2, p. 603).

b. Waqidi's mention of Khalid's traversing 'the land of Samawa', which is in the *Badiyat-ush-Sham* (the Desert of Syria), which covers Western Iraq and South-Eastern Syria (Waqidi: p. 14).

c. Tabari's statement that on completion of the march Khalid had left behind the frontiers of Rome and its garrisons facing Iraq.

d. Muthanna's accompanying Khalid up to Quraqir, which he would not have done if it were as far away as the border of North-Western Arabia.

Although the general alignment of the route can be deduced from historical accounts, no one can be absolutely certain about the exact route taken by Khalid. The route that I have here offered can only be generally correct, and may be out by several miles. Certain pointers used are explained below.

Arak is there even now in Syria, about 20 miles east-north-east of Palmyra. Suwa was a day's march from Arak, and the spring a day's march from Suwa. Thus the spring was about 50 miles from Arak, and this puts it at about the Bir Warid of today. Quraqir was five days from the spring which means between 100 and 150 miles away. This gives us a bel in which Quraqir was located. Of course, this Quraqir is not there today and no one knows its precise location, but according to Yaqut (Vol. 4, p. 49) it was a valley and a watering place of the Bani Kalb in Samawa; and the land of Samawa is part of the Desert of Syria, as stated above. Thus this Quraqir was certainly in Western Iraq and not in North-Western Arabia.

In Western Iraq exist the ruins of several ancient towns and castles which are clearly shown on an archaeological map of Iraq prepared by the Iraqi Directorate General of Antiquities. West of Muzayyah lie Qasr-ul-Khubbaz, Qasr Amij and Qasr Muheiwir, the last one being 120 miles from Muzayyah; and although the ruins relate to the Parthian period, they are believed to have existed as watering places for caravans until well into the Muslim period. This is the route which Khalid's movement must have followed before the start of the Perilous March; and I place Quraqir at or near Qasr Muheiwir, on the Wadi Hauran. This place is 120 miles from the spring, which is about the distance of a five day's march of that period.

Note 10: Battle of Marj-us-Saffar

Waqidi says nothing in his account about the Battle of Marj-us-Suffar. Yaqubi (Vol. 2, p. 139) and Balazuri (p. 125), mention this battle as being fought before the conquest of Damascus.

Waqidi, strangely enough, describes two sieges of Damascus, the first one being abandoned after a short time, while all other historians have mentioned only one siege-the successful one. According to Waqidi (p. 18), Khalid came to Damascus from the north-east, *after* the Battle of Busra, fought the battle against Azazeer and Kulus who came out of Damascus to meet him in the open, and then invested Damascus (p. 21). But soon after, he raised the siege and marched to Ajnadein to fight Wardan.

Gibbon accepts this version of Waqidi but it does not appear to be really acceptable. In the context of the mission given to Khalid by the Caliph, of taking command quickly of the Muslim forces in Syria and fighting the Romans (whose most threatening concentration was at Ajnadein) such an action does not make sense. It is unthinkable that Khalid should first approach Damascus from the north-east and raid Marj Rahit; then bypass Damascus, march to Busra and capture Busra; then again approach Damascus from the north-east, fight Azazeer and Kulus and invest the city; then raise the siege and march to Ajnadein.

But the events described by Waqidi in the action against Azazeer and Kulus are real and correct. They did take place. So I have assumed that they took place before the one and only siege, and this was the Battle of Marj-us-Suffar.

Note 11: Date of Conquest of Damascus

Most early historians and practically all later ones, have given the fall of Damascus as occurring in Rajab, 14 Hijri (September 635). Waqidi is the only one of the early writers who has placed it a year earlier, i.e. Rajab, 13 Hijri, and this I regard as correct. There are various pointers to this in the early accounts, most of which are linked with the death of Abu Bakr, which occurred on the 22nd of Jamadi-ul-Akhir, 13 Hijri, i.e. the month preceding the conquest of Damascus, and the removal of Khalid from the command of the army by Umar.

The letter which Umar wrote to Abu Ubaidah, appointing him the commander of the army, reached him while a great battle was in progress, and Abu Ubaidah withheld this information from Khalid until after the battle had been won. This battle could not have been that of Ajnadein, which was fought-and on this there is universal agreement-in Jamadi-ul-Awwal, 13 Hijri. Some writers have quoted sources saying that the event took place at Yaqusa. But this could not be correct either, as Yaqusa was just a one-day battle. Some who consider that the year of the Battle of Yarmuk was 13 Hijri have said that the letter reached Abu Ubaidah during this battle. This too is incorrect, as the Battle of Yarmuk was fought in 15 Hijri.

All the early historians have quoted some sources as saying that the letter reached Abu Ubaidah while the Muslims were besieging Damascus, and that he did not inform Khalid of this until after the surrender of the city. In my view this is correct; and if it is, then Damascus could not have fallen in Rajab, 14 Hijri, for this information regarding the change of command could not possibly have been concealed from the Muslims for a whole year. It could be concealed for a few weeks-certainly no more than a month-as I have assumed.

The pact with the Damascenes was signed by Khalid as the Muslim commander and witnessed by Abu Ubaidah. This pact was actually seen by Waqidi in the following century. There can be no question of this pact being signed by Khalid a year after his removal from command with Abu Ubaidah, the actual commander, only witnessing it.

According to Waqidi (p. 62), the oath of allegiance to Umar was taken by the Muslims at Damascus on the 3rd of Shaban, 13 Hijri, when they had finished with the siege of Damascus. And this was after the return of Khalid from his raid at Marj-ud-Deebaj; which came after the fall of Damascus.

Another point: The Battle of Fahl, according to every historian, was fought in Zu Qad, 13 Hijri, and all historians have quoted some sources as saying that this battle took place after the conquest of Damascus.

As for the duration of the siege of Damascus, historians disagree. It has been given variously as one year, six months, four months and 70 days. I put it at about a month. It could not have been very long, as then Heraclius would certainly have made more than one attempt to relieve the beleaguered garrison. If Heraclius could raise another large army for the battle of Fahl, fought in Zu Qad, 13 Hijri, he would certainly have used such an army for the relief of Damascus, had the siege continued till then. In any case, a garrison and a population so large could not have been victualled for a long siege in the three weeks which elapsed after the Battle of Ajnadein until the siege began. Here again Waqidi's version is the most sensible one.

Keeping all this in mind, we must conclude that Damascus fell on or about Rajab 20, 13 Hijri, after a siege of one month.

Note 12: Yarmuk-Opposing Strengths

There is a difference of opinion about the strength of both armies at the Battle of Yarmuk. As frequently happens in such cases, there has been a tendency to show ones own strength as less than it was and the enemy strength as more than it was.

Let us first take the Roman strength. Muslim historians assess it as follows:

a. Tabari, in one place, (Vol. 2, p. 598, where he gives his main account of the battle) shows it as 200,000 men. Elsewhere (Vol. 3, p. 74) he quotes Ibn Ishaq as saying that it was a 100,000 including 12,000 Armenians and 12,000 Christian Arabs.

b. Balazuri (p. 140) gives the Roman strength as 200,000.

c. Waqidi (p. 107) exaggerates it to a fantastic figure, but his estimate of the Roman who used chains (30,000, p. 139) seems very reasonable.

As for Western writers, Gibbon (Vol. 5, p. 325), taking his material from early Byzantine sources, gives the Roman strength as 140,000 including 60,000 Christian Arabs.

There is obvious exaggeration on both sides, but less so on the Western side, because the Byzantines would know their own strength better than their opponents would. We should dismiss the figure of 200,000 as incorrect. Such a vast army could not possibly have been assembled on one battlefield; and the problems of the concentration, movement, supply and feeding of such a force, with the relatively primitive communications of the time, would be such that any staff officer entrusted with the task would promptly resign his commission! This point will be more apparent to the trained military mind than to the civilian reader.

On the Western side, too, there is an attempt to minimise the Roman strength, especially the European part of it-partly perhaps for reasons of racial pride. It is absurd to say that the Arab section of the army amounted to 60,000 men. Just the Arabs of Syria could hardly have produced such a numerous army, when the entire Muslim State, which included Arabia, the Yemen, Iraq and Gulf States, could only produce 40,000. This is therefore probably nothing more than an attempt to pass the blame on to the Arabs. It is noteworthy that while Gibbon gives the Christian Arab strength as 40 per cent, Ibn Ishaq (a reliable source) gives it as only 12 per cent.

Allowing for exaggeration on both sides, I believe that the Roman army was 150,000 strong. It is impossible to say how strong each contingent was, but in the absence of any data, I have assumed that each of the five armies which comprised the Roman army at Yarmuk (including the Christian Arab army of Jabla) was roughly a fifth of the total Roman strength. There is, of course, the possibility of some error in this assumption. As for the Muslim strength, Tabari in one place (Vol. 2, p. 592) gives it as 40,000 plus a reserve of 6,000. Elsewhere (Vol. 3, p. 74), he quotes Ibn Ishaq as saying that the Muslims numbered 24,000-this against 100,000 Romans. Balazuri (p. 141) agrees with Ibn Ishaq, while Waqidi (p. 144) places the Muslim strength at 41,000.

I doubt that it could have been more than Waqidi's figure, and in order to accommodate Balazuri and Ibn Ishaq, have given the Muslim strength as 40,000, i.e. Tabari's figure without the reserve. This gives a ratio of roughly one Muslim to four of the opposition.

Note 13: Battlefield of Yarmuk

Early historians do not say precisely where the Battle of Yarmuk was fought, and as a result, a dispute has arisen among later writers on this point. Some have given the battle-front as the Yarmuk River itself; others have put it a little south of the Yarmuk; yet others have placed it east of the Yarmuk. We shall take up these theories in turn.

No battle could be fought across the Yarmuk River. I have described the ravine, and one look at it would convince any observer of this statement. Cavalry just cannot cross the gorge, and infantry movement is only possible in single file, with soldiers picking their way carefully over the sides, skirting the precipices. This is out!

A battle south of the river-on this grand scale that is-is out of the question. The terrain consists of spurs, flanked by steep *wadis*, running down to the Yarmuk, and nothing more than small-scale actions and skirmishes are possible in this terrain. This could not be the place which Khalid described as "a plain suitable for the charge of cavalry". Moreover, if the battle had been fought south of the Yarmuk, the Romans would have been pushed into the Yarmuk rather than the Wadi-ur-Raqqad, and this did not happen.

As for a battle east of the Yarmuk, the terrain permits a great battle here, as it consists of an open plain, north of Dar'a. But there is no indication anywhere in the early accounts of the battle being fought here. In fact the pointers are against it, for as indicated below, they favour the geography of the field I have described. Moreover, the Wadi-ur-Raqqad is too far from the plain east of the Yarmuk (over 20 miles from where the Roman left wing would be) for Khalid to turn the Roman, left, defeat the Roman army in position, herd the Romans into the Raqqad and then destroy them, all in a one-day counter-offensive.

Now for the battlefield I have reconstructed. There are the following clear proofs:

a. Khalid's advice to Abu Ubaidah-which was accepted-to put Azra *behind* him. This means facing roughly west.

b. Abu Ubaidah, after taking up positions, wrote to Umar that their army was deployed "on the Yarmuk, near Jaulan". (Waqidi.-p. 118).

c. The position of the Roman camp near Jaulan, which is between the Raqqad, Lake Tiberius and the area to the north. And this was 11 miles from the Muslim camp, putting the Muslim camp a little east of where I have placed the battle-line.

d. The Hill of Samein, 3 miles south-west of the present Nawa, is now known as the Hill of Jamu'a (gathering) because, according to local tradition, on it a part of Khalid's army was gathered. This is the clearest indication of the battle-front, as in most Arab countries these traditions have been passed down accurately from father to son since the earliest times.

e. The manoeuvre, as it was conducted, ending at the Wadi-ur-Raqqad in a one-day counter-offensive, could only be possible, in time and space, from the central or west-central part of the Plain of Yarmuk.

f. The front could not have been further east because the Muslims would not then have been on the Yarmuk, which starts at Jalleen, behind where I have shown the Muslim left wing. Moreover, Dhiraar's outflanking movement on the night before the last day of battle would not have been possible if the Muslim army had been 20 miles away.

Hence the battle-front could only have been more or less as it has been described in the chapter on Yarmuk. This should not be out by more than a mile or so, except for the Hill of Jamu'a, which was definitely in Muslim hands.

www.ingramcontent.com/pod-product-compliance
Lightning Source LLC
Chambersburg PA
CBHW081104080526
44587CB00021B/3441